ANCIENT TEXTILES SERIES 28

Gods and Garments

Textiles in Greek Sanctuaries in the 7th to the 1st Centuries BC

by
Cecilie Brøns

Oxford & Philadelphia

Published in the United Kingdom in 2017 by
OXBOW BOOKS
The Old Music Hall, 106–108 Cowley Road, Oxford OX4 1JE

and in the United States by
OXBOW BOOKS
1950 Lawrence Road, Havertown, PA 19083, USA

© Oxbow Books and the individual contributors 2017

Hardback edition: ISBN 978-1-78570-355-3
Digital Edition: ISBN 978-1-78570-356-0 (epub)

A CIP record for this book is available from the British Library

Library of Congress Cataloging-in-Publication Data

Names: Brøns, Cecilie, author.
Title: Gods and garments : textiles in Greek sanctuaries in the 7th to the 1st
 centuries BC / by Cecilie Brøns.
Other titles: Gods and garments (2017)
Description: Philadelphia : Oxbow Books, 2017. | Series: Ancient textiles
 series ; 28 | Based on the author's thesis (University of Copenhagen). |
 Includes bibliographical references.
Identifiers: LCCN 2016044906 (print) | LCCN 2016047159 (ebook) | ISBN
 9781785703553 (hardback) | ISBN 9781785703560 (epub) | ISBN 9781785703577
 (mobi) | ISBN 9781785703584 (pdf)
Subjects: LCSH: Textile fabrics, Ancient–Greece–Religious aspects. |
 Clothing and dress–Greece–Religious aspects. | Greece–Religion.
Classification: LCC BL65.C64 B76 2017 (print) | LCC BL65.C64 (ebook) | DDC
 292.3–dc23
LC record available at https://lccn.loc.gov/2016044906

All rights reserved. No part of this book may be reproduced or transmitted in any form or by any means, electronic or mechanical including photocopying, recording or by any information storage and retrieval system, without permission from the publisher in writing.

Printed in the United Kingdom by Short Run Press
Typeset in India by Lapiz Digital Services, Chennai

For a complete list of Oxbow titles, please contact:

UNITED KINGDOM	UNITED STATES OF AMERICA
Oxbow Books	Oxbow Books
Telephone (01865) 241249, Fax (01865) 794549	Telephone (800) 791-9354, Fax (610) 853-9146
Email: oxbow@oxbowbooks.com	Email: queries@casemateacademic.com
www.oxbowbooks.com	www.casemateacademic.com/oxbow

Oxbow Books is part of the Casemate Group

Front cover: Red-figure kylix by Makron. Staatliche Museen, Antikensammlung, inv. no. 2290. Photo: Johannes Laurentius.
Back cover: Mosaic with Nilotic scene, 2nd century BC. © Museo archaeologico Nazionale, Palestrina, inv. no. 149000907.

For my family

Contents

Acknowledgements vii
Abbreviations ix
Abstract xi

PART I: INTRODUCTION

Introductory framework 3

PART II: DEDICATIONS OF TEXTILES AND ACCESSORIES IN GREEK SANCTUARIES

1. Introduction: Textile dedications 21

2. The temple inventories: Written evidence for the dedication of textiles and accessories 33

3. Discussion: Textile dedications 145

PART III: CULT IMAGES AND DRESS

4. Introduction: Cult statues in ancient Greece 169

5. Iconographic evidence for the dressing of cult statues 183

6. Written evidence for the dressing of cult statues 239

7. Discussion: Dressing of cult statues 251

PART IV: SACRED DRESS CODES

8. Introduction: Sources and methodological discussion 269

9. Priestly garments 273

10. Iconographic evidence for the dress of sanctuary visitors 305

11. Clothing regulations in sanctuaries: The written sources 325

12. Discussion: Sacred dress-codes in sanctuaries 353

Conclusion...361

Appendix 1: The *peplos* of Athena at Athens..365
Appendix 2: Temple inventories. Greek texts and translations...................393
Appendix 3: Clothing regulations. Greek texts and translations.................401
Appendix 4: Dress-fasteners in sanctuaries 421
Bibliography..431

Acknowledgements

The current study has been carried out as a collaboration project between The Danish National Research Foundation's Centre for Textile Research (CTR) and the Department of Ancient Cultures of Denmark and the Mediterranean in the National Museum of Denmark. It was made possible by generous grants from these institutions as well as The Danish Ministry of Culture, for which I am very grateful.

I am grateful to more people than I can possibly name here, but some deserve special mention. First of all I am deeply grateful to Marie-Louise Bech Nosch, who oversaw this project from its inception as a doctoral thesis to its final publication, for all her invaluable guidance and support. This project has greatly benefited from her insightful comments, suggestions, and intellectual generosity. My deepest gratitude is also due to Bodil Bundgaard-Rasmussen and Lone Wriedt-Sørensen for all their kind support and advice. Furthermore, Mary Harlow, Tyler Jo Smith, and Jane Fejfer have all given invaluable advice on my dissertation and its revision into this publication.

I have had the great privilege of being affiliated with CTR, a research centre including the utmost experts in the field of textile research. I am grateful to the TEMA-group (Textile Economies in the Mediterranean Area) in particular for being so generous with their immense knowledge and for all our interesting discussions: it is definitely an inspirational environment that makes you grow as an academic scholar. I am especially indebted to Peder Flemestad for all his patient and generous assistance and guidance in epigraphic and literary matters. Without him, I would not have been able to perform this study. Warm thanks also to Giovanni Fanfani for help in various matters.

It has been an equally great privilege to work at the department of ancient cultures of Denmark and the Mediterranean at the National Museum of Denmark. Here I have been blessed with wonderful colleagues, who have always been generous with their immense knowledge and support and not least able to put a smile on my face.

I also wish to thank The Danish Institute at Athens, which provided me with the perfect place to stay in order to perform my research. Special thanks are due to Rune Frederiksen and Søren Handberg for kindly arranging my guest lecture at the Institute, as well as the staff at the Nordic Library in Athens for their kind assistance. My gratitude also goes out to the MacDonald Institute at the University of Cambridge for granting my stay as a guest scholar, especially to Margarita Gleba for arranging my visit as well as welcoming me and arranging my lecture at the university.

My warmest thanks are also due to Mark Nugent who did a wonderful job on editing my book for publication. My work has indeed benefited greatly from his expertise. I am also very grateful for all the help and assistance I have gotten from my friend and colleague Signe Skriver Hedegaard.

I would like to thank Jan Kindberg Jacobsen, Rune Frederiksen, Søren Handberg, Bjørn Lovén, and Kalliopi Sarri for their assistance in obtaining images for the book as well as Helle Horsnæs for providing excellent images from the Royal Collection of Coins and Medals in the National Museum of Denmark. Furthermore, I am grateful for the images of the artefacts in the Staatliche Museum zu Berlin, most kindly provided by Johannes Laurentius.

Great thanks are also due to Morten Svendsen and Werkstette for the excellent and professional work on the images for this book.

I am very grateful for the generous funding provided by Her Majesty Queen Margrethe and Prince Henrik's Foundation for the publication of this book.

I am also deeply indebted to my wonderful friends Luise Ørsted Brandt, Nora Petersen, and Dea Forchhammer for their friendship, eternal support, and – not least – their invaluable pep-talks. Finally, I wish to thank my family, especially my parents and my beloved sister Rebecca, for all their love and unfailing support – I could not have done this without you.

Abbreviations

AJA: American Journal of Archaeology
AM: Mitteilungen des Deutschen Archäologischen Instituts, Athenische Abteilung
ArchDelt: Archaiologikon Deltion
BCH: Bulletin de correspondance héllenique
BABesch: Bulletin antieke Beschaving
BICS: Bulletin of the Institute of Classical Studies
BMC 4: Gardner, P., *British Museum Coins. Seleucid Kings of Syria*, London 1878
BMC 8: Head, B. V., *British Museum Coins. Central Greece*, London 1884
BMC 16: Head, B. V., *British Museum Coins. Ionia*, London 1892
BMC 13: Head, B. V., *British Museum Coins. Caria and the Islands*, London 1960
BMC 17: Wroth, W. W., *British Museum Coins. Troas, Aeolis and Lesbos*, London 1894
BMC 19: Hill, G. F., *British Museum Coins. Lycia, Pamphylia and Pisidia*, London 1897
BMC 21: Hill, G. F., *British Museum Coins. Lycaonia, Isauria and Cilicia*, London 1960
BSA: Annual of the British School at Athens
CAJ: Cambridge Archaeological Journal
ClAnt: Classical Antiquity
CQ: Classical Quarterly
GRBS: Greek, Roman & Byzantine Studies
HarvTheolR: Harvard Theological Review
HSCP: Harvard Studies in Classical Philology
ID: Inscriptions de Délos, Paris 1926–1972
IG: Inscriptiones Graecae, Berlin 1873 –
JdI: Jahrbuch des Deutschen Archäologischen Instituts
JHS: Journal of Hellenic Studies
JÖIA: Jahreshefte des österreichischen archäologischen Instituts in Wien
LSCG: Sokolowski, F. *Lois sacrées des cités grecques*, Paris 1969
LSCGS: Sokolowski, F. *Lois sacrées des cités grecques, supplément*, Paris 1962
LSAM: Sokolowski, F. *Lois sacrées de l'Asie Mineure*, Paris 1955
LSJ: Liddell, H. G. & Scott, R. *A Greek-English Lexicon. Revised and augmented throughout by Sir H. S. Jones with the assistance of R. McKenzie*, Oxford 1940
MedelhavsMusfocus: Medelhavsmuseet. Focus on the Mediterranean
MusHelv: Museum Helveticum

OCD: *Oxford Classical Dictionary*, 3rd edition. Hornblower, S. & Spawforth, A. (eds.), Oxford 2003
OpAthRom: *Opuscula. Annual of the Swedish Institutes at Athens and Rome*
RE: *Pauly-Wissowa Real-Encyclopädie der Classischen Altertumswissenschaft. Begonnen von Georg Wissowa. Zweite Reihe (R-Z)*, Stuttgart 1939
RA: *Revue archéologique*
REG: *Revue des études grecques*
SCI: *Scripta Classica Israelica*
Schol. Callim.: Scholia in Callimachum, in Pfeiffer, R. *Callimachus*, Oxford 1949–1953
SEG: *Supplementum Epigraphicum Graecum*, Leiden & Amsterdam 1923 –
SIG^3: Dittenberger, W. et al. (eds.), *Sylloge Inscriptionum Graecarum*3, Leipzig 1915–1924
SNG: *Sylloge Nummorum Graecorum*
Suda: Adler, A. (ed.) *Suidae Lexicon*. Stuttgart 1928–1938
TAPA: *Transactions of the American Philological Association*
ThesCra I: *Thesaurus cultus et rituum antiquorum. I, Processions, sacrifices, libations, fumigations, dedications*, Los Angeles 2004
ThesCra V: *Thesaurus cultus et rituum antiquorum. V. Personnel of cult, cult instruments*, Los Angeles 2005
PAAH: *Πρακτικά της εν Αθήναις Αρχαιολογικής Εταιρείας* (1837–1926)

Abstract

Textiles comprise a vast and wide category of material culture and constitute a crucial part of the ancient economy. Yet, studies of classical antiquity still often leave out this important category of material culture, partly due to the textiles themselves being only rarely preserved in the archaeological record. This neglect is also prevalent in scholarship on ancient Greek religion and ritual, although it is one of the most vibrant and rapidly developing branches of classical scholarship.

The aim of the present enquiry is, therefore, to introduce textiles into the study of ancient Greek religion and thereby illuminate the roles textiles played in the performance of Greek ritual and their wider consequences. Among the questions posed are how and where we can detect the use of textiles in the sanctuaries, and how they were used in rituals including their impact on the performance of these rituals and the people involved.

This monograph adopts a broad, inclusive chronological scope to document the presence and ritual use of textiles in ancient Greek sanctuaries. The documentation includes source material primarily from the period from the 7th century until the end of the 1st millennium BC, with comparative archaeological evidence from other periods. Similarly, a broad geographical scope is adopted, thus providing useful comparanda. I do not confine myself to the area of modern Greece or the Aegean but, where relevant, include comparative material from a wider area across the Mediterranean from Southern Italy to the Near East, yet still with a primary focus on the Aegean.

The study is centred on three major themes: first, the dedication of textiles and clothing accessories in Greek sanctuaries is investigated through a thorough examination of the temple inventories, primarily from the late 5th to the 1st century BC. Second, the use of textiles to dress ancient cult images is explored. The examination of Hellenistic and Roman copies of ancient cult images from Asia Minor as well as depictions of cult images in vase-painting in collocation with written sources illustrates the existence of this particular ritual custom in ancient Greece. Third, the existence of dress codes in the Greek sanctuaries is addressed through an investigation of the existence of particular attire for ritual personnel as well as visitors to the sanctuaries with the help of iconography, such as sculpture and vase-painting and written sources.

By merging the study of Greek religion and the study of textiles, the current study illustrates how textiles are, indeed, central materialisations of Greek cult, by reason of their capacity to accentuate and epitomize aspects of identity, spirituality, position in the religious system, by their forms as links between the maker, user, wearer, but also as key material agents in the performance of rituals and communication with the divine. In sum, the present study demonstrates the importance of taking textiles into account in the study of ancient Greek religion.

Part I

Introduction

Introductory framework

Textiles are the widest imaginable category of material culture,[1] since they have a broad range of uses as a form of protection from environmental conditions, from clothing to shelter. Over time, textiles have been used as swaddling for babies, as shrouds for the dead, as bandages, as tents, blankets, curtains, and cushions, and also as containers for people, objects, or food. Textiles are also used in equipment such as fishing or hunting nets and sails. And, of course, they are used as clothing. The history of textiles is therefore vital to a greater understanding of the human experience, since textiles represent one of the earliest craft technologies and have always been part of human subsistence, economy, and exchange.[2] This centrality of textiles has caused Harlow and Nosch to argue for the inclusion of textiles among the 'big themes' of scholarship on the ancient Mediterranean, and for greater recognition of the fact that textiles were essential and ubiquitous in the past.[3]

Scholarship on Classical antiquity, however, still often overlooks this important category of material culture. This omission is, in part, connected with the fact that the textiles themselves are only very rarely preserved in the archaeological record. In the rare instances where textiles are preserved, it is due to exceptional climatic conditions, such as freezing, waterlogging, or desiccation – conditions that are not prevalent in ancient Greece. There are, therefore, only very few preserved textile remains from ancient Greece, and these rare examples are almost exclusively from burials.[4] Furthermore, textiles tend to be recovered either in the form of small fragments or in the form of so-called pseudomorphs, which means that, as a result of chemical interaction with metal objects, they are in a mineralised state.[5] A further reason for the neglect of textiles in Classical scholarship is the general underestimation of the importance of textiles in much modern scholarship, which is most likely rooted in the perceived gender divide in textile production. In other words, textiles were usually assigned to the realm of women, which was not deemed worthy of intense study in the predominantly male scholarship of the 20th century.[6]

Classical scholarship that has discussed textiles has tended to focus primarily on clothing. Over the past century, there has been a significant amount of scholarship on the appearance of ancient Greek dress,[7] but the earlier studies do not discuss other aspects of textiles or contexts. Within the last few decades, however, scholarship has moved beyond attempting to establish the appearance of Greek clothes to analysing the role of dress in signaling aspects of identity such as gender, status, ethnicity, and age.[8] Such studies have demonstrated the great importance of textiles in antiquity as well as their potential significance as a source of information about ancient societies. Furthermore, recent scholarship has established that textiles were a very important part of ancient economies.[9] The integration of textiles into the mainstream of current discussions in archaeological theory is therefore an important direction for research, since it has direct impact on basic theoretical assumptions and paradigms.[10] In neglecting textiles, we run the risk of developing a distorted view of ancient cultures.

Neglect of textiles is also prevalent in scholarship on ancient Greek religion and ritual, although it is one of the most vibrant and rapidly developing branches of Classical scholarship.[11] This neglect is related to the fact that classicists and philologists focus primarily on written sources, while archaeologists have focused mostly on architecture, iconography, and votive offerings that have survived in the archaeological record. Arguments have already been made, however, for the need of a better integration of archaeological material into historically oriented scholarship on ancient Greek religion, where material remains often serve as mere illustrations to aspects or features of Greek religion. Some scholars have proven to be important exceptions to this separation between written sources and the material record. Among the most important are Burkert and Ekroth,[12] who both have integrated different types of sources into their studies on ancient Greek religion and ritual, especially sacrifice. This focus on sacrifice in particular has become prevalent in studies of Greek ritual and religion, perhaps in part as a result of influence from studies in social anthropology, where such topics have been prominent in recent decades, and in part because these rites are recognisable in both the written and archaeological record, in the form of animal bones, vessels, altars, iconography, etc.

This brief introduction clearly illustrates that there is a gap in research on the role and importance of textiles in ancient Greek religion and ritual. My intention, therefore, is to introduce textiles into the study of ancient Greek religion in order to illuminate the roles that textiles played in the performance of Greek ritual and their wider effect on participants' experience of ritual. Among the questions that I plan to answer here are: how and where can we detect the use of textiles in sanctuaries? How were textiles used in rituals? What was the impact of textiles on the performance of these rituals and the people involved?

Chronological and geographical scope

I adopt a broadly inclusive chronological scope to document the presence and ritual use of textiles in ancient Greek sanctuaries. The documentation includes source material primarily from the 7th century until the end of the 1st millennium BC. In addition, comparative archaeological evidence from the surrounding periods is included. I do not confine myself to the conventionally defined chronological periods, because this historical or art-historical ordering of Greek antiquity into Archaic, Classical, and Hellenistic periods is not particularly well-suited to the study of cult and religion in ancient Greece.[13] I do, however, still employ these conventional period designations, since they are an established part of scholarly research and a useful tool in a study such as this. This long chronological view is not meant to give an impression of unbroken continuity, and I am aware of the possible pitfalls that may result, since this approach involves potentially severe methodological challenges, inasmuch as it is difficult to draw conclusions from material from very disparate periods. Nevertheless, the adoption of a wide time-span does allow for a broad perspective on what is not so clear in narrowly focused studies.

Although the Late Bronze Age falls well outside the parameters of this study, it should be noted that attestations of the ritual use of textiles is found in the Mycenaean Period, primarily in wall-paintings as well as in Linear B.[14] I am not necessarily arguing for continuity, but it must be said that what we know of Late Bronze Age ritual does seem to prefigure the later use of textiles in Greek cult and ritual.

Similarly, I adopt a broad scope with regard to geography, which allows for useful points of comparison. I do not confine myself to the area of modern Greece or the Aegean, but include relevant comparative material from the wider Mediterranean region, from southern Italy to the Near East, yet still with a primary focus on the Aegean. This geographical scope is meant to avoid an Atheno-centric study, and represents an attempt to balance the rich material from Athens with sources from the periphery of the Greek world.

Theoretical framework

As a discipline, Classical archaeology has always been separated from other archaeologies (which have sometimes evolved into or been absorbed as a branch of anthropology), and has instead been closely tied to classical philology and art history.[15] Until the 1970's, Classical archaeology focused primarily on the collection and classification of material remains, but did not tend to provide interpretation, and has therefore been considered more of a supplementary discipline.[16] Within the last few decades, however, Classical archaeology has been moving closer to ancient history, and has begun to employ theories developed in anthropology and sociology,[17] inspired by prehistoric archaeologists such as Renfrew. This has meant that Classical archaeology has slowly moved into more immaterial fields and consequently has provided more theoretically informed and wide-ranging analyses, which is also the aim of the current study.

Terminology: religion, ritual, and cult

The study of ancient Greek religion involves a number of terms that are very difficult to define: especially *'religion'*, *'ritual'*, and *'cult'*. This is not made any easier by the fact that the Greeks did not have any terms for either *religion*, *ritual*, or *cult*.[18] This means that these terms are not reflections of reality, but etic scholarly constructs not immediately recognisable to the ancient Greeks. Instead, they reflect the observer's point of view, not that of the actor.[19] We often take the meanings of these terms for granted, but, in fact, people tend to attach different meanings to them, and they sometimes appear to be used interchangeably. This is, to a large extent, a result of the lack of consensus on the definition of these terms. This particular subject is enormously complex, however, and it is not my goal to discuss or define these terms in depth, but rather to offer some observations.

Different definitions of religion have been proposed, e.g. by Durkheim, who defines religion as:

> 'a unified system of beliefs and practices relative to sacred things, that is to say things set apart and forbidden – beliefs and practices which unite into one single moral community called a Church, all those who adhere to them.'[20]

Spiro defines religion as:

> 'an institution consisting of culturally patterned interaction with culturally postulated superhuman beings.'[21]

In sum, the essence of religion can therefore be considered to be some framework of beliefs. These beliefs are not necessarily simply philosophical beliefs about the world or its origin, but must relate to transcendental or supernatural forces to warrant the term religion.[22] Thus religion can be understood as consisting of a binary opposition of myth/beliefs and ritual,[23] as well as the involvement of supernatural beings.

As with the term 'religion,' different definitions of the term 'ritual' have been proposed. Rappaport defines ritual as:

> 'the performance of more or less invariant sequences of formal acts and utterances not entirely encoded by the performers'.[24]

Kyriakidis defines ritual as:

> 'an etic category that refers to set activities with a special (not-normal) intention-in-action, and which are specific to a group of people.'[25]

He argues that this definition clarifies the frequent association of religion with ritual, since it is common in the Western world to consider religious practice as special, sometimes even irrational. Furthermore, religious practice is always associated with a specific group of people. He therefore concludes that religious practice can always be seen as ritual. This does not imply, however, that all religious practice is ritual, nor that all ritual is religious (a modern day example is civil weddings, which are ritual, but not religious).[26] This means that rituals can be defined as acts performed on particular occasions by a particular group of people.

There are differences, however, in how scholars conceive of the relationship between ritual and religion. Some scholars are inclined to see ritual as an exclusively religious activity,[27] whereas others are of the view that ritual can be religious as well as non-religious.[28] According to Renfrew, most analyses of archaeological evidence for religion or religious ritual have not made sufficient distinctions between evidence of ritual practice in general and that of specifically religious ritual practice.[29] There are thus many cases of ritual practices that we need not situate in a context of religious belief or practice, and it has been argued that there is no necessary separation between religious and secular in most ancient societies. According to Renfrew, we should therefore emancipate ritual from an automatic association with religion.[30]

Since archaeology is largely the study of material remains and of material culture, it is considered not to have direct access to faith or beliefs. Kyriakidis argues that an archaeology of religion cannot have as great a scope as an archaeology of ritual, which is more accessible to us, if one takes ritual to be a practice or action, since practice has a direct effect on material, whereas belief or faith has an effect on material only through practice. He therefore argues that we need to make fewer deductions for the study of practice than for a study of belief.[31] This means that it is ideal – or at least easier – for archaeology to focus on ritual rather than belief or religion, which explains the many studies that adopt this particular approach.

Whereas both religion and ritual have been subjects of numerous theoretical discussions in religious studies, the study of cult has been somewhat overlooked.[32] A body has been developed for the study of religion and ritual, which seem universally applicable, but cult studies appear to have been replaced by ritual analyses. Christensen argues that the difference between the two fields of study lies in their use of social versus model-orientated modes of criticism. She argues that cult studies focuses on social, historical, political, and economic aspects of worship (the role of diverse people, priests, expenses, and prayers) and may be focused in the 'special case', while ritual analyses are model-orientated and focus on the synchronic and general laws of the rites in question (the elements, stages, and classification of rituals). In this way, cult is somewhat similar to ritual, in that it forms a pair with myth as constituent parts of religion.[33] There is a tendency, however, among Classical archaeologists and philologists to use the word 'cult' synonymously with 'religion', rather than with ritual, and this general preference for 'cult' rather than 'religion' allows for the problem of beliefs to be left out of the equation. Thus, discussion of cult in Classical studies is also indicative of the fact that the focus is on action rather than on beliefs or immaterial aspects of religion.[34]

Studies of religion and material culture

In *The Elementary Forms of the Religious Life* (1915), Durkheim argued that religion is at the base of all social and cultural behaviour. In this work he also posited the idea that the dichotomy between the sacred and the profane is the central characteristic of religion:

> 'all known religious beliefs, whether simple or complex, present one common characteristic: they presuppose a classification of all the things, real and ideal, of which men think, into two classes or opposed groups, generally designated by two distinct terms which are translated well enough by the words profane and sacred.'[35]

He considers this division of the world into two domains – one containing all that is sacred, the other all that is profane – as the distinctive trait of religious thought. By the sacred, he means not only gods or spirits, but also material objects and rites – anything can be sacred. He defines the profane in the negative: it is that which is not sacred.[36] According to Durkheim, there is thus nothing inherent or intrinsic about an object that makes it sacred. An object only becomes sacred when a community invests it with that meaning. Durkheim goes on to define religion as 'a unified system of beliefs and practices relative to sacred things, that is to say, things set apart and forbidden'.[37] Religion, then, to Durkheim, is a social institution involving beliefs and practices based on recognising the sacred and maintaining a distance between these two realms or entities.

This perception of a sharp division between the sacred and the profane has been highly influential in scholarship on ancient Greek religion. This is reflected – to put it simply – in the usually unstated assumption that religious activities took place in sanctuaries, whereas the remaining community is considered to be profane. Recent scholarship, however, considers this division of the world into two essentialist spheres to be conceptually false and methodologically misleading.[38] Even so, this dichotomy often forms the basis for the differentiation between the ritual and the domestic in much archaeological work. Scholars such as Droogan argue that, instead, we should move away from such polarities such as 'sacred' and 'profane', and 'ritual' and 'domestic', since religion can be embedded in all aspects of past life.[39] We should not, then, limit our consideration of ancient Greek religion to sanctuaries, and we should consider the fact that religious activity could also occur in the domestic sphere, the agora, or anywhere else in the public domain. Conversely, we can still assume the 'sacredness' of the sanctuary, even though we cannot exclude the possibility that profane/secular acts occurred therein.

This divide between the sacred and the profane is also reflected in other fields of academic scholarship, where the focus differs greatly depending on the subject and discourse. Thus, in the study of religion, the 'material' and the 'spiritual' have often been presented as having radically differing, even opposing, natures – what appears to be a polarisation between the two. This alludes to what Droogan calls the 'deep and long-lasting character of the quasi-Gnostic split between the transcendent religious and the immanent material in a number of Western discourses'.[40] Meyer and Houtman describe further oppositions – introduced by Protestantism in particular – that privilege spirit above matter, belief above ritual, content above form, and mind above body, which has resulted in a devaluation of religious material culture, and materiality at large, as lacking serious theoretical interest.[41]

These dichotomies are deeply entrenched in the thought processes of modernity, and are reflected not only in Christian discourse, but also in much academic study of religion.[42] As Whitehead argues, enlightenment rationality has enabled and contributed to the development of influential ideas that have turned the focus away from the material world, and especially the roles that religious materiality plays.[43] Thus, the academic study of religion in the modern West has been shaped primarily by the idea that a religion consists of a discrete, subjective experience concerning such things as the origin of the cosmos, the existence of deities, or the purpose of life.[44]

Other theoretical perspectives have caused a further divide: for example, scholarship influenced by a Marxist perspective has generally treated religion as a secondary belief or emotion arising out of the harsh material conditions of existence, and other philosophical and religious traditions have argued that religiosity is an existential state, separate from the world of material things.[45] This issue is also discussed by Whitehead, who argues:

> 'The problem of materiality can be simplified and considered in two veins: there is the problem of materiality in material cultural studies, and there is the problem of materiality and how it is understood in the study of religions. In terms of religious materiality, objects are often relegated to statuses of "representation" or "symbolic", which means that they point towards something other than themselves. In terms of material culture studies, objects are often referred to as "things", commoditized, and discussed in terms of their circulation, exchange or as representational means to the expected ends of social constructions and engagements. While the paths of these disciplines intersect at points, they often have different expected outcomes and are framed against different backdrops.'[46]

Thus, it has been claimed that the discipline of archaeology has in the main ignored religion, and the discipline of religion has largely ignored archaeology, predominantly because there has been little specific theorisation of the materiality of religion. This has often caused religion to be considered absent from the archaeological record, or reduced to a secondary consideration that is derivative of greater social realities.[47] Even though the situation is perhaps not so drastic, there is no doubt that there has been a clear and different focus – more or less conscious – in the two disciplines.

This view, however, has changed, and scholarly studies now recognise the significance of the material in the experience and understanding of religion. One scholar who advocates for the inclusion of the material in the study of religion is Morgan. He argues that scholars seeking to understand a religion have tended to inquire about its teachings, a focus that derives from the creed-based tradition of

Christianity, which was intensified by Protestantism.[48] Arguing against this tendency, he contends that scholars should not regard belief (i.e. religion) as a universal mental or inner state. Instead, he proposes another way of thinking about belief:

> 'what if belief were about more than faith in things unseen, trust in divine promises, or the declaration of the truth of certain teachings? What if believing was not fundamentally different from seeing or smelling or dressing or arranging space?'[49]

Morgan thus argues that we should look for religion in a broader range of manifestations than creedal utterances, and for ways of believing that engage more of a human being than discursive performance alone. This means that we should examine how people behave, feel, intuit, and imagine as ways of belief.[50] Morgan argues, therefore, that, instead of asking what a religion teaches, we should focus on the social and interpersonal relations that characterise practitioners of a religion, such as what it is that people teach their children. A better approach might thus investigate how, when, and where people teach their children what they choose to impart. This moves the inquiry to the register of material culture by examining the conditions that shape the feelings, senses, spaces, and performances of belief – that is, the material coordinates or forms of religious practice.[51]

This discourse is what academic scholarship terms 'material religion' or religious materiality. Whitehead, very much in line with Morgan, defines material religion as:

> 'anything that can be seen, touched, smelled, felt, tasted and heard in multitudes of forms of religious expression, and like religion generally, it is inescapably entangled with material culture'[52]

whereas religious materiality can be understood as the medium through which religion 'happens'. Religion and material culture are mutually reinforcing: religion cannot occur without material culture any more than material culture can emerge without the ideas that shape and inspire its creation.[53] The study of material religion, however, is not only an examination of how material things reflect or are shaped by lived religion, or how religious people understand 'religious things', such as iconography, relics, art, etc. Instead, it entails a recognition of what images or objects or spaces themselves do: how they engage believers and what powers they possess.[54] In studying religion, one should therefore take a close look at what people do, the objects that they exchange, use, and display, and the spaces in which they perform. The material record is thus of great importance to the study of religion, since it can provide information about other aspects of the practice of religion than is conveyed in written texts or spoken words. The material record is consequently crucial to the study of ancient religious practices, since we not always have texts to rely on, and, if we do, they do not necessarily provide us with the full picture. The theoretical concept of material religion is thus central to the current study, since it means that we can infer aspects or expressions of religion through the material objects in the archaeological record.

Methodological framework
Source material
Textiles have largely disappeared from the archaeological record, and no textiles have so far been recovered from any Greek sanctuary. A study of ancient Greek textiles must therefore be based, to a large extent, on a range of secondary sources. My aim is therefore to present the best preserved evidence from the widest range of sources in order to construct as complex as possible a portrait of textiles in sanctuaries. The sources include archaeological objects, iconographic representations, as well as literary and epigraphic sources. These sources are all crucial for our knowledge of ritual textiles, since they each contribute different perspectives and types of data, and focusing on only one type of evidence would result in skewed results.

As Roland Barthes states in *The Fashion System*, it is possible to distinguish between three garments: image-clothing, written clothing, and real clothing.[55] In this study, I consider all three types of clothing: dress that is represented, described, named, counted, used and re-used, illustrated, and worn and discarded, as attested in different source materials.

The archaeological material gathered in this study includes:

- Iconography
- Dress-fasteners and other textile ornaments
- Textiles preserved in other archaeological contexts

Iconography is a crucial source of information about ancient textiles. The study includes a wide range of object groups, such as vase-painting, sculpture, figurines, reliefs, mosaics, and coins.

A second important source of information about Greek dress is the fibulas and pins used to fasten the garments of both men and women. These have been recovered in large numbers in certain Greek sanctuaries and thus may provide indirect evidence for the dedication of garments. Metal textile ornaments, originally sewn onto garments, are also included so as to illuminate the variety and decoration of garments.

Finally, textiles preserved in other archaeological contexts and geographical areas are included as comparative evidence for the range of possible fibres, colours and dyes, types of weave, decoration, techniques, etc. Although preserved textiles from Greece are rare, a vast corpus exists from Italy, Egypt, the Levant and the ancient Near East, and from Russia and Ukraine as well as from Northern Europe. Even though this material should be used as comparative evidence for Greek textiles only with the greatest care, it nonetheless provides us with an excellent opportunity to confirm which textile types existed. Some textiles, however, have been recovered from Greek contexts, but all derive from burial contexts, and are thus not direct sources of information about ritual textiles in sanctuaries. Like the textiles from other geographical areas, they do, however, provide an idea of the possible types and materials of textiles.

Written sources constitute another major source of information about the study of ancient textiles. The two types of written sources included in the current study are:

- Epigraphy (inventories, sacred laws, and decrees)
- Literary sources (e.g. historical texts, geographical descriptions, lexical works, drama, poetry, epigrams, and medical texts)

Inscriptions are one of the richest sources of information about ancient textiles used in Greek sanctuaries, since they provide us with otherwise unavailable insight into the realities of cult organisation and the importance of textiles in Greek religion. The study includes inscriptions that testify to the ritual use of textiles in sanctuaries, especially the so-called temple inventories, the clothing regulations for entering specific sanctuaries, inscriptions describing priestly garments, and, not least, the inscriptions concerning the famous *peplos* of Athena in Athens. It is not within the scope of this study, however, to provide a complete corpus of all epigraphic evidence for the ritual use of textiles, but only to provide a large survey and to choose the most salient examples and important evidence that allows us to acquire considerable insight into the subject.

Ancient Greek and Roman literary sources are also included to illuminate further the ritual use and meaning of textiles in sanctuaries. Like the inscriptions, they contribute information not otherwise provided in the archaeological material. Literary sources are particularly helpful in reconstructing the religious sentiments behind the practice of worshipping the gods, since they offer longer narratives than do other sources.[56]

Methodology and structure

Researching textiles in antiquity presents particular methodological challenges. As Harlow and Nosch argue, for textile research to be really effective it needs to combine the approaches of academic disciplines often kept separate in university departments.[57] Thus, because of the different types of source material employed, the study assumes an interdisciplinary approach in that it encompasses Classical archaeology, philology, and ancient history, and includes perspectives from the history of religions and anthropology. Although these disciplines are related lines of research, mostly belonging to the field of 'classics', they represent different lines of inquiry, methods, and theoretical implications. The diversity of the source material therefore requires a diversity of strategies.

Rather than treating each type of source material sequentially in separate analyses – that is, dividing the evidence into archaeological material, iconography, and epigraphic and written sources[58] – I treat the material in three separate themes, which each feature the relevant sources:

- Dedications of textiles and accessories
- Cult statues and dress
- Sacred dress-codes

Part 2, *Dedications of textiles and accessories in Greek sanctuaries*, is divided into three chapters. Chapter 1 introduces the theoretical discussion and praxis of votive offerings and the iconographic evidence for the dedication of textiles. Chapter 2 gives an in-depth analysis of the temple inventories that record textiles or dress-fasteners. For the analysis, I present all of the relevant inscriptions in an appendix of the inventories, which includes the Greek text and an English translation as well as several tables that provide an overview of all of the relevant information. Archaeological material and/or information from textile research is combined with the texts, so that the most complete possible conclusions can be reached. Finally, Chapter 3 discusses the wider implications and significance of the dedication of textiles for our understanding of Greek society and religion through discussion of gift theory, especially the theories formulated by Marcel Mauss and employed in more recent scholarship.

Part 3, *Cult images and dress*, examines the evidence for the use of textiles to dress cult images. It includes four chapters: Chapter 4 reflects on the archaeological and literary evidence for ancient Greek cult images. Chapter 5 analyses the iconographic evidence for cult statues and dress, on the basis of depictions in sculpture and vase-painting. Chapter 6 examines the literary sources that testify to the dressing of cult statues. Chapter 7 concludes part 3 with a summary and a discussion that draws upon agency theory – especially as formulated by Alfred Gell – with the aim of illustrating the far-reaching impact of the use of textiles in ritual.

Part 4, entitled *Sacred dress codes*, examines the evidence for the use of special types of attire in the performance of ritual acts. It is divided into five chapters. Chapter 8 discusses the terminology, taxonomy, and hierarchies of Greek priests and priestesses. Chapter 9 deals with the iconographic and written evidence for the existence of priestly garments. Chapter 10 examines the iconographic evidence for the dress of sanctuary visitors, and Chapter 11 analyses the inscriptions that record specific clothing regulations for sanctuary visitors. Part 4 concludes with Chapter 12, which discusses the evidence for and significance of dress in ancient cult and ritual. This chapter employs theory on dress and the relationship that dress has with identity as well as the effect that it has on human behaviour.

Appendix 1 is a case-study of the famous *peplos* of Athena, in which all relevant material is included. The evidence for this particular ritual is substantial and has had a profound influence on scholarly interpretations of textiles in Greek ritual. For this reason, a thorough re-examination is in order. **Appendixes 2** and **3** contain the Greek text and English translations of relevant inscriptions. **Appendix 4** includes a case-study of dress-fasteners in Greek sanctuaries.

Methodological discussions

Iconography is an important source of information about certain parts of ancient society. Making sense of images can be difficult, but we tend to take the process of interpreting images very much for granted.[59] A constant problem for anyone interpreting ancient iconography is therefore determining *how* we interpret what we see.[60] The most important problem for the use of iconography as a source on ancient Greece is the extent to which iconography corresponds to reality. There is no doubt that images are not simple replicas or photographs of reality, and that they were not produced to document or to provide information.[61] Instead, iconography may draw on and select elements from the surrounding world that were 'recognisable' to the ancient audience. Thus, in many cases, the images reflect the perceptions, ideologies, and ideas of the society in which they were produced.[62] Furthermore, reading images is not just a question of decoding a single meaning, since the interpretations of images changes in different contexts, with different viewers and different expectations.[63] In addition, images could be read on several levels at the same time: for example, an image on a vase of a departing warrior can also depict a mythological figure such as Hektor departing for battle. This also implies that a strict division of the scenes into 'everyday life' and 'myth' is arbitrary.[64]

With regard to depictions of dress, the underlying assumption of most scholarship on Greek dress appears to be that the garments represented on artefacts accurately reflect contemporary garment types and modes of dress. There is no *a priori* reason, however, that this should be the case,[65] and we cannot be entirely sure that works of art reproduce clothing from the place and time of the artist, since he might have copied an exotic or earlier type of dress.[66] This is especially the case with regard to depictions of divinities, which are often rendered in archaising types of dress. It thus appears that what is often depicted is ideal dress rather than real dress. Furthermore, as Harlow and Nosch argue, the context and genre of any piece of visual culture creates a particular visual message, and clothing is often used as a direct signifier of the modest wife, the courtesan, or the goddess with special attributes.[67] Thus, the iconographic wardrobe forms part of the identifying features of the individual depicted, and it may be exaggerated in the presentation of certain elements.[68] In sum, depictions of clothing do not map easily onto clothing worn in real life.[69]

Depictions of clothing can present challenges of interpretation too, in that articles of clothing are often superimposed on the body, and it can be difficult to separate the different garments, and determine where one article of clothing begins and another ends. Moreover, it is important to keep in mind that depictions usually omit an important element: colour. Even so, iconographic depictions are still an invaluable source of information about dress, and there is certainly a substantial degree of correspondence between images and real life.[70] A final note on iconography: although chronological developments can be important, in this study it matters very little whether an object can be dated to 450 or 430 BC. I therefore use datings provided by scholars in the relevant fields or give more general datings to periods. I am not concerned with the attribution of vase-paintings, sculpture, etc. to specific artists, since this is not of crucial importance to the topic under investigation here. I am primarily focused more on what the depictions show, less on who produced them.

When dealing with Greek inscriptions there are a number of factors that one needs to consider. First of all, all written sources are selective.[71] This means that inscriptions represent decisions about inclusion and exclusion, and that we cannot expect them to tell the full story.[72] Another potential problem facing the scholar dealing with inscriptions is restorations conducted by epigraphists. As Badian demonstrates, there is a peculiar brand of historical fiction created by those who construct far-ranging theories on the basis of words or phrases inserted by epigraphists between square brackets in a fragmentary text.[73] These restorations should of course always be included with caution, since they might reflect the scholar's preconceived ideas rather than the actual ancient text.[74] For this very reason, the restored parts of the inscriptions included in this study are reproduced in the appendixes.

The translations of inscriptions can be problematic as well, since the translators rarely have specific knowledge of ancient Greek dress. Most garment terms are therefore usually translated with generic terms such as 'tunic', 'robe', or 'cloak'. Such generic translations, however, obscure the important nuances and meanings that the particular choice of terminology reflects. I therefore employ the Greek garment terms wherever possible in the analysis and interpretation of the Greek texts in order to avoid misunderstandings and generalisations.[75] The same applies to literary sources, for which reason I often give the passage in English, but also provide the Greek garment term. This practice, in combination with the appendixes of the temple inventories and clothing regulations, which supply both the Greek texts and English translations, will make the monograph accessible for readers who are not experts in ancient Greek. Furthermore, it allows for a discussion of the problematic translations of garment terms.

Literary sources are not without bias either, and the pitfalls are legion.[76] With regard to dress, literary sources are constrained by genre in much the same way as visual media, since each type of literature will privilege particular and often contradictory images of the clothed body and the use of textiles. Furthermore, the authors were usually upper-class men writing for their peers, and did not have much interest in the details of clothing. Instead, descriptions of dress served other functions: e.g. as a shorthand for character, gender, or ethnicity.[77] Sometimes, therefore, it can be impossible to ascertain how much of the rhetoric was simply literary fiction used for effect.[78] Literary sources do, however, still provide a great deal of important information, such as descriptions of how dress was used, by whom, and for what purposes. Literary texts are, therefore, included in this study, although mostly on a comparative basis, in order to provide a more complete picture of the ritual use of textiles.

Due to the lack of textile finds – at least of entire garments – from Greece, the reconstruction and interpretation of Greek dress is based on the juxtaposition of three primary sources: epigraphy, literary sources, and iconographic depictions.[79] These three media serve different purposes and offers different data. Images of dress provide information where written and archaeological evidence have gaps: only from images can we reconstruct how clothing items might have been assembled, draped, and worn on the body, and how they were combined with particular accessories.[80] Inscriptions, on the other hand, mention garments or other textile terms that we cannot identify in the archaeological or iconographic record. Yet, as Harlow and Nosch argue, it is often not clear to the modern reader what type of garment is being described or mentioned in the written sources, or why rectangular garments with essentially the same shape and function should have different names (e.g. *chlamys, chlanis, himation*). A question that arises is thus whether these terms refer to the way that the garment is worn, the material from which it was made, its decoration, the wearer's identity, or perhaps a combination of these elements.[81] Sometimes it is possible to illuminate the meaning of specific words on the basis of their derivation or etymology or through comparison with related forms, but this is not always the case. Besides, many ancient Greek garment terms are poorly attested, which makes interpretation even more difficult.

The nomenclature of ancient Greek dress is, moreover, far from secure, although the names of ancient garments are rarely questioned. As Lee has already argued, it may be impossible and perhaps even undesirable to create a new system of dress-terminology, but it is still important to note that many words for ancient Greek garments have been erroneously identified, and that we use many of these terms with false authority, since their use is a product of scholarship, not ancient Greek nomenclature.[82]

Another problem that becomes evident when juxtaposing written sources and iconography is that the depictions of dress often appear fairly standardised and represent limited modes of dress, whereas the written sources reflect a much more varied garment terminology.[83] The relatively few garment types depicted in iconography appear to have caused a common perception that Greek dress consisted primarily of the *chitōn, peplos, himation*, and *chlamys*. This means that we run the risk of attempting to map evidence onto preconceived ideas of Greek dress consisting of only a few garment types.

In sum, the juxtaposition of these three types of sources is not unproblematic, and linking words and images is an extremely difficult, and sometimes even an impossible, task. I therefore do not attempt to devise definite identifications of several of the more elusive garment types examined in this study, since this can be a very precarious undertaking. It is still possible, however, to identify certain garment terms on the basis of iconographic depictions, and I therefore employ many of the more established Greek garment terms in my descriptions and interpretations of depictions of dress.

Another methodological challenge that must be addressed is the degree to which our sources are representative of the ancient reality. For one thing, the evidence is uneven and dependent on chances of survival, and the situation may change with new discoveries. Furthermore, some archaeological material is better preserved in the archaeological record than others, and iconographic analyses are therefore based especially on vase-painting and stone sculpture. There is also a geographical bias, since we have a wealth of Athenian source material, written as well as archaeological, which might cause us to see ancient Greece with 'Athenian eyes'. It should also be kept in mind that the sources primarily deal with the elite, and the social upper-classes are heavily overrepresented.[84] Trying to access everyday clothing of other social groups is often difficult. Only rarely do we see or hear of people in 'ordinary dress' or in rags. Instead, some groups appear as genre pieces: e.g., the elderly, the poor, and labourers, who are often depicted in short tunics worn over only one shoulder.[85] Thus, our interpretation is constrained by the evidence.

Notes

1. Schneider 2006, 203.
2. Gleba & Mannering 2012, 1.
3. Harlow & Nosch 2014, 3–4.
4. For textiles from ancient Greece, see e.g. Spantidaki & Moulhérat 2012.
5. Moulhérat & Spantidaki 2009a; 2009b.
6. Harlow & Nosch 2014, 3.
7. Bieber 1928; Johnson 1964; Losfeld 1991; Marinatos 1967; Studniczka 1886.
8. Cleland, Harlow & Llewellyn-Jones 2002; Llewellyn-Jones 2003; Tellenbach et al. 2013.
9. E.g. Droß-Krüpe 2014; Nosch 2014; Liu 2012.
10. Good 2001, 201.
11. Kindt 2011, 696.
12. Ekroth & Wallensten 2013; Ekroth 2009b; Burkert 1983; 1985.
13. Günther 2013, 245.
14. See Nosch 2000 and Nosch & Perna 2001 for the use of wool and textiles as votive offerings, and sheep under the protection of the gods in the Bronze Age Aegean.
15. Kindt 2011, 698.
16. Kindt 2011, 699.
17. E.g. Humphreys 1983.
18. Bremmer 1998, 12; Christensen 2009, 21.
19. Bremmer 1998, 12, 31.
20. Durkheim 1995, 44.
21. Spiro 1966, 96.
22. Renfrew 2007, 113; Spiro 1966, 91.
23. Christensen 2009, 13.
24. Rappaport 1999, 24.
25. Kyriakidis 2007, 294.
26. Kyriakidis 2007, 294.
27. E.g. Bell 2007; McCauley & Lawson 2007.
28. E.g. Kyriakidis 2007; Renfrew 2007; Elsner 2012, 8.
29. Renfrew 2007, 114.
30. Renfrew 2007, 110.
31. Kyriakidis 2007, 298.
32. Christensen 2009, 14.
33. Christensen 2009, 13, 16.
34. Christensen 2009, 21.
35. Durkheim 1995, 37.
36. Durkheim 1995, 37.
37. Durkheim 1995, 44.
38. Droogan 2012, 19.
39. Droogan 2012, 19.
40. Droogan 2012, 3.
41. Houtman & Meyer 2012.

42. Droogan 2012, 3.
43. Whitehead 2013, 179.
44. Morgan 2010, 1.
45. Droogan 2012, 3.
46. Whitehead 2013, 24.
47. Droogan 2012, 4.
48. Morgan 2010, 1.
49. Morgan 2010, 5.
50. Morgan 2010, 5.
51. Morgan 2010, 6.
52. Whitehead 2013, 23.
53. Whitehead 2013, 97.
54. Droogan 2012, 9.
55. Barthes 1990, 3–5.
56. Connelly 2007, 10.
57. Harlow & Nosch 2014, 1.
58. Nor would it make sense to divide the material on the basis of sanctuaries, since many objects and inscriptions cannot be attributed to a specific sanctuary. The same applies to a division on the basis of divinities, since it can be difficult to determine whether an object is for a specific god or goddess.
59. Beard 1991, 12.
60. Ekroth 2009a, 91; Bérard & Durand 1989, 23.
61. Ekroth 2011, 7.
62. Ekroth 2011, 11.
63. Beard 1991, 13.
64. Beard 1991, 21.
65. Lee 2005, 58.
66. Riis 1993, 151.
67. Harlow & Nosch 2014, 7.
68. Harlow & Nosch 2014, 7.
69. For art historical perspectives on the understanding of the relationship between real clothes and their representation in art, see Hollander 1978.
70. Ekroth 2009, 91. See also Harlow 2004, 205 and Geddes 1987, 308.
71. Osborne 2011, 97.
72. Osborne 2011, 99.
73. Badian 1989, 59.
74. For specific examples of wrongly restored inscriptions, see Badian 1989.
75. I have transcribed the Greek terms, using 'y' for upsilon, 'ch' for chi, ō for omega, and ē for eta.
76. For the methodological problems of written sources, see Hansen 1999, 11–19.
77. Harlow & Nosch 2014, 8.
78. Harlow & Nosch 2014, 12.
79. See Harlow 2004, 205 for the methodology on interpreting female dress in Late Antiquity.
80. Harlow & Nosch 2014, 8, 11.
81. Harlow & Nosch 2014, 9.
82. Lee 2004, 221, 224.
83. This might indicate the existence of a public and private wardrobe, Harlow & Nosch 2014, 12.
84. Lee 2012b, 181.
85. Harlow & Nosch 2014, 12.

Part II

Dedications of textiles and accessories in Greek sanctuaries

Part II

Chapter 1

Introduction: Textile dedications

The Ancient Greeks sought to interact or communicate with the gods and goddesses in order to enter into and sustain a good personal relationship with them. This could be achieved in three main ways: via the dedication of votive offerings, the performance of sacrifices, and choruses/prayers.[1] Dedication and sacrifice were both a form of gift, but, whereas the sacrifice went up to the gods as smoke or was consumed by the community, a dedication survived as physical object in the sanctuary.[2] Or, in the formulation of van Straten, sacrifices are occasions when the offered object is intended for consumption (human or divine), while votive offerings are basically durable.[3]

A votive offering can be defined as a gift for a deity deposited in a sacred place.

Virtually any object could be used as a dedication, and, in many respects, an account of Greek offerings is also an account of all Greek arts and crafts:[4] bronze and terracotta figurines, reliefs, statues and sculptures, *pinakes*, jewellery, wreaths, war trophies such as arms and armour, tripods and cauldrons, sickles, strigils, athletic equipment, toys, money, clay models of body parts, miniatures, furniture and vessels, and of course textiles.[5] Organic and degradable items could also function as votive offerings, and we know of the dedication of flowers, fruits, food, cakes, bread, etc. It was thus not the type of object in itself that made it suitable as a votive offering. Offerings generally fall into two categories: purpose-made and secular, the latter being used in non-sacred contexts before their dedication.[6] It is, however, often difficult to discern between the two, since determining whether an object has been in use before ending up in a sanctuary can prove to be a challenge - except in the case of e.g. terracotta figurines, which appear to have been made specifically for dedication.

A further problem is that it can be difficult, if not impossible, to distinguish between votive offerings dedicated to a deity and ritual equipment. One way around this problem is to consider everything in a sanctuary to be an offering to the deity, including the temple and the cult image as well as ritual equipment.[7]

Since the textiles have disappeared from the archaeological record, they are usually not included in scholarly studies, and if so, only briefly mentioned. For example, among the scholarly works on votive offerings and their significance is Rouse's *Greek Votive Offerings* (1922), which aims to 'collect and classify those offerings which are not immediately perishable.'[8] Another is Baumbach, who provides a thorough study of a large number and variety of votive offerings in these specific sanctuaries, with a focus on preserved archaeological artefacts.[9] Thus, no one has so far attempted to conduct a more thorough study of the presence and use of textiles as votive offerings in Greek sanctuaries. Furthermore, scholarly research on votive offerings tends to focus either on literary and epigraphic aspects (for example, dedicatory inscriptions)[10] or on a specific type of votive object (such as figurines) – the latter type of study often focusing on typological or art-historical aspects, and often comprising a separate part of a publication on a specific sanctuary.[11] Only rarely are the two merged.[12] Generally, votive offerings thus appear to have been analysed primarily from an art historical perspective, or in order to perform typologies, while the study of Greek religion traditionally has focused on literary and epigraphic evidence.[13]

In the following section, different sources attesting to the use of textiles as votive offerings – both written and archaeological – will be examined to shed light on this practice. First, iconographic evidence for the use of textiles as votive offerings will be investigated; second, written evidence consisting of the temple inventories will be analysed and compared with archaeological evidence and textile research. Literary evidence will be included where relevant. Finally, the implications of this evidence for our knowledge of Greek votive practices, textiles, and women's roles in these practices will be discussed.

Iconographic evidence for the dedication of textiles

Although not preserved in the archaeological record, the use of textiles as votive offerings is attested in iconographic sources. The most famous example of the collective dedication of a garment is the offering of a *peplos* to Athena on the Athenian Acropolis, described in both epigraphic and literary sources and depicted on the east frieze of the Parthenon.[14] Besides the example from the Parthenon frieze, however, iconographic depictions of the offering of textiles in Greek sanctuaries are extremely rare. There are, though, a few exceptions, which will be examined below.

One possible example of a depiction of a textile dedication is a Boeotian relief *pithos* dated to the 7th century BC (Fig. 1).[15] On the shoulder of the *pithos* is a scene of five mounted warriors riding in procession towards the right, but of particular interest is the scene on the neck of the *pithos*, which depicts five women walking in procession towards the left. All five women are clad in garments that reach to their feet and are decorated with circular, rosette-like patterns; the woman leading the procession also wears a decorated mantle and carries a long staff – possibly an indication of her status as a priestess. The vase is fragmented, but at least two women each carry a square object with exactly the same patterns as those on their own garments, which makes the conclusion that the objects they carry are folded textiles attractive. It has been suggested that the scene represent a mythical or epic scene of Hekabe or the priestess Theano and the Trojan women carrying textiles to be dedicated to Athena,[16] although the Homeric interpretation is now rarely advanced.[17] Perhaps the scene rather depicts women walking towards a sanctuary with offerings of textiles – but, of course, the textiles might also be wedding or burial gifts.

1. Introduction: Textile dedications

Fig. 1. Boeotian relief pithos, 7th century BC. © Museum of Fine Arts, Boston, inv. no. 99.506.

Another example is a votive relief from Echinos, Central Greece (Fig. 2). The relief is relatively large (1.21 m long and 68 cm high) and has been dated to the end of the 4th century BC.[18] The relief is framed by two pillars and a peristyle, indicating that the scene takes place inside a temple. To the far right stands a tall goddess with a long torch in her right hand, and before her is a small altar, in front of which is a small male figure holding a sacrificial victim. Behind him stands a young woman holding out an infant towards the goddess, and behind her is a smaller woman or girl. Finally, to the far left stands a tall woman with a mantle over her head. But, most interestingly, on the wall behind the worshippers hang several clothing items – votive offerings to the goddess residing in the temple. It is important to note, however, that these clothing items are already in the temple and are not a focal point of the scene, but are rather props serving perhaps to indicate the nature of the deity in question. Dakoronia and Gounaropoulou identify the textiles (from left to right) as a pair of shoes, 'a short garment with short sleeves', two bed sheets, and a belted *peplos* with *apoptygma*.[19] They are doubtlessly right in their identification of the shoes and the short sleeved garment, which might be a so-called *chitōniskos cheiridotos* (sleeved *chitōniskos*). I have doubts, however, about their identification of the two bed sheets. These items, in fact, represent two rectangular pieces of textile decorated with fringes at the border. There is nothing to indicate that these are actually sheets, and they are more likely to be garments, perhaps mantles or veils, or possibly a *kalasiris*, a garment of Egyptian type with fringes at the hem. Next to the rectangular fringed textiles is a long ribbon, perhaps a belt or girdle. The textile to the far left is more difficult to identify, and Dakoronia's and Gounaropoulou's identification of it as a *peplos* with overfold is rather unsubstantiated. At present, it can only be defined as an unknown textile type with long fringes.

The relief is usually interpreted as a thank offering for a successful child birth and for the continued protection of the child.[20] Some scholars identify the woman to the far left as the mother of the newborn child, who is being presented by the younger woman at the altar,[21] while others identify the mother as the younger woman holding the child and the woman to the left as either the mother or the mother-in-law of the woman who has given birth.[22] According to Dillon, the goddess should be identified as Artemis, in part because of the garments hanging in the background and the torch, but in part because it appears that she originally held a quiver of bronze (now missing) in her left hand.[23] Dakoronia and Gounaropoulou, on the other hand, suggest that the lost object in her hand was a second torch, an arrow, or a poppy head. Due to the torch and the general appearance of the goddess, the occasion for the offering, and the garments in the background, they conclude, like Dillon, that the goddess depicted is Artemis.[24] Dakoronia and Gounaropoulou further consider this relief evidence for and a source of information about the Artemis cult in Central Greece. The authors themselves concede that there is no solid evidence for the veneration of Artemis in this part of Central Greece, but they argue that this gap is mitigated by this particular relief, which depicts Artemis in her function as Eileithyia and *kourotrophos*.[25] Although this might very well be the case, it is also possible that the goddess in question is Demeter, since an inventory

Fig. 2. Votive relief from Echinos, end of the 4th century BC. Archaeological Museum of Lamia, inv. no. AE 1041.© Hellenic Ministry of Culture and Sports/Archaeological Receipts Fund.

from Tanagra testifies to the dedication of garments in her temple, and she is often depicted carrying a torch. The veneration of this goddess in Boeotia is attested in several sources, written and archaeological.[26]

Another example of iconographic depictions of the dedication and ritual use of textiles is a group of votive terracottas from the Doric Greek colony of Lokroi Epizephyrioi in Southern Italy. This city developed a distinctive pantheon with Persephone and Aphrodite as the key deities. Demeter was worshipped there, too, in a typical Thesmophorion, but her role was overshadowed by that of her daughter. According to Larson, the Persephone of Lokroi Epizephyrioi received the prenuptial

Fig. 3. Woman placing a folded textile in a chest. Pinax from Locris, ca. 500–450 BC. National Museum of Reggio Calabria, inv. no. 2866, 57414, 61251. © Soprintendenza Archeologica della Calabria.

sacrifices known as *proteleia* because she served many of the functions related to female maturation, marriage, and childbirth – roles that were primarily fulfilled by Artemis and Hera for the mainland Greeks.[27]

At the Sanctuary of Persephone, a large number of mould-made votive terracotta plaques or *pinakes* decorated in relief with different scenes was uncovered. The majority of the plaques measure ca. 26–28 cm square and date to the first half of the 5th century BC.[28] The plaques have holes and were placed in the sanctuary, either attached to the walls of the temple, hung from trees in the sanctuary, or possibly even attached to cult images.[29] The main types include: scenes of Persephone's abduction by Hades; wedding ritual scenes, such as libations, processions, and women packing wedding gifts; an enthroned goddess receiving visitors or worshippers; and scenes with Aphrodite sometimes accompanied by Hermes.[30] MacLachlan argues that the plaques clearly portray cult activities and mythical narratives, as is evident from the presence of libation vessels, fillets, etc.[31] Some of the scenes are of special interest, since they involve textiles. According to Prückner, three groups of scenes involve textiles: the storage of textiles, the transport of textiles, and the dedication of textiles.[32] The scenes of textile storage depict a woman placing a folded textile in a chest (Fig. 3).[33] These scenes take place indoors, since there is furniture, and different items – such as *kalathoi*, mirrors, *kantharoi*, and *lekythoi* – can be seen hanging on the wall. The scenery has been interpreted as a sanctuary, although it could also be a representation of the female quarters of a private household. The ritual nature of the event depicted, however, is clear.[34] The scenes of textile transport can be divided into two main types: women carrying a folded textile on a sort of tray, balanced on the head ('Gewandprozession') (Fig. 4); and processions interpreted as *peplophoria* (Fig. 5).[35] The *peplophoria* scenes consist of about four women walking in procession, carrying a large piece of unfolded spotted textile. The procession is either led or followed by a differently dressed woman, with a mantle covering her hair, who carries a lustration bowl and an aspersion rod. She is usually interpreted as a priestess.[36] In rare instances, the robe is not carried by women, but by men.[37] The scene could either be a ritual honouring a goddess or it could be a procession connected with a marriage ceremony.[38] Occasionally, the women carrying folded textiles on their heads are also accompanied by a priestess with bowl and aspersion rod. The last type of scene related to the textile dedications depicts a woman standing in front of a seated goddess (Fig. 6). In between the two figures is a table with offerings, on which a folded piece of textile has been placed.[39] It has been suggested that the scene depicts Persephone receiving the *peplos* during the *peplophoria*.[40] It is usually assumed that the textiles in these scenes, whether folded or not, are *peploi*. In fact, though, we cannot determine this with certainty; we cannot even conclude that they are garments, only that they are textiles. Nevertheless, whether *peploi*, other garment types, or simply large pieces of textile, these votive plaques testify to the ritual use of textiles and their dedication in sanctuaries.

This raises the question of the receiving goddess' identity. According to Prückner, the goddess is Aphrodite, but others argue that the goddess is Persephone.[41] As

Fig. 4. 'Gewandprozession'. Pinax *from Locris, ca. 500–450 BC. National Museum of Reggio Calabria, inv. nos. 57416, 57419, 57421. © Soprintendenza Archeologica della Calabria.*

MacLachlan argues, though, deciding whether it is one goddess or the other exclusively is missing the point, since they might overlap and both represent marital or nuptial aspects.[42] In fact, it seems that both goddesses can be identified on the plaques: in some scenes Persephone is depicted, e.g. in scenes where she is enthroned, sometimes together with Hades, and in scenes depicting her rape by Hades; and in other scenes Aphrodite is depicted, e.g. in scenes representing her birth.[43] It does seem, however, that the scenes including textiles primarily involve Persephone. No scholar has yet considered the possibility that the goddess might be Demeter. As we shall see in the later chapters, textiles and garments appear to have been important in the cult and rituals of this particular deity.

It may appear initially that these four examples of iconographic representations of textile dedications provide a small sample. As compared with identifiable

Fig. 5. Peplophoria. Pinax *from Locris, ca. 500-450 BC. National Museum of Reggio Calabria, inv. no. 57482. © Soprintendenza Archeologica della Calabria.*

representations of other votive gifts, though, this does not seem so. How often are we actually capable of identifying a votive offering of e.g. a statue? A weapon? Jewellery? Generally, these scenes are difficult to identify in iconography. Among the exceptions are, for instance, the *korai*, who present small votive gifts, such as wreaths, birds, or pomegranates, but even so none of these are depicted bearing more 'material' offerings, only degradable objects. This is also the case for terracotta figurines, which often hold a small animal or ritual equipment. Generally, sacrificial scenes are recognisable in vase-painting and on the famous wooden plate from Pitsa, and thus it is possible to identify offerings in the form of sacrificial animals, as well as offerings of food items, such as cakes.[44] Possibly this is connected with the fact that the objects bearing these scenes – such as the *pinakes* from Lokris – are votive offerings in themselves. The depictions thereupon can thus be considered examples of tautology: a gift featuring a depiction of another gift.

We should be careful, however, in interpreting these depictions as valid and true representations of ritual and specific votive practices. As Gaifman argues, dedications featuring images of rituals and dedicatory inscriptions are expressions of an individual's piety rather than reflections of actual cult practices.[45] She argues,

Fig. 6. Dedication of textiles, woman standing in front of a seated goddess. Pinax from Locris, ca. 500-450 BC. National Museum of Reggio Calabria, inv. no. 28272, 57798. © Soprintendenza Archeologica della Calabria.

further, that dedications, such as votive reliefs, are visual constructs related to cultic realities, but not direct reflections of actual rituals. Instead, they illustrate a selection of elements that can be recognised from reality as well as imagined modes of interactions between humans and divinities (e.g. epiphanies).[46]

The scenes examined here can be divided into two overall types: textiles as votive offerings, as on the relief from Echinos and some of the *pinakes* from Lokris, and processions with textiles, as on the Boeotian *pithos* and some of the *pinakes* from Lokris. None of the examples fits with the more or less standardised depictions of offering scenes on e.g. votive reliefs, which depict processions of dedicants carrying offerings and leading a sacrificial animal towards a deity, in this way blending sacrifice (the depicted animal) and dedication (the dedicated object – the relief). According to Gaifman, such a dedication may have served as a substitute for the performance of a real sacrifice, but it may also have accompanied it.[47] This is also the case for these depictions, which may not refer to a 'real' offering, but instead may be intended to recall a previous offering or, possibly, to represent a generic ritual scene. These depictions, even though few in number, furthermore clearly establish that textiles did play a part in votive practices, and that these offerings primarily – if not exclusively – involved women and female deities.

Notes

1. Van Straten 1981, 65.
2. *ThesCra* I, 270.
3. Van Straten 1981, 66.
4. *ThesCra* I, 282.
5. See *ThesCra* I, 283–316.
6. Baumbach 2009, 206.
7. *ThesCra* I, 282.
8. Rouse 1902, 1.
9. Baumbach does include textiles and clothing, although only briefly.
10. E.g. Jim 2012.
11. Terracotta figurines published as separate binds in sanctuary publications: e.g. Merker 2000 (sanctuary of Demeter, Corinth); Huysecom-Haxi 2009 (Artemision, Thasos).
12. E.g. Gaifman 2008.
13. Baumbach 2009, 205.
14. This example is examined in Appendix 1.
15. Museum of Fine Arts, Boston, inv. no. 99.506.
16. *LIMC* Hekabe no. 12; Caskey 1976, 33.
17. Burgess 2001, 226, note 162.
18. Dillon 2002, 231. Archaeological Museum of Lamia, inv. no. AE 1041.
19. Dakoronia & Gounaropoulou 1992, 223.
20. Dakoronia & Gounaropoulou 1992, 218.
21. Dillon 2002, 232.
22. Dakoronia & Gounaropoulou 1992, 222.
23. Dillon 2002, 232–233.
24. Dakoronia & Gounaropoulou 1992, 218.
25. Dakoronia & Gounaropoulou 1992, 224.
26. E.g. Paus. 9.3.4; 9.4.3; 9.16.5–6; 9.25.5–6. Hdt. 9.57; 9.65; 9.69. *SNG* Copenhagen 380, 383. *BMC* 77.
27. Larson 2007, 83.
28. Dillon 2002, 222.
29. Dillon 2002, 222; Larson 2007, 84; see also *ThesCra* I, 293–295.
30. Larson 2007, 84.
31. MacLachlan 1995, 218.
32. Prückner 1968, 39.
33. Prückner type 14–15.
34. MacLachlan 1995, 214.
35. Prückner types 16–28.
36. Dillon 2002, 227; Prückner 1968, 42; MacLachlan 1995, 213.
37. Dillon 2002, 227.
38. MacLachlan 1995, 213.

39. Prückner type 29.
40. MacLachlan 1995, 215.
41. Dillon 2002, 225–227; Prückner 1968, 45, 46.
42. MacLachlan 1995, 218.
43. MacLachlan 1995, 211–217.
44. Examples include the *Lēnaia* vases depicting cakes for Dionysos. See Chapter 5.
45. Gaifman 2008, 85, 86.
46. Gaifman 2008, 99.
47. *ThesCra* I, 279.

Chapter 2

The temple inventories: Written evidence for the dedication of textiles and accessories

The study of textile dedications in sanctuaries presents obvious challenges, inasmuch as the textiles themselves have disappeared today, and one must therefore seek other sources of information. One such potential source is literary and epigraphic evidence. For example, Homer describes the offering of a garment to a deity in the *Iliad*, when Hecabe goes to the Temple of Athena in order to give the goddess the largest and most beautiful *peplos*, which she places on the knees of the cult statue.[1]

As far as the epigraphic record is concerned, the so-called temple inventories, in particular, have great potential to increase our knowledge of the dedication of textiles to various deities. Throughout the Greek world, the sacred property of the gods was kept in treasuries, either in a separate building or in a part of the temple reserved for the storage of these offerings.[2] Although the Greek temple treasures are no longer directly accessible to us, we still have evidence from which to reconstruct them in the form of temple inventories inscribed on stone *stelai* that suggest their possible extent, monetary value, and arrangement.[3] These temple inventories reflect what was perceived as the most important objects, and some confirm the presence of textiles, their type, fibres, colours, and decoration, and the gods to whom they were dedicated. So even though the textiles and garments donated at the sanctuaries have physically disappeared, they occasionally survive in the lists preserved on stone.[4]

The temple inventories

The term 'temple inventories' generally refers to lists of votive offerings that were kept in the temple treasuries. They have been found over large parts of the Greek world (Map 1), and generally span a rather long period, from the 5th century BC to the 2nd century AD,[5] but the large majority of inventories with references to textiles belong to the Hellenistic Period, while a few can be dated to the late Classical Period. It is important to note that written administration and record keeping had a strong

Sites with temple inventories recording textiles and/or dress-fasteners.
textiles dress-fasteners textiles and dress-fasteners

1. Athens
2. Brauron
3. Tanagra
4. Thebes
5. Delos

6. Heraion Samos
7. Miletos
8. Oia, Aegina
9. Perga, Pamphylia

tradition in Ancient Greece and was neither an invention nor a prerogative of the sanctuaries.

Most temple inventory lists are single finds, and, apart from Athens and Delos, few Greek cities seem to have regularly published their temple records on stone.[6] The absence of inventories among the epigraphic evidence does not necessarily mean, however, that they did not exist, but can possibly be due to the chance of survival. Perhaps records were also kept on perishable materials such as wood or parchment or even textiles,[7] and not inscribed on stone.[8]

2. The temple inventories

Most inventories appear to list the offerings in the sanctuary in a kind of topographical order, building for building and room for room. Yet it seems unlikely that the treasurers, when preparing the inventory, checked the items against previous lists. If this had been the case, the lists of two consecutive years ought to follow each other closely, but the lists are usually not bound by prior ones as to arrangement, vocabulary, or selection of items. Furthermore, the narrative form of many of these texts, with all entries following in a continuous flow, often of 200 lines or more, prevents consultation for verification purposes. According to Linders, their purpose is therefore not to serve as a practical instrument for checking temple treasures and individual votive gifts;[9] instead, the majority of inventories are primarily records of *paradoseis* – transactions in which (an) outgoing official(s) personally inspected and handed over the treasures to his successors.[10] She argues that *temple* inventories and, to a lesser extent, *temple* accounts reflect the personal encounters of the administrators with other officials, especially members of the Council, during the *paradosis* procedure at the end of their term, rather than being continuous and regular book-keeping records. Besides the *paradosis* inventories, two other types of procedures, *katairesis* and *exetasmos*, are recorded in the inventories. *Katairesis* was the listing of metal objects being removed from the sanctuary to be melted and re-cast into new objects for the sanctuary,[11] while *exetasmos* was a kind of total inventory, a list of all items in a building or sanctuary.[12] Linders stress the symbolic purpose of the inventories, since they were dedications to the gods in themselves and testified to the piety of the officials and/or donors and of the city.[13] It is likely, however, that they served a dual purpose, practical and symbolic, and, as Scott emphasises, one should not distinguish between a functional and symbolic role, since the difference between the two in this instance is unclear, and the lists must have provided some usable information.[14] Yet a third purpose of the inventories could have been to encourage a certain form of behaviour, since the inventory lists create normative patterns of dedicatory practice and thus encourage people to behave and dedicate in a specific way similar to that of previous visitors.[15] Also, the motivation of the benefactors to have their names recorded and publicly displayed is another important factor when discussing the purpose of the inventories. Thus, further offerings were encouraged, which would increase the property of the sanctuary.[16]

In the following section, temple inventories testifying to the presence of textiles in the sanctuaries will be examined in order to shed light on the use of textiles as votive gifts. The inventories themselves will be only briefly presented; for more thorough studies, I refer to the respective epigraphic publications.[17] Subsequently, the evidence will be discussed thematically, and relevant source material from archaeology, textile research, and literary sources will be included in order to shed light on the nature of the dedicated textiles.

Temple inventories recording textiles

The Brauron catalogues[18]
The inventories from the Sanctuary of Artemis Brauronia on the Athenian Acropolis are among the central documents for the study of Greek clothing and its use as offerings.[19] This centrality is justified by their sheer size and unusual exclusivity, and the clothing section of the votive records is the single most significant body of inscriptional evidence for Greek clothing of the Late Classical Period.[20] The inventories consist of 13 fragments originally belonging to six separate *stelai*, and record offerings made to Artemis. According to the traditional interpretation of these lists, they record dedications made at the Sanctuary of Brauron, and it has been argued that the same records were inscribed and displayed both at Brauron and at Athens.[21] Exactly equivalent inscriptions found in the sanctuary complex at Brauron seem to establish this beyond reasonable doubt.[22] Thus, the fragments published in *IG* II² come from duplicate *stelai* set up on the Athenian Acropolis.[23] The date is easily established, since the lists are organised by year of dedication. As preserved, they relate to dedications made in the years 349/8, 348/7, 347/6, 346/5, 345/4, 344/3, 343/2, 342/1, 338/7, 337/6, and 336/5.[24] The inventory of votive gifts was begun long before the 4th century BC, however, and continued as long as offerings were dedicated to Artemis Brauronia.[25] These inventories also record metal objects, which are not however included in this study. Long lists of textile items are recorded afterwards, and the two are thus kept separate.

The inventories of the Treasurers of Athena[26]
Inventories were also kept in the Parthenon – a building that was not a place for congregational worship, but rather a house for Athena and a location for her possessions to be kept.[27] The building was dedicated in 438/7 BC, and yearly inventory lists were produced from 434/3 BC until the end of the 4th century BC.[28] The lists record the official act of one board of treasurers passing the responsibility for the contents of the temple along to their successors. This passing along (*paradoseis*) appears to have taken place at the end of the Panathenaia, after objects carried in the Panathenaic procession were returned to the temple.[29] The objects are divided up on the lists by their location. For the first 40 years of the inventories, separate *stelai* were used for each of the main three rooms of the Parthenon: the Pronaos, Hekatompedon, and Parthenon; these were later consolidated into one single stele.[30] According to Harris, the inventories of the 5th century BC indicate that there were four separate rooms that contained the treasures of Athena: the Pronaos, Hekatompedon, Parthenon, and Opisthodomos. The location of these rooms has been a subject of debate for well over a century, but, according to Harris, these terms refer to the four rooms of the temple that we today call the Parthenon: the Pronaos is the east-end portico, the Hekatompedon the eastern cella with the statue, the Parthenon the western room, and the Opisthodomos was the western portico.[31] The terms 'Hekatompedon'

and 'Parthenon', however, were also used for the building as a whole.[32] The rooms appear to have contained different objects: the Pronaos seems to have contained only vessels; the Opisthodomos, which seems to have functioned more as a bank, held mostly undecorated gold and silver vessels; and the Hekatompedon and the Parthenon contained more varied objects.[33]

The inventories of the Parthenon temple (not just the single room) include not only various items such as arms and weapons, containers, coins, figurines and statues, furniture, jewellery, vessels, wreaths (the most numerous category), but also items of clothing: a garment called a *xystis* dedicated by a man named Pharnabazos,[34] a coarse linen *chitōn* (*chitōn stuppinos*),[35] a purple dyed *chitōniskos*,[36] something made of fine cloth (*sindōn*),[37] corselets/cuirasses (*thōrakes*),[38] belts/girdles (*zōnia*),[39] as well as an *apoptygma*,[40] which is somewhat strange, since this is the overfold of a garment, and in principle it should not be possible to separate it. Two boxes with footwear are also recorded,[41] a purple *mitra*,[42] and an inventory from 320 BC lists *himatia*, *chitōniskoi*, *chitōnes*, *kekryphaloi*, and grey and white wool in baskets.[43] These items are not recorded together as a list of garments, but rather individually among other objects, such as jewellery.

Inventory of a temple at Tanagra (Appendix 2.1)
A marble stele from Tanagra provides further information about the offering of textiles in sanctuaries. The stele has inscriptions on both sides.[44] Side A bears an inscription with a sacred regulation concerning a sanctuary of Demeter and Kore at Tanagra. Nevertheless, in view of the similarities between this inventory and the inventories of Artemis Brauronia, Schachter suggests assigning the texts on both sides of the stele to the Sanctuary of Artemis Aulideia rather than that of Demeter and Kore, an attribution he bases on the fact that there are so far no other attested inventory list of textiles dedicated to Demeter and Kore.[45] Since, however, side A clearly states that the city consulted the oracle concerning the Sanctuary of Demeter and Kore, it is reasonable to assume that side B also relates to this sanctuary. Furthermore, there are examples of inscriptions from sanctuaries of Demeter that regulate the dress and adornment of visitors, which indicates that garments were of importance to this goddess.[46] Below the regulation on side A is a list of women who contributed financially to the construction of the temple, while side B carries an inventory of individuals (primarily women) who donated garments to the sanctuary. Every entry records the name of the donor and the type of garment given. The two sides are not of the same date, but both inscriptions can probably be dated to ca. 230–200 BC.[47]

Inventory of the possessions of an unknown deity at Thebes[48] (Appendix 2.2)
A marble slab from the first half or middle of the 3rd century BC, probably from the Sanctuary of the Kabeiroi located 8 km west of Thebes, bears an inscription with an inventory list.[49] The inscription is fragmented, and the right and bottom parts are missing.[50] The inventory records garments dedicated to a deity whose name is

unfortunately not preserved, but according to Günther it could possibly be Artemis Eukleia, who was venerated in Boeotia.[51] The inscription is not long, but records several garments donated in the sanctuary by women.

The Delian records of the hieropoioi and of the Athenian temple-administrators[52]
Extraordinarily rich epigraphic evidence has been recovered at Delos. Around 2800 inscriptions have been recovered on the island, among which about half are either an account, or a year's list of expenditures and income of the *hieropoioi*, or an inventory of one of the island's many temples.[53] The temple accounts begin in 433 BC and continue until shortly after 166 BC,[54] while the inventories of votive gifts cover the period ca. 367 to the 130s.[55] The Delian history and inventories are conventionally divided into three chronological phases: (Athenian) Amphictyonic, Independence, and Athenian.[56] The Delian Period of Independence is the only of the three historically defined periods with firm dates (314–166 BC). During this time, the island was free of Athenian control, and the *hieropoioi* were the chief religious officials. These elected *hieropoioi* were in charge of all the sanctuaries and responsible for all their contents, and the accounts and inventories are the records of their activity.[57] In the third period (from 166 BC), Athens regained control of the island, and the sanctuaries were now administered by Athenian temple administrators.[58]

The island of Delos had temples dedicated to Apollo, Leto (the Letoon), Artemis, Hera (the Heraion), Zeus, Athena, Herakles, and Asklepios, as well as a temple to the Olympian gods (the Dodekatheon) and to the Egyptian gods Isis and Serapis, many of which appear to have included textiles in their inventories. Thus, the textiles recorded in the Delian inventories do not form a long, continuous list, as at for example Brauron or Tanagra, but are recorded separately according to their presence in this or that sanctuary, e.g. in the Artemision or the Iseion. They are often recorded among other precious objects, such as jewellery. This gives the impression that the textiles were scattered here and there, and not gathered in the same place.[59] In other instances, the textiles are described as being used to dress the cult images.[60]

The temple inventory of Samos[61] *(Appendix 2.3)*
A marble stele, which can be dated to 346/45 BC, bearing an inscription with an inventory of the Sanctuary of Hera, was recovered from the Heraion on Samos.[62] The inventory demonstrates that the statue of Hera was dressed in costly raiments, and numerous ceremonial garments and materials are listed as 'belonging to' the goddess under the rubric 'jewellery of the goddess'. Hera was originally the sole deity of the sanctuary, but later other gods appeared on the scene. For example, the inventory lists two Hermes statues, one of which is thought to have stood in the Temple of Aphrodite. Furthermore, the multitude of buildings (such as temples and treasuries) that have been excavated suggests that there were other gods in the sanctuary, but precisely which of the existing temples belonged what god is still unresolved. The inventory gives a long list of different textiles donated in the sanctuary. In contrast

to, for example, the inventory from Tanagra, there is no record of gender for either garments or donors, and only one donor is named – a man by the name Diogenes.

Inventory of the Temple of Artemis Kithōnē at Miletos[63] *(Appendix 2.4)*
An inscription from Miletos provides an inventory of textiles stored in a temple that was perhaps dedicated to Artemis Kithōnē ('Artemis clothed-in-a-tunic').[64] There is, however, uncertainty regarding the goddess for whom the dedications were meant. Only about half of the white marble stele has survived intact.[65] The top and bottom of the stele are broken as a result of a later reuse of the stone. Accordingly, an important part of the inventory is missing – most importantly, the name of the deity and sanctuary in question. Judging from the writing, the inscription can be dated to the 2nd century BC.[66] The first five preserved lines of the inventory contain the end of a catalogue of metal objects. The remaining part of the inscription (lines 5–24) contains a long list of textile products, methodically recorded: first larger clothing items are listed, then pieces of fabric of smaller dimensions, and lastly clothing for children. The individual pieces vary greatly from one another, and are classified and recorded according to the type and quality of the fabric. The detailed descriptions illustrate different garments, their cut, colour, and decoration, and the inventory is thus of inestimable value for the study of Ancient Greek costume and costume terms.[67]

Inventories with only a few records of garments
Not all temple inventories record textiles, and some include just a very few garments. Among the lists with only a few records of textiles is the Lindian Temple Chronicle,[68] a monumental inscription erected in 99 BC to commemorate the legendary treasures and the four epiphanies of Athena Lindia on the island of Rhodes.[69] Among the gifts are examples of garments: the fine linen cuirass (*thō[rak]a lineon*), each thread of which has 360 strands,[70] donated by the Egyptian king Amasis,[71] the pair of trousers (*anaxyridas*) presented by the Persian general Darius,[72] and the royal robe (*basilikan stolan*) bequeathed by the demos.[73]

Another example is the ca. 200 inscriptions from the Asklepieion in Athens, ranging in date from ca. 400 BC to ca. 250–300 AD.[74] There are nine inventories preserved, which record (fully or partially) 1347 dedications to the sanctuary,[75] but only one of them records garments.[76] This particular inventory is dated to 329/8 BC, and has some records of garments and clothing items, such as a *chlamys*,[77] a grey *chlamys*,[78] three pairs of women's sandals (*hypodēmatōn gynaike*),[79] and a broken sequence recording the dedication of clothing or similar soft goods, followed by a phrase stating that it was ornamented with silver.[80]

Inventories recording dress-fasteners
Other inventories do not record textiles, but instead different types of dress-fasteners (Table 1). Different terms are employed to denote these, especially *porpē* (plur. *porpai*),

peronē (plur. *peronai*), and *peronis*. The word *porpē* derives from the verb *peirō* – to pierce[81] and is usually translated as 'brooch' or 'clasp for fastening dresses'[82] and commonly interpreted as denoting a fibula.[83] It is often semantically confused with the term *peronē*. The term *peronē*,[84] as well as the much rarer *peronis*,[85] are also primary derivatives of the verb *peirō*.[86] It is normally translated as 'pin', 'brooch' or 'buckle', but can also mean 'rivet', 'pivot' or 'a small arm bone' (radius).[87] Prêtre suggests a translation as 'pin' (fr. *épingle*).[88] Deonna emphasises a distinction between *peronē* – pin (fr. *épingle*) and *porpē* – fibula.[89] Modern Greek provides a clear distinction between *peronē*, a pin that is simply attached to a garment, and *porpē*, a fibula which is used to fasten garment.[90] This implies that *peronai* were of a more decorative character, while the *porpai* were of a more functional character. Yet connecting the term with an exact type of archaeological artefact is difficult, and one should obviously be wary of considering *peronai* to be solely decorative and *porpai* to be merely functional. This confusion of the terms is often reflected in scholarly literature, where the two words are sometimes translated as 'fibula' and sometimes as 'pin', 'clasp' or 'brooch', yet the logic behind these different translations is seldom explained. Especially 'clasp' and 'brooch' are very vague and appear to be used as more generic terms. This confusion emphasises all the more the importance of employing the Ancient Greek terms in order to avoid confusion and preconceptions. Yet perhaps this confusion in the ancient sources concerning what is a pin and what is a fibula reflects the multiple functions of the objects, and it may also indicate that it was not of great importance for the ancient author or reader to specify whether it was a pin or a fibula, but only to note that a metal dress-fastener was used. This inference is supported by the fact that a passage from Pollux is the only ancient text in which *peronē* and *porpē* are distinguished.[91]

Inventory of the Temple of Mnia (Damia) and Auzesia (Auxesia) at Oia, Aegina[92]
The inventory of the Temple of the goddesses Mnia and Auzesia records many dedicated dress-fasteners. Mnia and Auzesia were the Aiginetan titles for the goddesses Damia and Auxesia, which were the names used by Pausanias in relation to their Troizenian cult.[93] They were deities of fertility, and possibly associated with Demeter and Kore/Persephone.[94]

The temple of the two goddesses has not been discovered, but Herodotos locates it in the middle of the island of Aegina ca. 20 stades from a town called Oia.[95] He also mentions the tradition of dedicating dress-fasteners in the sanctuary:

> 'As for the Argives and Aeginetans, this was the reason of their passing a law in both their countries that brooch-pins (*peronai*) should be made half as long as they used to be and that brooches (*peronai*) should be the principal things offered by women in the shrines of these two goddesses.'[96]

Godley's translation clearly illustrates the confusion over the connection of these different terms with archaeological material that was touched upon above, since he translates the same term (*peronai*) as brooch-pins as well as brooches.

2. The temple inventories

Table 1. Dress-fasteners

A. Dress-fasteners

Greek term	Translation	Material	No.	Site	Reference	Date
peronai	dress-fasteners		2	Athens	IG II², 1388, line 20	398/7 BC
peronai	dress-fasteners			Athens	IG II², 1424a, line 55	369/8 BC
peronai	dress-fasteners			Athens	IG II², 1425, line 54	368/7 BC
peronai	dress-fasteners			Athens	IG II², 1425, line 58	368/7 BC
peronides	dress-fasteners			Athens	IG II², 1425, lines 60-61	368/7 BC
peronai	dress-fasteners			Athens	IG II², 1428, line 16	367/6 BC
peronides	dress-fasteners			Athens	IG II², 1428, line 22-23	367/6 BC
peronai	dress-fasteners		2	Athens	IG II², 1428, line 33	367/6 BC
peronai	dress-fasteners	bronze	2	Perge	IvPerge 11:99,2 =IK Perge 10, line 26	Hellenistic
peronai	dress-fasteners	gold	4	Perge	IvPerge 11:99,2 =IK Perge 10, lines 28-29	Hellenistic
peronai	dress-fasteners	iron	120	Oia	IG IV, 1588, lines 10-11	5th century
peronai	dress-fasteners	iron	5	Oia	IG IV, 1588, lines 12-13	5th century
peronai	dress-fasteners	iron	6 (fragments)	Oia	IG IV, 1588, line 14	5th century
peronai	dress-fasteners	iron	22	Oia	IG IV, 1588, line 27	5th century
peronai	dress-fasteners	iron	180	Oia	IG IV, 1588, lines 35-36	5th century
peronai	dress-fasteners	iron	8	Oia	IG IV, 1588, lines 40-41	5th century
peronai	dress-fasteners	iron	5 (fragments)	Oia	IG IV, 1588, lines 42-43	5th century
porpa	dress-fastener	gold		Thebes	IG VII, 2420, lines 7-9	3rd century
porpa	dress-fastener	gold		Thebes	IG VII, 2420, lines 17-18	3rd century

B. Dress-fasteners, Delos

Greek term	Translation	Material	No.	Site	Reference	Date
peronē	dress-fastener	silver, gilded		Delos	IG XI,2, 161B, line 62	278 BC
peronē	dress-fastener	silver, gilded		Delos	IG XI,2, 162B, line 50	278 BC
porpē	dress-fastener	silver, gilded		Delos	ID 407, line 8	ca. 190 BC
porpē	dress-fastener	silver, gilded		Delos	ID 442B, line 198	179 BC
porpē	dress-fastener	silver, gilded		Delos	ID 443Bb, line 123	178 BC
porpē	dress-fastener	silver, gilded		Delos	ID 444B, line 42	177 BC
porpē	dress-fastener	silver, gilded		Delos	ID 461Bb, line 30	169 BC
porpē	dress-fastener	gold		Delos	ID 439, line 77	181 BC
porpē	dress-fastener	gold		Delos	ID 442B, line 85	179 BC
porpē	dress-fastener	gold		Delos	ID 443Bb, line 123	178 BC
porpē	dress-fastener	gold		Delos	ID 448B, line 7	175 BC
porpē	dress-fastener			Delos	ID 461Bb, line 47	169 BC
porpion	dress-fastener			Delos	ID 1416A (I), line 3	156/5 BC
porpion	dress-fastener			Delos	ID 1417 (II), line 162	155/4 BC
porpion	dress-fastener			Delos	ID 1452A, line 5	after 145 BC
porpion (restored)	dress-fastener			Delos	ID 1430a, line 4	153/2? BC
porpion (restored)	dress-fastener			Delos	ID 1439Abc (I), line 78	166–140/39 BC
porpion	dress-fastener			Delos	ID 1441A (I), line 86	ca. 150 BC
porpion (restored)	dress-fastener			Delos	ID 1450A, line 58	140–39 BC
porpas	dress-fastener	gold		Delos	ID 1428 (II), line 56	166 BC
porpas (restored)	dress-fastener	gold		Delos	ID 1429B (II), line 3	155/4? BC
porpas	dress-fastener	gold	2	Delos	ID 1449Ba (I), line 8	after 166 BC
porpas	dress-fastener	gold		Delos	ID 1433, line 5	153/2 BC
porpion	dress-fastener			Delos	ID 1442A, line 53	146/5–145/4 BC

2. The temple inventories

C. Dress-fasteners, Delos

Greek term	Translation	Material	No.	Site	Reference	Date
peronai	dress-fasteners	gold		Delos	ID 103, line 20	372/67–364/3 BC
peronai	dress-fasteners	gold		Delos	ID 104, line 80	364/3 BC
peronai	dress-fasteners	gold		Delos	ID 104–2B, line 2	434–315 BC
peronai	dress-fasteners			Delos	ID 104–11B, line 19	353/2–352/1 BC
peronai	dress-fasteners	gold		Delos	ID 104–12, line 56	353/2–352/1 BC
peronai	dress-fasteners			Delos	ID 104–21Ba, line 10	364/3? BC
peronai	dress-fasteners			Delos	ID 104, line 51	364/3 BC
peronai (restored)	dress-fasteners			Delos	ID 104–12, line 36	353/2–352/1 BC
peronai	dress-fasteners	gold		Delos	ID 104, line 146	364/3 BC
peronai (restored)	dress-fasteners	gold		Delos	ID 104–21Ba, line 6	364/3? BC
peronai	dress-fasteners	gold		Delos	IG XI,2, 161B, line 41	278 BC
peronai	dress-fasteners	gold		Delos	IG XI,2, 162B, line 33	278 BC
peronē	dress-fastener	silver, gilded		Delos	IG XI,2, 161B, line 62	278 BC
peronē	dress-fastener	silver, gilded		Delos	IG XI,2, 162B, line 50	278 BC
peronai	dress-fasteners			Delos	IG XI,2, 164A, line 65	276 BC
peronai	dress-fasteners			Delos	IG XI,2, 199B, line 45	273 BC
peronai	dress-fasteners			Delos	IG XI,2, 203B, line 64	269 BC
peronai (restored)	dress-fasteners			Delos	IG XI,2, 208, line 29	314–250 BC
peronai	dress-fasteners			Delos	IG XI,2, 223B, line 23	262 BC
peronai (restored)	dress-fasteners			Delos	IG XI,2, 247, line 13	265–255 BC
peronai	dress-fasteners			Delos	IG XI,2, 287B, line 23	250 BC
peronai	dress-fasteners			Delos	ID 300B, line 6	250–166 BC
peronai (restored)	dress-fasteners			Delos	ID 298A, line 159	240 BC
peronai	dress-fasteners	bronze		Delos	ID 298A, line 179	240 BC
peronai	dress-fasteners	bronze		Delos	ID 457, line 18	174? BC
peronai	dress-fasteners	gold		Delos	ID 399Ba, line 154	192 BC

An inscription preserves an inventory list from the sanctuary.[97] It can be dated to the 5th century BC,[98] and, according to Figueira, it is unlikely that the inscription can date much after 431;[99] Polinskaya, however, dates it to 431–404 BC.[100] The inscription records different items, but the dominant type of object is the dress-fasteners (*peronai*), of which there are 346, including fragments: 131 in the Sanctuary of Mnia and 193 in the Sanctuary of Auzesia. Scholars commonly interpret the *peronai* recorded in this inventory as pins – not fibulas.

The recorded groupings of dress-fasteners, comprising 5–180 specimens, are, according to Jacobsthal, not original dedicated units, but were formed in antiquity for the purpose of display and inventory.[101] According to Dunbabin, most of them were probably individual dedications, while a few were pinned to *peploi*.[102] A word of caution is necessary, however, since he bases this assumption on the wording of the inscription, which he interprets as indicating that the dress-fasteners, not the *peploi*, were most important.

With regard to placement in the temple, the inscription records that the back chamber (*hypisthodomos*) held 120 dress-fasteners and that next to the *peploi* were five complete dress-fasteners and six fragments. Line 26 mentions a door or entrance to the Temple of Auzesia, above which were placed 22 *peronai*. Line 28, referring to the Temple of Auzesia, records 180 arranged in a group by themselves. Furthermore, according to the inscription, eight whole and five broken dress-fasteners were present/placed in front of a *peplos*.[103] Jacobsthal suggests that there was a row of *peploi* displayed on the wall, and that the dress-fasteners were fixed near the first of them.[104] It seems that there are only few *peploi* in relation to the dress-fasteners recorded in the inventory, and the two clearly do not correspond in number. This has caused Polinskaya to conclude that the dress-fasteners must have been votive, and that the *peploi* are therefore 'best understood not as personal garments dedicated by female worshippers, but rather as ritually crafted offerings for a deity, perhaps presented and replaced on a periodic basis',[105] just as the *peplos* for Athena on the Athenian Acropolis. This is perhaps taking the interpretation too far, but what the inventory clearly shows is that dress-fasteners could be dedicated alone, without garments.

All the specimens are in iron. Iron pins became rare after the Proto-Geometric Age.[106] Jacobsthal therefore suggests that, if the 346 *peronai* in this sanctuary were indeed pins, the *peploi* and pins could thus be relics of a Proto-Geometric sanctuary, e.g. an *oikos* or a grove, and, when this was replaced by a temple, the votive offerings were not buried but transferred into the new building. Alternatively, they were never worn, but were contemporary to the inscription and made for the goddesses in an archaic material.[107]

The Delian inventories
Dress-fasteners are recorded in several of the Delian inventories: *porpai* are recorded in the inventories of the Artemision, the Temple of Apollo, the Sanctuary of the

Egyptian Gods, and in the Temple of Isis. In the Artemision, a *porpē* of gilded silver is recorded several times, while in the Sanctuary of the Egyptian Gods a *porpion*, probably a diminutive of *porpē*, is recorded, and in the Temple of Isis there is a *porpion* with a silver boss. In the Temple of Apollo there are several records of *porpai*. There is a *porpē* of gold, a *porpion*, and gold *porpas*. The *peronai* are almost exclusively recorded from the Artemision: several are specified to have been made of gold, but there is also one silver gilded example, and others are undescribed. Besides those in the Artemision, gold *peronai* are recorded in the Temple of the Delians, and the inventory of the Building of the Andrians records copper *peronai*. Thus, both *porpai* and *peronai* are recorded on Delos, but never in the same inventory. There seems to be a chronological difference, since *porpai* are recorded in lists dating to the 2nd century BC, while *peronai* are listed in earlier inventories dating to the 4th and 3rd centuries BC.[108] Yet it is difficult to determine whether this merely reflects a change in terminology or whether this implies a change in the type of dress-fastener.[109] According to Prêtre, it is only *porpai* in gold or silver that are recorded.[110] She suggests, therefore, a distinction between the pin *peronē*, which, according to Prêtre, can be equated with an ordinary object in ordinary metal, and the fibulas, *porpai*, which were jewellery intended exclusively for the divine garments.[111] Still, this is not an undisputed distinction, since there are also records of gold *peronai*.

Inventories with few records of dress-fasteners
As shown above, some inventories record many textiles, while others mention only very few; this is also the case for inventories recording dress-fasteners. For example, the inventories of the Athenian Hekatompedon record *peronai* as being present in the temple in 398/7, 369/8, 368/7, and 367/6 BC.[112] They are recorded either in association with an overfold (*apoptygma*)[113] or a corselet/cuirass,[114] indicating that dress-fasteners were used with these types of attire. IG II² 1425 records *peronai* just after two ribbons. Furthermore, an inventory of the Temple of Artemis Pergaea at Perga, Pamphylia,[115] dating to the Hellenistic Period, records a pair of bronze dress-fasteners (*peronai*), dedicated by Ra[..]gas, daughter of Erymneus from Aspendos (line 26), and four gold dress-fasteners (*peronai*) (line 28), but no textiles.[116]

A final example is the inventory of the Temple of the Kabeiroi/Kabiri near Thebes.[117] It dates to the 3rd century BC, and contains lists of three successive years, each headed by the names of the archon and the two priests of the Kabeiroi.[118] The inventory lists items of gold, silver, and jasper, but, according to Schachter, there must have been a much longer list of offerings, to which this inventory was attached, and a place to keep them.[119] There are two gold dress-fasteners (*porpai*) among the votive offerings, one of which was dedicated by two women. One of them is very light (the gold *porpa* of Autarxia has an estimated weight of 1.93 g, and the gold *porpa* dedicated by Danatria and Satyra has an estimated weight of 8.6 g), which indicates that at least the lightest one was possibly a token dedication.[120] These two gold dress-fasteners are, however, insignificant in comparison to the dominant offerings in the sanctuary.[121] If the recovered and recorded offerings are reliable indicators of the offerings given, there is thus not much to indicate that sacred textiles played any role at this sanctuary.

Conclusions on the dress-fasteners
Dress-fasteners are recorded in at least five inventories. Some list only a few, as for example the inventories of the Hekatompedon, the Sanctuary of the Theban Kabeiroi, and the Temple of Artemis Pergaea at Perga; but at Delos and especially the Sanctuary of Mnia and Damia many are recorded. Two terms are used to denote dress-fastener: *peronē* and *porpē*. Yet, as already discussed in connection with Delos, the two terms are never used together: *peronai* are recorded in lists dating to the 5th to the 3rd century BC, while *porpai* are recorded in lists dated to the Hellenistic Period, with the notable exception of the inventory from Perga, which records *peronai*, but is dated to the Hellenistic Period. There thus appears to be a chronological difference in the use of these terms. With regard to material, *porpai* and *peronai* are both described as being made of gold or gilded silver, while only *peronai* are described as of bronze or iron. One should be wary, however, of drawing the conclusion that *porpai* were generally of a more precious character, since the material in many cases is not specified.

It is interesting to note, though, that, despite the large numbers of garments recorded in the Brauron catalogues, there is no mention of dress-fasteners. This is also the case for the other inventories recording textiles. This could be an indication that these items not were dedicated together, that they had gone out of use/fashion, or that they belong to two different ritual/cultic practices or traditions (see Appendix 4).[122] But what does this mean in case of the inventories that do record dress-fasteners? Several possible explanations come to mind: one is that these lists record old pins and fibulas, dedicated a long time ago, but registered in the lists because they are still present in the sanctuaries. Another possible explanation is that people still dedicated dress-fasteners to the deities residing in these sanctuaries, even though 'regular people' had stopped wearing them, and that they thus represent an archaic type of offering. Finally, it is of course also possible that people still wore these items and that they thus represent offerings of contemporary clothing accessories. With the possible exception of the inventory of Mnia and Auzesia, which specifically record dress-fasteners in bronze and iron, possibly indicative of more old-fashioned items, the inventories probably reflect the third explanation, since there is nothing to indicate either an archaic custom or that these items could not have been worn.

The evidence of textiles from the temple inventories

As shown, several Greek temple inventories record the donation of textiles. Other temple inventories do not, however, mention textiles, e.g. the inventory of the Temple of Aphaia on Aegina,[123] the inventory of the Theban Kabeiroi,[124] or the more than 50 inventory lists from Didyma.[125] This indicates that textiles were not systematically donated in all sanctuaries. It is possible that they were considered appropriate as offerings only for certain deities, or that they were connected to certain rites or regions. The possibility cannot, of course, be excluded, though, that textiles were simply not consistently recorded.

A comment should be made on the chronology of these inventories, since they span a rather long period of time. The earliest inventory is that from Oia (5th century BC). The inventories from the 4th century BC include those of the Treasurers of Athena (ca. 400–315 BC), of Samos (346/5 BC), and of the Asklepieion (329/8 BC), while the recording of textiles on Delos is attested from ca. 340 (ca. 340–140 BC). Dating to the 3rd century BC there are the inventories from Tanagra (ca. 250 BC), from Thebes (ca. 250 BC), and of the Theban Kabeiroi. From the 2nd century BC there is the inventory from Miletos, and from the 1st century BC there is the Lindos Chronicle (99 BC). Thus, the inventories from Delos and most other cities are primarily from the 4th century BC and the Hellenistic Period. Only the Acropolis of Athens offers several early inventories, mainly from 434/3 BC on, but here they disappear from the end of the 4th century BC.[126] This is of course important to keep in mind when comparing the evidence from these inventories.

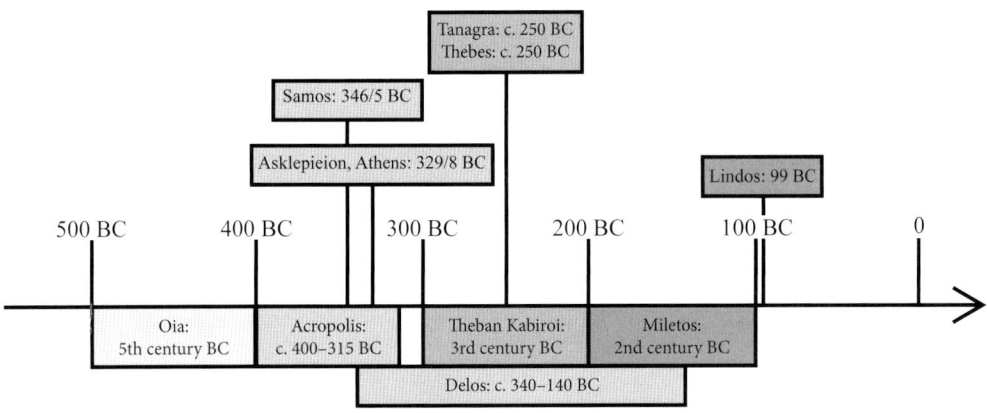

Deities

According to the inventories discussed here, several divinities received textiles. There is evidence for the donation of textiles to Athena, Artemis, Leto, Demeter and Kore, Hera, Eileithyia, Dionysos, Hermes, and Asklepios. On closer inspection, however, it appears that the offering of textiles in the case of certain deities in these specific sanctuaries or temples was unsystematic or inconsistent. This is the situation for the Athena sanctuary at Lindos and the Asklepieion in Athens, for example. It seems that, at these sanctuaries, garments were dedicated primarily because of their great economic value or because of their association with famous donors, such as the Egyptian pharaoh Amasis, who dedicated the fine linen cuirass, and the Persian general

Darius, who dedicated a pair of trousers at Lindos. These garments were not related to any particular ritual, but were merely an expression of piety towards the gods and a display of wealth. Yet, at Lindos, they also served as a way for the sanctuary to illustrate its importance via the powerful people who had visited it and donated offerings. And this is not to omit that, at Lindos, not only deities, e.g. Herakles, gave offerings, but also mythical persons such as Menelaos and Helen, which places the votive offerings in a special sphere.

At the sanctuaries on Delos and Samos there are records concerning the adorning of the cult statues in new garments. This is the case for Leto, Artemis, Hera, and Demeter and Kore (the latter goddesses receive garments, but whether these are used to dress the statues is uncertain). The cult statues of Dionysos and Hermes also wear garments, but these are said to be the old garments of, respectively, Artemis and Hera. Textiles were thus not directly offered to these male gods, who instead 'inherited' the garments of the goddesses. Such garment types must therefore be suitable for both sexes, or at least would not appear ridiculous on a male statue.[127]

In sum, the inventories do not provide solid positive evidence for male gods receiving textiles as offerings. This leaves us with the goddesses Artemis, Athena, Hera, and Demeter and Kore. As discussed above, the Brauron catalogues provide ample evidence for the offering of textiles to Artemis. This is supported by the inventory from Miletos, which is possibly meant for Artemis as well, while the inventory list from Tanagra, on the other hand, most likely testifies to the offering of textiles to the goddesses Demeter and Kore.

Garments (Tables 2–12)
Many different garment types appear in the inventories. Both inner and outer garments were dedicated, as well as items such as veils, belts, and loincloths. Some garment types are much more common than others, though, and a few only occur once or twice. The descriptions of dedicated garments are not 'technical descriptions', that is, they do not always accurately or completely describe the actual attire, and it has been argued that the primary purpose of describing the garments was to distinguish them from each other, rendering them recognisable and significant to users of the catalogue.[128] As such, the records provide us with valuable information about the different types of garments donated in the sanctuary.

Chitōn, chitōnion, chitōniskos, and peplos (Tables 2-4)
Three types of inner-garments dominate in the inventories: the *chitōniskos*, the *chitōnion*, and the *chitōn*, which probably reflect three sizes of the same type of garment.[129] The *chitōn* is generally understood as a full-length, sleeved garment, usually worn with a belt and consisting of one or two pieces of textile sewn together to make a tube. Sleeves were created either by adding buttons or dress-fasteners along the top edge of the garment or by sewing.[130] The *chitōniskos* was a shortened version of the

chitōn.[131] The *chitōn* is by far the most commonly recorded in the inventories. It occurs in the inventories of the Parthenon, the Sanctuary of Demeter and Kore at Tanagra, on Delos and Samos, and at Thebes and Brauron. These descriptions show that the most emphasised and therefore important property of the *chitōn* was its quality or fibre, but at Tanagra the *chitōn* is often described in gendered terms (i.e., for a girl or a boy), which tells us that the *chitōn* was for both genders as a rule, but could somehow be made gender specific, perhaps via decoration. Several of the *chitōnes* are described as possessing coloured borders, especially at Tanagra and Samos. Besides the coloured borders, however, only one *chitōn* is described as having decoration (a *chitōn katastiktos* at Samos), which indicates that this garment type (at least for dedicatory purposes) was usually undecorated.

The *chitōnion* is only recorded at Brauron and Delos. Indeed, in Brauron it is one of the most common types of inner garment. Just as in the case of the *chitōn*, the *chitōnion* is often described in relation to fibre and/or quality (*amorginos, stuppinos*), and only in one instance is it described as having decoration (*katastiktos*).

Table 2. Chitōn

Greek term	No.	Colour	Decoration/fibre	Site	Reference
chitōn			amorginon	Brauron	IG II², 1514, line 10
chitōn			amorginon	Brauron	IG II², 1514, line 22
chitōn				Brauron	IG II², 1514, line 66
chitōn				Brauron	IG II², 1517B, line 174
chitōn				Brauron	IG II², 1517B, line 139
chitōn				Brauron	IG II², 1517B, line 139
chitōn			liton, amorgi	Brauron	IG II², 1524B, line 211
chitōn			amorgi	Brauron	IG II², 1524B, line 216
chitōn				Brauron	IG II², 1521B, line 69
chitōn				Brauron	IG II², 1522, line 18
chitōn				Brauron	IG II², 1523, line 1
chitōn			stuppinos	Brauron	IG II², 1523, line 12
chitōn			stuppinos	Brauron	IG II², 1523, line 12
chitōn			stuppinos	Brauron	IG II², 1523, line 17
chitōn			amorginon	Brauron	IG II², 1523, line 21
chitōn			amorginon	Brauron	IG II², 1528, line 19
chitōn				Brauron	IG II², 1528, line 21
chitōn			amorg.	Brauron	IG II², 1529, line 1

(Continued)

Table 2. Chitōn (Continued)

Greek term	No.	Colour	Decoration/fibre	Site	Reference
chitōn		with purple borders	amorg.	Brauron	IG II², 1529, line 7
chitōn			stuppinos	Brauron	IG II², 1529, line 15
chitōn			stuppinos	Brauron	IG II², 1529, line 16
chitōn				Brauron	IG II², 1529, line 20
chitōn				Brauron	IG II², 1529, line 31
chitōn			amorg.	Brauron	IG II², 1530, line 4
chitōn			stuppinos	Athens	IG II², 1414, line 26
chitōn				Athens	IG II², 1469, line 127
chitōn		purple	boy's	Tanagra	SEG 43: 212B, line 6
chitōn			girl's, amorgine	Tanagra	SEG 43: 212B, line 7
chitōn		purple	linen	Tanagra	SEG 43: 212B, line 9
chitōn			woman's, with a border	Tanagra	SEG 43: 212B, line 14
chitōn		yellow	with 7 buttons	Tanagra	SEG 43: 212B, line 17
chitōn			girl's, finely worked	Tanagra	SEG 43: 212B, line 21
chitōn		with a dark border	girl's	Tanagra	SEG 43: 212B, line 40
chitōn		with a dark border	girl's, uninscribed	Tanagra	SEG 43: 212B, line 42
chitōn		with a border	boy's	Tanagra	SEG 43: 212B, line 43
chitōn		with a purple border	linen	Tanagra	SEG 43: 212B, line 44
chitōn		with a dark border	girl's	Tanagra	SEG 43: 212B, line 45
chitōn			girl's, amorgine	Tanagra	SEG 43: 212B, line 45
chitōn		with a dark border	girl's	Tanagra	SEG 43: 212B, line 47
chitōn		with a dark border	girl's	Tanagra	SEG 43: 212B, line 48
chitōn		with a dark border	girl's, tarantine	Tanagra	SEG 43: 212B, line 49
chitōn		with a purple border	boy's	Tanagra	SEG 43: 212B, line 50

(Continued)

Table 2. Chitōn (Continued)

Greek term	No.	Colour	Decoration/fibre	Site	Reference
chitōn		with purple borders		Thebes	*IG* VII, 2421, line 5
chitōn		apple/quince-coloured, with purple borders		Thebes	*IG* VII, 2421, line 6
chitōnes				Delos	*IG* II,2, 287A, line 121
chitōn		white	woollen	Delos	*ID* 1412A, line 35
chitōn		white	uninscribed	Delos	*ID* 1417A, I, line 7
chitōn		white		Delos	*ID* 1417A, I, line 28
chitōnes	3			Delos	*ID* 1417A, I, lines 28–32
chitōn			linen	Delos	*ID* 1417A, I, line 100
chitōn		white	woollen	Delos	*ID* 1417A, II, line 22
chitōn			linen	Delos	*ID* 1425, II, lines 14–17
chitōn			linen	Delos	*ID* 1425, II, lines 17–19
chitōn				Delos	*ID* 1444Aa, line 38
chitōn				Delos	*ID* 1450A, line 200
Lydian chitōn		dark blue edges		Samos	*IG* XII 6, 1, 261, line 13
Lydian chitōn		purple fringe		Samos	*IG* XII 6, 1, 261, line 13
Lydian chitōn		purple fringe		Samos	*IG* XII 6, 1, 261, line 14
Lydian chitōn		purple fringe		Samos	*IG* XII 6, 1, 261, line 15
chitōn			katastiktos	Samos	*IG* XII 6, 1, 261, line 16
Lydian chitōn		with white edges		Samos	*IG* XII 6, 1, 261, line 16
chitōnos			stuppinos	Samos	*IG* XII 6, 1, 261, line 20
Lydian chitōnes		with purple edges		Samos	*IG* XII 6, 1, 261, line 27
chitōnes				Samos	*IG* XII 6, 1, 261, line 28
chitōnes	2			Samos	*IG* XII 6, 1, 261, line 29
chitōnes	2			Samos	*IG* XII 6, 1, 261, line 31
chitōnes	38			Samos	*IG* XII 6, 1, 261, line 31
chitōnes	2			Samos	*IG* XII 6, 1, 261, line 37

Table 3. Chitōnion

Greek term	No.	Colour	Decoration/fibre	Site	Reference
chitōnion			amorginon	Brauron	IG II^2, 1514, line 51
chitōnion			hemihyphes	Brauron	IG II^2, 1514, line 59
chitōnion			amorginon	Brauron	IG II^2, 1514, line 61
chitōnion			amorginon	Brauron	IG II^2, 1514, line 63
chitōnion			isoptyches, amorginon	Brauron	IG II^2, 1514, line 65
chitōnion			isoptyches, diploun, uninscribed	Brauron	IG II^2, 1514, line 66
chitōnion				Brauron	IG II^2, 1516, line 47
chitōnion				Brauron	IG II^2, 1516, line 48
chitōnion			woman's	Brauron	IG II^2, 1524B, line 131
chitōnion			for a child/boy, diploun	Brauron	IG II^2, 1524B, line 133
chitōnion			amorginon	Brauron	IG II^2, 1524B, line 120
chitōnion			stuppinon	Brauron	IG II^2, 1524B, line 125
chitōnion			amorginon	Brauron	IG II^2, 1518B, line 65
chitōnion			stuppinon	Brauron	IG II^2, 1518B, line 66
chitōnion			amorginon	Brauron	IG II^2, 1518B, line 69
chitōnion			amorginon, uninscribed	Brauron	IG II^2, 1518B, line 70
chitōnion				Brauron	IG II^2, 1524B, line 210
chitōnion			isoptyches	Brauron	IG II^2, 1524B, line 228
chitōnion		thapsinos with purple borders		Brauron	IG II^2, 1522, line 24
chitōnion			stuppinos, katastiktos	Brauron	IG II^2, 1523, line 27
chitōnion			stuppinos	Brauron	IG II^2, 1528, line 14
chitōnion			stuppinos	Brauron	IG II^2, 1529, line 17
chitōnion			amorginos	Brauron	IG II^2, 1529, line 18
chitōnion			peripoikilos	Brauron	IG II^2, 1529, line 5
chitōnion			linen, amorginon	Delos	ID 104(26bis), C, line 7
chitōnion		with purple edges		Delos	ID 104(26bis), C, line 8

2. The temple inventories

Table 4. Chitōniskos

Greek term	No.	Colour	Decoration/fibre	Site	Reference
chitōniskos			peripoikilos, ktenotos	Brauron	IG II², 1514, line 7
chitōniskos		purple	peripoikilos	Brauron	IG II², 1514, line 12
chitōniskos			poikilos	Brauron	IG II², 1514, line 14
chitōniskos		purple?	periegetos, women's	Brauron	IG II², 1514, line 21
chitōniskos			purgōtos	Brauron	IG II², 1514, line 25
chitōniskos			paideion, uninscribed	Brauron	IG II², 1514, line 28
chitōniskos			ktenotos	Brauron	IG II², 1514, line 29
chitōniskos			ktenotos, peripoikilos	Brauron	IG II², 1514, line 41
chitōniskos			ktenotos, periegetos	Brauron	IG II², 1514, line 43
chitōniskos			ktenotos	Brauron	IG II², 1514, line 45
chitōniskos		with with purple borders	purgōtos, parakumatios	Brauron	IG II², 1514, line 46
chitōniskos			ktenotos	Brauron	IG II², 1514, line 51
chitōniskos			periegetos	Brauron	IG II², 1514, line 52
chitōniskos			hemihyphes	Brauron	IG II², 1514, line 53
chitōniskos			haplous	Brauron	IG II², 1514, line 55
chitōniskos		yellow	paidiou, agraphos	Brauron	IG II², 1514, line 58
chitōniskos			ktenotos	Brauron	IG II², 1516, line 50
chitōniskos				Brauron	IG II², 1517B, line 169
chitōniskos		yellow	amorginos	Brauron	IG II², 1524B, line 132
chitōniskos			poikilos	Brauron	IG II², 1517B, line 122
chitōniskos		white		Brauron	IG II², 1517B, line 123
chitōniskos		with purple borders		Brauron	IG II², 1517B, line 126
chitōniskos			stuppinos	Brauron	IG II², 1517B, line 127
chitōniskos			ktenotos	Brauron	IG II², 1517B, line 129
chitōniskos			for a man	Brauron	IG II², 1517B, line 129

(Continued)

Table 4. Chitōniskos (Continued)

Greek term	No.	Colour	Decoration/fibre	Site	Reference
chitōniskos			for a man	Brauron	IG II², 1517B, line 130
chitōniskos			peripoikilos	Brauron	IG II², 1517B, line 131
chitōniskos		with purple borders	uninscribed	Brauron	IG II², 1517B, line 132
chitōniskos		white	purgōtos	Brauron	IG II², 1517B, line 136
chitōniskos		frog-coloured		Brauron	IG II², 1517B, line 137
chitōniskos				Brauron	IG II², 1518B, line 52
chitōniskos		with purple borders		Brauron	IG II², 1518B, line 61
chitōniskos			for a woman	Brauron	IG II², 1518B, line 61
chitōniskos		with purple borders	uninscribed	Brauron	IG II², 1518B, line 62
chitōniskos			periegetos	Brauron	IG II², 1524B, line 208
chitōniskos			ktenotos, ptergas	Brauron	IG II², 1524B, line 212
chitōniskos			hemihyphes	Brauron	IG II², 1524B, line 213
chitōniskos			periegetos	Brauron	IG II², 1524B, line 215
chitōniskos		white	parapoikilos	Brauron	IG II², 1522, line 21
chitōniskos				Brauron	IG II², 1523, line 6
chitōniskos		white		Brauron	IG II², 1523, line 14
chitōniskos		frog-coloured		Brauron	IG II², 1523, line 14
chitōniskos		with purple borders		Brauron	IG II², 1523, line 16
chitōniskos		bluish-grey (glaukeios)		Brauron	IG II², 1523, line 18
chitōniskos		with purple borders	parapoikilos	Brauron	IG II², 1523, line 18
chitōniskos			periegetos, cheiridotos	Brauron	IG II², 1523, line 22

(Continued)

Table 4. Chitōniskos (Continued)

Greek term	No.	Colour	Decoration/fibre	Site	Reference
chitōniskos		frog-coloured	peripoikilos	Brauron	IG II², 1523, line 24
chitōniskos			for a man	Brauron	IG II², 1523, line 26
chitōniskos				Brauron	IG II², 1528, line 6
chitōniskos				Brauron	IG II², 1528, line 7
chitōniskos		yellow		Brauron	IG II², 1528, line 22
chitōniskos			ktenotos	Brauron	IG II², 1529, line 6
chitōniskos				Brauron	IG II², 1529, line 9
chitōniskos			cheirido = cheiridotos?	Brauron	IG II², 1529, line 10
chitōniskos		with purple borders		Brauron	IG II², 1529, line 11
chitōniskos			periegetos, ktenotos	Brauron	IG II², 1529, line 12
chitōniskos		purple		Athens	IG II², 1475, line 7
chitōniskoi				Athens	IG II², 1469, line 125
chitōniskos		with a dark border	for a girl	Tanagra	SEG 43: 212B, line 41
chitōniskos				Delos	ID 1428, II, line 53
chitōniskos				Delos	ID 1433, line 3
chitōniskos				Delos	ID 1442A, lines 52–54
chitōniskos		with a purple fringe	linen	Samos	IG XII 6, 1, 261, line 15
chitōniskos			chrysōi peripoikilmenos	Samos	IG XII 6, 1, 261, line 17

The *chitōniskos* occurs at the Parthenon, Delos (for Leto), Samos, Tanagra, and Brauron, where it is by far the most commonly recorded garment type. It is very often described as decorated or coloured (purple, yellow, white, frog-coloured, bluish-grey, or with a dark border).

Although these three garment types are lexically similar, as Cleland has also demonstrated, each garment type displays distinct characteristics based on the descriptions in the catalogues: the *chitōn* and the *chitōnion* are characterised by their fibre/fabric quality, while the *chitōniskos* is distinguished by its decoration.

These garment types are generally the most common in the catalogues. In most of the inventories, however, the *chitōn* is the most commonly recorded, while at Brauron it is the *chitōniskos*, second-most the *chitōnion*, and third-most the *chitōn*.

The *peplos*, on the other hand, is only rarely recorded. The *peplos* is conventionally identified as a rectangular piece of (woolen) cloth draped around the body and fastened at the shoulders by pins or fibulas. The open vertical edge is sometimes sewn together, but is usually left open, while the top part of the garment could be folded to create an overfold. The garment could be belted at the waist.[132] The garment type occurs in a fragmented state pe[plos] in the Sanctuary of Delos (for Demeter and Kore),[133] and several *peploi* are recorded in the inventory from Oia.

Mantles (Tables 5–7)
The inventories record a large number of mantles, but the most common type is the *himation*, worn by men and women alike.[134] It should be kept in mind, however, that the term *himation*, in its plural form *himatia*, also became a general Greek term for clothing, but one that increasingly came to mean 'a large and voluminous oblong cloth diagonally draped across the torso, wrapped around the body, supported on one shoulder or arm'.[135] It is recorded in all the inventories, except that from Tanagra.

The *chlamys* is recorded at the Athenian Asklepieion, Tanagra, and Delos, while *chlamydes ephebikai* are only recorded at Miletos. Such ephebian *chlamydes* are claimed to have been the 'uniform' of Greek male youths, and according to literary and epigraphic sources were donated as offerings at the end of military service.[136] Generally, the *chlamys* was a garment worn throughout the Greek world by a wide range of men: horsemen, foot soldiers, heralds, travellers, and ephebes; it was also used as a royal mantle and as a tragic costume.[137] In appearance it was a short woollen mantle of rectangular or semi-circular shape.[138]

The *chlanis* was a mantle worn by both sexes, but is interpreted as a garment made of finer wool, worn on festive occasions and as a wedding mantle.[139] It is recorded at Brauron, Tanagra, Samos and Miletos. Several diminutives such as the *chlanidion/chlandion* and the *chlanidiskion*, usually translated as shawls or 'women's mantles',[140] are also recorded in the inventories that record *chlanides*, but there are no records of the *chlaina*.

The *tribōn* was a short or small mantle, and it has been claimed that it was characteristic of Spartan men and philosophers.[141] It is recorded at Brauron and in the inventory from Tanagra, where it is specifically denoted as for a man (*tribōn andreios*).

Other types of mantles are more elusive, and we cannot be exactly certain of their appearance. They are often interpreted as 'wraps' or 'shawls'.[142] One is the *epiblēma*, which is a garment type only recorded at Brauron, another is the related *periblēma*, which is only recorded once in the inventory from Samos. Both these terms have the general meaning of 'that which is thrown over, around or about'.[143] These garments appear to have been large pieces of cloth which generally were used as mantles, cloaks, veils or shawls.[144] The *ampechonon* is similar, and comes from the word *ampechō*,

2. The temple inventories   57

Table 5. Himation

Greek term	No.	Colour	Decoration/fibre	Site	Reference
himation gynaikeion		white		Brauron	IG II², 1514, line 16
himation gynaikeion		purple border	perikumation	Brauron	IG II², 1514, line 17
himation		white		Brauron	IG II², 1514, line 20
himation		white		Brauron	IG II², 1514, line 27
himation andreion				Brauron	IG II², 1514, line 47
himation		white with purple borders		Brauron	IG II², 1514, line 69
himation				Brauron	IG II², 1516, line 51
himation				Brauron	IG II², 1517B, line 163
himation paideion				Brauron	IG II², 1517B, line 124
himation				Brauron	IG II², 1517B, line 138
himation		purple borders		Brauron	IG II², 1518B, line 60
himation			hemihyphes gynai	Brauron	IG II², 1518B, line 67
himation		white		Brauron	IG II², 1524B, line 205
himation		white		Brauron	IG II², 1524B, line 210
himation		purple borders		Brauron	IG II², 1522, line 11
himation			chrysa grammata	Brauron	IG II², 1529, line 14
himation		purple borders		Brauron	IG II², 1529, line 16
himatia	2			Athens	IG II², 1469, line 124, 129
himatia	5			Thebes	IG VII, 2421, line 10
himatia	2	grey		Delos	ID 104(26bis), C, lines 3–4
himation	1	white		Samos	IG XII 6, 1, 261, line 27
himatia	48			Samos	IG XII 6, 1, 261, line 32
himation	1	purple border		Miletos	Milet VI, 3, 1357, line 6
himatia	3	purple		Miletos	Milet VI, 3, 1357, line 9

Table 6. Chlamys, chlanis

Greek term	Translation	Type	No.	Colour	Decoration/fibre	Site	Reference
chlamys	chlamys	mantle				Athens, Asklepieion	IG II², 1533, line 8
chlamys	chlamys	mantle		grey		Athens, Asklepieion	IG II², 1533, line 18
chlamys	chlamys	mantle				Delos	ID 1442A, lines 52–53
chlamydiskan	small chlamys	mantle				Tanagra	SEG 43: 212 (B), line 36
chlamydes ephebikai	ephebic chlamydes	mantle	4			Miletos	Milet VI, 3, 1357, lines 11–12
chlanis	chlanis	mantle			agraphos parabolon	Brauron	IG II², 1514, line 39
chlanida andreian		mantle				Brauron	IG II², 1517B, line 128
chlanida		mantle				Brauron	IG II², 1523, line 7
chlaniskion	small chlanis	mantle		white		Brauron	IG II², 1514, line 40
chlaniskion	small chlanis	mantle				Brauron	IG II², 1518B, line 55
chlanida	chlanis	mantle				Tanagra	SEG 43: 212 (B), line 12
chlanida	chlanis	mantle				Tanagra	SEG 43: 212 (B), line 13
chlanidiskan	small chlanis	mantle		white		Tanagra	SEG 43: 212 (B), line 33
chlanidas	chlanis	mantle	2			Tanagra	SEG 43: 212 (B), line 38
chlanides	chlanis	mantle	2	purple		Samos	IG XII, 6, 1, 261, line 36
chlanidion		mantle		purple		Samos	IG XII, 6, 1, 261, line 30
chlanides	chlanis	mantle	3			Miletos	Milet VI, 3, 1357, line 8
chlandion paidika	child's chlanis	mantle		purple		Miletos	Milet VI, 3, 1357, line 22
chlandion paidika	child's chlanis	mantle		with purple border		Miletos	Milet VI, 3, 1357, line 23

Table 7. Other mantle types

Greek term	Type	No.	Colour	Decoration/fibre	Site	Reference
epiblēma	mantle				Brauron	*IG* II², 1514, line 33
epiblēma	mantle	2		poikilos	Brauron	*IG* II², 1529, line 19
epiblēma	mantle	4		poikilos	Brauron	*IG* II², 1514, line 31
enkyklon	Shawl/wrap			poikilos	Brauron	*IG* II², 1514, line 48
enkyklon	Shawl/wrap	2	white		Brauron	*IG* II², 1524B, line 206
enkyklon	Shawl/wrap	3			Brauron	*IG* II², 1529, line 6
enkyklon	Shawl/wrap	2	white		Brauron	*IG* II², 1524B, line 223
periblēma	Shawl/wrap	1		linen	Samos	*IG* XII, 6, 1, 261, line 18
proslēmma	Shawl/wrap		with purple edges		Samos	*IG* XII, 6, 1, 261, line 20
proslēmma	Shawl/wrap			linen	Samos	*IG* XII, 6, 1, 261, line 26
ampechonon	Shawl/wrap				Brauron	*IG* II², 1514, line 18
ampechonon	Shawl/wrap			hieron epigegraptai	Brauron	*IG* II², 1514, line 34
ampechonon	Shawl/wrap				Brauron	*IG* II², 1514, line 36
ampechonon	Shawl/wrap				Brauron	*IG* II², 1514, line 50
ampechonon	Shawl/wrap				Brauron	*IG* II², 1524B, 218
ampechonon	Shawl/wrap				Brauron	*IG* II², 1522, line 17
ampechonon	Shawl/wrap		white		Tanagra	*SEG* 43: 212 (B), line 11
tribōna andrion	mantle				Tanagra	*SEG* 43: 212 (B), line 32
tribōnia	mantle	2			Brauron	*IG* II², 1514, line 22
emphiesmena linois	Shawl/wrap			linen	Delos	*ID* 1424B, line 3–4
emphiesmena linois	Shawl/wrap			linen	Delos	*ID* 1417A, II, line 22
emphiesmena linois	Shawl/wrap			linen	Delos	*ID* 1417A, II, line 52
emphiesmenon linōi	Shawl/wrap			linen	Delos	*ID* 1417A, I, lines 100–103
emphiesmenon linōi	Shawl/wrap			linen	Delos	*ID* 1425, II, lines 17–19
emphiesmenon linōi	Shawl/wrap			linen	Delos	*ID* 1426B, I, lines 25–30
emphiesmena linois	Shawl/wrap			linen	Delos	*ID* 1426B, II, line 22
emphiesmena linois	Shawl/wrap			linen	Delos	*ID* 1442B, lines 44–45

which translates as 'surround, enclose'[145] or 'to have around' and the middle form *ampechomai*, which means 'to drape' or 'to wear'.[146] This particular garment type is recorded primarily in Brauron, but also once in the inventory from Tanagra. The *enkyklon* is another similar garment type only recorded at Brauron. It was, like the above garments, probably used as a mantle or 'an encircling'.[147] The *proslēmma* most likely belongs in the same category.[148] It is only recorded in the inventory from Samos. Finally, the inventories from Delos use the expression *emphiesmenon linois*, which possibly can be translated as 'wrapped in linen'. Yet, again, we do not know whether this reflects a specific garment type or way of draping it. It is interesting to note that these different terms for mantles, shawls or wraps usually only occur in inventories from one site (with the exception of Tanagra and Brauron, which overlap, since both inventories record the *ampechonon* and the *tribōn*). But possibly this simply reflects local fashions or terminologies – perhaps what was termed a *proslēmma* on Samos might have been termed an *ampechonon* in Brauron.

Non-Greek garments (Table 8)
The inventories also testify to the presence of non-Greek garments in the sanctuaries, although they are quite rare. One such non-Greek garment is the *kalasīris*, a long fringed garment of Egyptian or Persian origin.[149] The garment is described by Herodotos as a linen *chitōn* adorned with tassels or fringes worn by Egyptians.[150] In the sacred regulations of Andania,[151] where the *kalasīris* was prescribed for ceremonial worship, this particular garment type might thus have had ritual connotations.[152] It

Table 8. Non-Greek garments

Greek term	Translation	No.	Colour	Decoration/fibre	Site	Reference
kalasiris					Miletos	Milet VI, 3, 1357, line 5
kandys				poikilos	Brauron	IG II², 1514, line 19
kandys			purple		Brauron	IG II², 1524B, line 217
kandys				linon, poikilos	Brauron	IG II², 1524B, line 219
kandys					Brauron	IG II², 1524B, line 204
kandys				with gold pasmatia	Brauron	IG II², 1523, line 8
kandys					Brauron	IG II², 1523, line 27
anaxyrides	Persian trousers				Lindos	Blinkenberg, Lindos II, 2, C, lines 65–68
xenikē	foreign garment		purple		Brauron	IG II², 1514, line 49
endyma phrygion	Phrygian garment				Delos	ID 1442 B 57, lines 56–57

is only recorded in the inventory from Miletos, where it is described as having gold edgings and a vertical stripe down the middle in bluish grey.[153]

Another non-Greek garment dedicated in sanctuaries is the *kandys*, which in Greek literature denotes a Persian garment: a coat with sleeves, usually worn over a tunic and *anaxyrides* (trousers) by Persian noblemen and the Great King, and usually slung over the shoulders.[154] Pollux, though, describes the *kandys* as purple, sleeved, sometimes made of leather, and fastened from the shoulders.[155] The *kandys* is also mentioned by Xenophon, who tells us that Cyrus wore Median dress consisting of a purple *chitōn* and a *kandys*.[156] In Attic sculpture and vase-painting of the late 5th and early 4th century BC, however, Greek women and children are sometimes depicted wearing the *kandys*, indicating that the garment type had been widely adopted by these groups (Fig. 7).[157] Therefore, the recording of the *kandys* in the inventories does not necessarily indicate an imported male garment item from Persia, but could just as well be a locally produced garment for women or children. It should be noted that Miller observes that the use of the *kandys* had come to an end within a century of its introduction to Athens, and claims that there is no evidence for Athenians wearing this type of garment in the Hellenistic Period.[158] The *kandys* is only recorded in the Brauron catalogues, where it occurs six times. Some are described as e.g. 'patterned' (*poikilos*), 'with gold *pasmatia*', or 'patterned all over' (*peripoikilos*), and one is also described as being of linen.[159]

Another non-Greek garment type is the Persian or Median trousers (*anaxyrides*), which were dedicated at Lindos.[160] *Anaxyrides* were baggy trousers, sometimes made of leather and usually patterned, but the term also refers to the tighter trousers of the Amazons depicted in e.g. vase-painting, as well as the trousers with a coloured rhombic pattern belonging to the Persian rider from the Athenian Acropolis (Fig. 8).[161] Also, the Brauron catalogues record a *xenikē*, which can be interpreted as a garment of foreign type, but we do not know its appearance.[162] Finally, on Delos there is a record of a Phrygian garment (*endyma phrygion*), which indicates that it was something particular from this area, perhaps a garment of a certain type or made from a specific fibre, in a specific technique or with specific decoration.

Luxurious garments (Table 9)
Other garment terms refer more to the textiles' luxuriousness and value rather than their actual type and appearance, as for example the *thryphēma*, a garment named after the term *tryphē*, meaning 'softness', 'delicacy' or 'daintiness'.[163] The term is recorded seven times in Brauron.

Another garment type that can be considered luxurious is the *xystis*, a full-length robe of rich soft fabric,[164] which is recorded in the inventory of the Parthenon.[165] Three garments are similarly described with the term *xystidotos* in the Brauron catalogues. Lee considers the *xystis* a version of the *chitōn*, and identifies it as a long, sleeveless garment worn by charioteers in vase-paintings and sculpture, such as the famous Motya charioteer from the 5th century BC. She further argues that the *xystis* was

Fig. 7. Kandys. Attic grave relief of the girl Myttion, 4th century BC. The J. Paul Getty Museum, inv. no. 78.AA.57. Digital image courtesy of the Getty's Open Content Program.

never worn by Greek women.[166] The royal robe (*basilika stolē*) from Lindos probably also belongs with the luxurious garments. We do not know the appearance of any of these three garment types, only that they were luxurious and thus of a very high value.

Fig. 8. Persian rider from the Athenian acropolis, ca. 520–510 BC. © Acropolis Museum, Athens, inv. no. Acr. 606. Photo: Socratis Mavrommatis

Not clearly defined garment types[167] (Table 10)
The inventories from Brauron record several garment terms about which we do not know much. One is the *lēdion*, a light summer garment,[168] another the *lasion*, translated by Cleland as a 'shaggy garment', and another is the *dipterygon*, which literally means that it has two *ptera*, feathers or wings, and has been interpreted by Cleland as a mantle with two fluttering corners.[169] The inventory from Tanagra records a *gada*[170] *paillos*, a type of child's garment, and a woman's robe (*symmetria gynaikeia*).[171] The inventories from Tanagra and Thebes both record an open garment (*schistos*): one is yellow and has six buttons, the other has purple borders. Yet not one of these garments can be identified in the archaeological record or in iconography, so we are, for now at least, at a loss as to their appearance.

Table 9. Luxurious garments

Greek term	Translation	No.	Colour	Decoration/fibre	Site	Reference
thryphēma	luxurious garment			katastiktos	Brauron	*IG* II², 1514, line 71
thryphēma	luxurious garment		yellow (krokōtos)		Brauron	*IG* II², 1517B, line 162
thryphēma	luxurious garment			peristiktos	Brauron	*IG* II², 1525, line 4
thryphēma	luxurious garment				Brauron	*IG* II², 1517B, line 136
thryphēma	luxurious garment				Brauron	*IG* II², 1518B, line 69
thryphēma	luxurious garment			amorginon	Brauron	*IG* II², 1523, line 21
thryphēma	luxurious garment				Brauron	*IG* II², 1523, line 25
xystis					Athens	*IG* II², 1412, line 11
xystis					Athens	*IG* II², 1421, line 118
xystis					Athens	*IG* II², 1428, line 143
xystidōtos				katastiktos	Brauron	*IG* II², 1514, line 11
xystidōtos				katastiktos	Brauron	*IG* II², 1524B, lines 208–209
xystidōtos				katastiktos	Brauron	*IG* II², 1523, line 10
basilikan stolan					Lindos	Blinkenberg, Lindos II, 2, C, lines 85–89
stolide		2			Athens	*IG* II², 1428, lines 30–31

The Lindos Chronicle and the inventories from the Parthenon and Brauron record *thōrakes*, which can be translated as 'corselet', 'scale armour', or 'linen jerkin', but 'breastband' is also a possibility.[172] Such linen corselets were in use from the Archaic Period through the Hellenistic Period, and perhaps date back as far as the Greek Bronze Age.[173] In the inventories from the Parthenon, they are often recorded together with belt/girdles and dress-fasteners, which might indicate that these items were worn together. Literary sources also attest to the dedication of linen corselets in sanctuaries: for example, Pausanias reports that Gelon of Syracuse dedicated three linen corselets at Olympia,[174] and Herodotos records the dedications of linen corselets by Amasis at Lindos and Sparta.[175] According to Gleba, linen corselets were of composite type and

Table 10. Not clearly defined garment types

Greek term	Translation	No.	Colour	Decoration/fibre	Site	Reference
lēdion	a light summer garment				Brauron	IG II², 1514, line 45
lasion	a shaggy garment				Brauron	IG II², 1524B, line 130
dipterygon	mantle with 2 fluttering corners			katastiktos	Brauron	IG II², 1514, line 38
dipterygon	mantle with 2 fluttering corners			amorgi	Brauron	IG II², 1524B, line 214
schistos	open garment		yellow	with 6 buttons	Tanagra	SEG 43: 212B, line 30
schistos	open garment		purple borders		Thebes	IG VII, 2421, line 8
symmetria gynaikeia	woman's garment				Tanagra	SEG 43: 212B, line 10
gada paillos	child's garment		silver		Tanagra	SEG 43: 212B, line 18
thōrax	Corselet/cuirass				Athens	IG II², 1388, line 19
thōrax	Corselet/cuirass				Athens	IG II², 1424a, line 55
thōrax	Corselet/cuirass				Athens	IG II², 1425, line 54
thōrax	Corselet/cuirass	16			Athens	IG II², 1425, line 256
thōrax	Corselet/cuirass				Athens	IG II², 1428, line 16
thōrax	Corselet/cuirass				Athens	IG II², 1428, line 30
thōrax	Corselet/cuirass				Athens	IG II², 1428, line 205
thōrax	Corselet/cuirass			katastiktos	Brauron	IG II², 1523, line 20
stolion	Robe			linen	Delos	ID 1442A, lines 52–54

made of many rectangular sections of thick and stiffened linen fabric, sewn together, while the skirt had one or two rows of flaps, termed *pteryges*, or a thick fringe of plaited cords.[176] Linen corselets are not represented in the archaeological record - with the exception of the linen *pteryges* adhering to the composite armour recovered at Masada.[177] The example made of iron and gold from the tomb of Philip II at Vergina is thought, however, to be a translation of a linen corselet into metal, which gives us an idea of its appearance.[178] Furthermore, linen corselets are represented in Etruscan wall-paintings, where they are recognisable due to their white colour (Fig. 9),[179] and in the famous Alexander mosaic, where Alexander himself wears a corselet of cloth and metal.[180]

Fig. 9. Linen corselet represented in Etruscan wall-painting. Tomba del Orco II at Tarquinia, 4th century BC (after Steingräber 1968, pl. 129).

Finally, some terms simply have the general meaning of garment, and we do not know what they look like. One of these terms is *esthēs* (garment), which is only recorded at Delos; another is the term *endyma* (garment), also recorded at Delos.[181] According to Lee, however, *endymata* are undergarments, that is, the first layer of garments worn directly on the body. She thus considers the *chitōn* and the *peplos* to be types of *endymata*.[182]

Veils and headdresses (Table 11)
According to Llewellyn-Jones, there are three Ancient Greek terms for veil: *krēdemnon*, *kalymma* and *kalyptrē*.[183] The latter term does not occur in the inventories investigated here. We do not know how these items looked or how they differed from each other – whether by size, material, decoration, or colour, or if they simply reflect variations in use.[184] The *krēdemnon* is only recorded at Samos. Its literal meaning is 'head-binder', and, according to Llewellyn-Jones, it appears to have been an especially fine or luxurious garment, and is best regarded as a head-veil that hung from the back part of the head and covered the back and shoulders of the wearer, reaching either to shoulder-length or, more likely, floor-length.[185] In connection with the *krēdemnon*, the *epikrēnon* (literal meaning: 'about the head') needs to be addressed, since this garment should perhaps be equated with the *krēdemnon*.[186] *Epikrēna* of silk and linen are recorded only at Miletos. Günther associates these garments with a type of headgear depicted on east Greek funerary *stelai* from the 2nd century BC, which is fastened to the hair at the back of the head and falls down over the shoulders.[187] This identification, however, is unsubstantiated. The *kalymma* (covering) is recorded only at Brauron. Its appearance and usage is ambiguous. In epic poetry it is twice alluded to as black or dark, which might imply a use in mourning. Other sources, however, describe the *kalymma* as golden.[188] In Aeschylos, a bride wears a *kalymma*,[189] and in Euripides a bridal veil is also referred to as a *kalymma*.[190] Llewellyn-Jones therefore argues that, by the Classical Period, the *kalymma* had lost any precise meaning and had come to refer to a standard type of veil.[191] The inventories from Brauron do not assist much in the interpretation of this term, since there are no descriptions of colour, fibre, or decoration, although one is 'inscribed'.

There are also other types of veils or headdresses, such as the *kekryphalos*, which has been interpreted as 'a net-like cap or woven snood used to keep the hair in order and particularly well-suited to the major female hairstyle of the day, the chignon', sometimes covering the hair completely and sometimes leaving it visible over the forehead.[192] *Kekryphaloi* are, like the veils, only recorded at Brauron and Samos. We know from both iconography and the archaeological record that women wore hair nets.[193] Such hair nets might have been woven or made of sprang – a technique that we know was in use in Ancient Greece, due to the depictions of sprang looms on Greek vases.[194] There are no preserved sprang fragments from Ancient Greece, but a piece of woollen sprang dating to the second half of the 1st millennium BC was recovered in a burial in Kertch,[195] and a linen hair net, dated to

Table 11. Veils and headdresses

Greek term	Translation	No.	Colour	Decoration/fibre	Site	Reference
krēdemnon	veil				Samos	IG XII 6, 1, 261, line 36
krēdemna	veil	7			Samos	IG XII 6, 1, 261, line 21
prosōpidia	face veil	4		of silk	Miletos	Milet VI, 3, 1357, lines 12–13
prosōpidion	face veil	2		of wool	Miletos	Milet VI, 3, 1357, line 13
prosōpidia or linen pieces	face veil	12		of linen	Miletos	Milet VI, 3, 1357, line 14
epikrēnon	veil	1		of linen	Miletos	Milet VI, 3, 1357, line 14
epikrēnon	veil	2			Miletos	Milet VI, 3, 1357, line 15
epikrēnon	veil	1			Miletos	Milet VI, 3, 1357, line 15
epikrēnon	veil	1		silk	Miletos	Milet VI, 3, 1357, line
epikrēnon	veil	1		silk	Miletos	Milet VI, 3, 1357, line
kekryphalos	hair veil	1			Athens	IG II², 1469, line 125
kekryphalos	hair veil		purple		Samos	IG XII 6, 1, 261, line 22
kekryphalos	hair veil		white		Brauron	IG II², 1525, line 4
kekryphaloi	hair veil	3			Brauron	IG II², 1522, line 18
kalymma	veil				Brauron	IG II², 1524B, 204
kalymma	veil				Brauron	IG II², 1529, line 9
kalymma	veil				Brauron	IG II², 1529, line 13
kalymma	veil			agraphos	Brauron	IG II², 1529, line 14
kalymma	veil				Brauron	IG II², 1514, line 3
kalymma	veil			synerrammenon	Brauron	IG II², 1517B, line 133
trichapton	veil				Samos	IG XII 6, 1, 261, line 37
trichapton	veil				Brauron	IG II², 1524B, line 177
trichapton	veil				Delos	IG XI, 2, 287A, line 53
tegidion	face veil		white		Tanagra	SEG 43: 212(B), line 38

the 1st centuries around the birth of Christ, has been preserved from the Judean Desert Caves (Fig. 10).[196]

Another type of veil is the *tegidion*, which is recorded only once, in the inventory from Tanagra. Llewellyn-Jones identifies the *tegidion* as a face veil, created by cutting eyeholes into a single rectangular cloth, which was sometimes edged with a fringe, and was bound around the head with a fillet and occasionally fastened

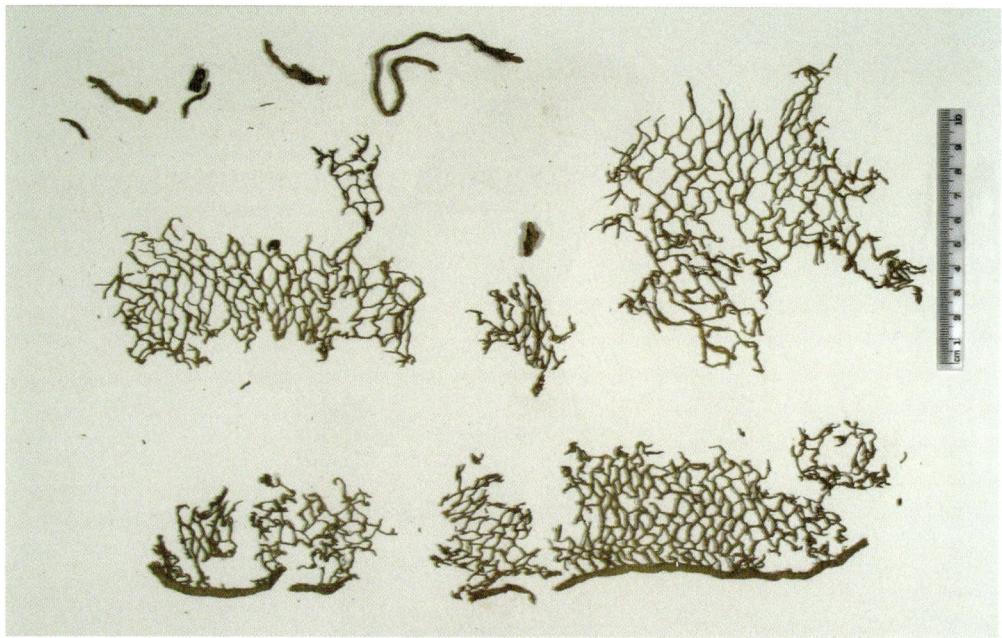

Fig. 10. Linen hairnet, 1st centuries BC/AD. © The Israel Antiquities Authority, inv. no. IAA 577048. Photo: Clara Amit.

over the forehead with a brooch.[197] The *tegidion* is represented in art, the most famous example probably being a Hellenistic bronze figurine representing a female dancer (Fig. 11).[198] Her body and head is wrapped in a *chitōn* and a long mantle/veil, which she draws over her lower face. Her *tegidion* is visible beneath, and covers her entire face except the eyes. The *tegidion* enters the sources in the late 4th and disappeared in the 1st century BC; it was worn in Boeotia, Macedonia, Asia Minor, Egypt, and possibly Attica, but it does not seem to have been used in the Peloponnese.[199]

Related to or possibly the same as the *tegidion* is the *prosōpidion*, which can be translated as 'little mask' or 'face cover' (*prosōpos*, 'face').[200] It is recorded in the inventory from Miletos, where it is of either silk, wool, or linen – an ambiguity which demonstrates that this garment type was not defined by fibre. According to Güther, *prosōpidia* were a kind of accessory that women wore when outside the house to protect themselves from dust and heat.[201] Thanks to Heraclides, we have a fairly accurate idea of how this item was worn.[202] His description is similar to that of the costume of Theban women, who wore the veil like a mask that covered the entire face except the eyes. It is thus likely that the *prosōpidia* are veils of the same type as the *tegidion*. Literary sources also mention this veil-type: Antipater of Sidon tells of a young woman named Herakleia wearing a *kalypteira prosōpou* 'veil for the face', which she dedicates to Aphrodite immediately prior to her wedding.[203]

Fig. 11. Bronze figurine, woman wearing tegidion, *3rd–2nd century BC. © Metropolitan Museum of Art, inv. no. 1972.118.95.*

A final clothing item to be discussed in connection with the veils is the *trichapton*, which is recorded at Brauron, Delos, and Samos. The adjective *trichaptos* is translated by *LSJ* as 'plaited' or 'woven of hair',[204] and the substantive has been identified as 'a fine veil of hair'.[205] There is nothing, however, in the constituents of the word (hair [*thrix*], fasten [*aptō*]) to confirm that it was woven or plaited. Instead, we might simply interpret it as a clothing item used to fasten the hair, such as a veil.

Girdles, belts, loincloths, and ribbons (Table 12)
Belts and girdles were also donated in the sanctuaries and thus recorded in the inventory lists. They are denoted by different terms, primarily *zonē* (pl. *zonai*),[206] but also the variants *zōma* and *diazōma*. It is difficult to determine whether these items denote belts or girdles, and one could argue that these items are the same,

Table 12. Girdles, belts, loincloths, and ribbons

Greek term	Translation	No.	Colour	Decoration/fibre	Site	Reference
perizōma	loincloth		purple	poikilos	Samos	IG XII 6, 1, 261, line 22
diazōma	belt/girdle			epichrysos, wool	Miletos	Milet VI, 3, 1357, line 19
diazōma	belt/girdle			linen	Miletos	Milet VI, 3, 1357, line 20
zōnion	belt/girdle				Athen	IG II², 1424a, lines 55
zōnion	belt/girdle				Athen	IG II², 1425, line 55
zōnion	belt/girdle				Athen	IG II², 1428, line 16
zōnion	belt/girdle			diachrysos	Delos	ID 1428, II, lines 58
zōnion	belt/girdle			diachrysos	Delos	ID 1429B, lines 4
zōnion	belt/girdle			diachrysos	Delos	ID 1433, lines 7
zōnion	belt/girdle			diachrysos	Delos	ID 1450A, lines 200–201
zōnarion	belt/girdle				Delos	ID 1442A, lines 52–54
zōnai	belt/girdle				Miletos	Milet VI, 3, 1357, line 21
zōnai	belt/girdle				Miletos	Milet VI, 3, 1357, line 22
zōma	belt/girdle				Brauron	IG II², 1514, line 15
zōma gynaike	belt/girdle				Brauron	IG II², 1518B, line 58
zōma	belt/girdle		white		Brauron	IG II², 1523, line 16
mitra		1	purple		Athens	IG II², 1448, line 4
mitrē		1		stuppeiou	Samos	IG XII 6, 1, 261, line 17
mitrē		1		paraulos	Samos	IG XII 6, 1, 261, line 18
mitrē		1			Samos	IG XII 6, 1, 261, line 36
mitrai		2		stuppinai	Samos	IG XII 6, 1, 261, line 36

(Continued)

Table 12. Girdles, belts, loincloths, and ribbons (Continued)

Greek term	Translation	No.	Colour	Decoration/fibre	Site	Reference
spleniskos				hypogegrammenon hippea	Samos	*IG* XII 6, 1, 261, line 24
spleniskos				of linen & wool	Samos	*IG* XII 6, 1, 261, line 25
sphendonai		2		of linen	Samos	*IG* XII 6, 1, 261, line 21
lemniskoi		2	green (prasinos)		Miletos	Milet VI, 3, 1357, line 17
lemniskos		1	scarlet		Miletos	Milet VI, 3, 1357, lines 17–18
strophion					Athens	*IG* II², 1388, line 19
strophion					Athens	*IG* II², 1428, lines 30–31
strophion					Delos	*ID* 1442 B 57, lines 56–57
strophoi		2		epichrysoi	Miletos	Milet VI, 3, 1357, line 18
strophos		1	spandikinos	chrysopoikilos	Miletos	Milet VI, 3, 1357, line 19
pezidion					Brauron	*IG* II², 1525, line 4
pezidion					Brauron	*IG* II², 1524B, lines 178–179
pezidion					Brauron	*IG* II², 1522, line 15
pezidion					Brauron	*IG* II², 1522, line 28
anadema	hairband			poikilos	Brauron	*IG* II², 1523, line 15
katōride					Athens	*IG* II², 1425, line 58

the girdle simply indicating a female association. Both terms are therefore, in this context, translated as 'belt/girdle' in order to avoid misconceptions. Belts/girdles (*zonai*) are recorded in the inventories from Athens, Delos, and Miletos, while the *zōma* is recorded only at Brauron and the *diazōma* only at Miletos – a fact that again illustrates the local differences in terminology. The loin cloth (*perizōma*) is recorded only once, at Samos.

A related term that may also denote a belt/girdle is the *mitra*, which is recorded in Athens and on Samos. According to Papadopoulou, the *mitra* is synonymous with *tainia* (band), *diadēma* (diadem), and *zōnē* (belt),[207] but it can also denote a kind of headdress for both men and women. In both cases, it expresses the idea of 'wearing something, usually of cloth, which goes around some part of the person, forming a circle.'[208] Thus, the objects labelled as *mitra* come in many shapes and sizes, can be worn in different ways around different parts of the body, and are often used in texts in lieu of synonymous terms.[209] In later poets, the term can denote a 'maiden's girdle',[210] but is also used to describe a girdle worn by wrestlers, and as a headband or snood.[211] The word is specifically used of a 'victor's chaplet at the games',[212] of a headband worn as a sign of rank at the Ptolemaic court, of an oriental headdress,[213] and of the headdress of the priest of Herakles on Cos.[214] Finally, the term *mitra* was also used to describe a piece of armour in the form of a metal guard worn around the waist.[215] It seems quite clear, though, that the *mitrai* recorded in the lists are made from textiles, whether they are headdresses or belt/girdles, because they are listed among other textile items.

Finally, *strophiones/strophoi* are recorded at Athens, Delos, and Miletos. These are interpreted either as a girdle, a breastband, or as a headdress worn by priests.[216] In later literature the meaning of *strophion* varies. According to Stafford it sometimes means 'headband', being explicitly linked with head or hair, but in Aristophanes it appears to refer to a breastband.[217]

It must be remarked, though, that relatively few belts/girdles are recorded in the inventories, which is somewhat surprising given that literary sources suggest that it was common to dedicate belts, especially to Artemis and/or Eileithyia, either in connection with marriage or childbirth.[218] Among the literary sources, Pausanias (2.33.1) records that girls at Troizen dedicated a girdle to Athena before marriage, and, in Euripides' *Heracles*, the hero dedicated Hippolyta's girdle in the Temple of Hera at Mycenae.[219] Moreover, the *Palatine Anthology* contains several epigrams attesting to the dedication of belts and girdles to goddesses.[220] The scarcity of these clothing items in the inventories, however, can probably be explained as a reflection of their low value in comparison to such garments as *chitōnes* and *himatia*.

Related to the belts/girdles are the ribbons or textile strips recorded in the inventories, and denoted by different terms. On Samos two *spleniskoi* are recorded, and in Miletos two *lemniskoi*, both to be translated as ribbons. On Samos two *sphendonai* are also recorded, perhaps a form of headband, and in Brauron an *anadema* is recorded, possibly a type of hairband.[221]

Footwear

There is thus a wealth of different garments recorded in the inventories. But there are also clothing items that are not so widely attested. For example, with the exception of the pair of boots (*hypodēmatōn koilōn zeugos*) dedicated to Leto on Delos, and the two boxes of shoes (*opisthokrēpides*) from the Athenian Acropolis, there are no records of footwear, even though we know from iconographical and literary sources that many different types of shoes, boots, and sandals were available to ancient men and women.[222] Furthermore, epigrams in the *Palatine Anthology* describe the offering of footwear to different deities.[223] The lack of a record for these items in the inventories can therefore be explained either as a reflection of the fact that, although they were dedicated, they were not considered important enough to be recorded, or simply as a result of the fact that they were not dedicated in these particular sanctuaries. Alternatively, footwear was not considered appropriate to bring into a sanctuary (see Chapter 11).

Textile furnishings

The inventories almost exclusively record different types of garments, and furnishing textiles are quite rare. The inventory from Samos records textile furnishings consisting of cushions (*hypokephalaia*), two of linen and one embroidered/patterned, a ragged table cover (*katapetasma tēs trapezēs rhakinon*), two foreign patterned curtains (*parapetasmata barbarika poikila*), and two curtains.[224] Furnishing textiles also occur in the inventories from Delos, which record cushions (*proskephalaia*), some of linen.[225] A linen curtain (*parapetasma linoun*) is recorded among the furnishings of the Temple of Kynthus,[226] and another inventory mentions a man who was paid for mending another curtain, which was placed on the portrait Queen Arsinoe in another part of the sanctuary.[227] Finally, eight carpets (*tapides*) are recorded in the inventory of the Temple of Apollo.[228]

The presence of furnishing textiles in temples is also attested by Pausanias:

> 'In Olympia there is a woollen curtain, adorned with Assyrian weaving and Phoenician purple, which was dedicated by Antiochus, who also gave as offerings the golden aegis with the Gorgon on it above the theatre at Athens. This curtain is not drawn upwards to the roof as is that in the temple of Artemis at Ephesus, but it is let down to the ground by cords.'[229]

And, in Euripides' *Ion*, we hear of the banquet that Xuthos, Ion's father, ordered to be laid out in his absence. Ion erected a spacious tent, made from a framework of poles on which were hung tapestries with figural decoration taken from the sacred treasury of Delphi.[230]

Furthermore, there are also iconographic depictions of textiles being used as furnishings to supplement architecture in sanctuaries. An example, although somewhat later, is a late 2nd century BC mosaic from Palestrina depicting a Nilotic scene, in which a huge undecorated sunshade/sunshield hangs in front of a tetrastyle temple, fastened at the top of the pitched roof, providing cover for Roman soldiers who are gathered to perform a religious ceremony (Fig. 12).[231] A further example is a marble relief from Greece dated to ca. 200 BC, which depicts a scene from a sanctuary (Fig. 13).[232] The scene takes place outside. In the centre, there is an altar; on the right, are a god and goddess; and on the left is a group of worshippers. On the far left is a large tree around which is tied a textile; a large textile, either a sunshield or a curtain, also hangs from the branches of the tree.[233] It is possible, moreover, that wall-paintings developed from the use of wall-hangings, as a cheaper version of the very costly tapestries.

The use of textiles for architectural interiors is also attested in wall-painting. For example, in the Capitolium in Brescia, northern Italy, the two cellae on the external side have a painted dado covered with representations of drapery attached to the walls by rings, which most probably imitate a practice of covering the walls with drapery.[234] The use of textiles as wall-hangings and as backdrops for statues may also be confirmed by finds from the so-called Hall of the Colossus (*Aula del Colosso*) in the Forum of Augustus in Rome. The hall's back wall, in front of which the colossal

Fig. 12. Mosaic with Nilotic scene, 2nd century BC. © Museo archaeologico Nazionale, Palestrina, inv. no. 149000907.

Fig. 13. Marble relief, ca. 200 BC. © Staatliche Antikensammlungen und Glyptothek München, inv. no. 206. Photo: Christa Koppermann.

statue stood, was faced with rectangular white marble slabs at least as high as the 11–12 m statue.²³⁵ The marble panels are slightly curved, imitating the folds of a decorated fabric, which formed the background to the colossal statue. Examinations of the fragments have shown that the decoration consisted of different decorative red bands on a blue background.²³⁶

Unfortunately, in terms of real textiles, there are almost no preserved wall-hangings, curtains, tents, etc. from the 1st millennium BC Mediterranean, and it is therefore difficult to say much about their appearance and use in sanctuaries. Instead, such items are primarily preserved from the Late Antique Period onwards. One rare example, though, is the famous textile found in The Tomb of the Seven Brothers on the Taman peninsula, southern Russia: a resist-dyed wool coverlet or hanging bearing mythological scenes (Fig. 14).²³⁷ Despite the lack of finds, textile furnishings are usually considered to have been widespread in antiquity, and were

Fig. 14. Resist-dyed wool sarcophagus cover with mythological scenes. From Semibratny Barrow no. 6, main sepulchre. Early 4th century BC. © State Hermitage Museum, St. Petersburg, inv. no. Sbr. VI-16. Photo: Vladimir Terebenin, Leonard Kheifets, Yuri Molodkovets, Aleksey Pakhomov.

used for various purposes, such as doorway covers, room subdividers, curtains between columns in arcades, wall-hangings, and awnings.[238] According to Schrenk, wall-hangings were widespread and were often very prestigious items, even though they did not attract the interest of ancient artists to the extent that they represented them in wall-paintings, mosaics,[239] or other works of art as much as they represented curtains, sun shields, and other furnishing textiles.[240]

Curtains are known to have been used in religious buildings to protect against the sun and the dust, to protect and adorn entrances and doorways, and to separate areas into a public and a more private sphere or for men and women, initiates and non-initiates. Furthermore, curtains can not only mark and protect a very special place, but also serve decorative purposes.[241] Such curtains are known from the Bible and Talmudic literature: there is the *kila* (a curtain or canopy), which is defined as the

Biblical parochet on the outside or inside the Ark, and the *gulta*, an additional covering placed beneath the *kila*, covering the scrolls.[242] They are also known from iconography: for example, curtains in synagogues are depicted in mosaics.[243] Curtains generally carry great importance in Hebrew, Christian, and Islamic ritual, an importance that arises from their role of concealing what is most sacred. Such curtains were thus highly significant coverings in basilicas, synagogues, and in the holy Kaaba at Mecca, where a new *kiswa* is wowen at each Haaj.[244]

Cushions are also often depicted in art: many occur in red-figure vase-painting along with chairs and *klinai*. Many of them have inwoven decoration, usually in the form of different patterns or stripes, as shown on a *kalyx krater* from the 4th century BC, where a woman is placing a cushion decorated with flower-like motifs and a wave pattern on a chair (*klismos*) (Fig. 15).[245] It is thus interesting

Fig. 15. Woman with cushion and chair. Kalyx krater by the Kadmos painter, 4th century BC. © State Hermitage Museum, St. Petersburg, inv. no. UO-28. Photo: Vladimir Terebenin

that, even though different forms of architectural textiles or furnishing textiles are known to have been in use in Ancient Greece, they are only rarely recorded in the temple inventories, which instead are primarily concerned with garments. Of course, since the majority of Greek garments were rectangular in form, and many of them heavily decorated, they could double as tapestries that were hung on walls or as curtains.[246]

Fabrics and fibres (Tables 13–19)

Animal fibres

Terms for different fibres are rarely recorded in the inventory lists, and some of them are mentioned only in a single inventory. Whether this means that the remaining garments whose fibre is not specified were in fact all of the same fibre remains unclear, but it is possible that for the ancient reader it was obvious that a certain fibre was implied.[247] This implied fibre could possibly be sheep's wool, which is known to have been a common textile fibre.[248] This assumption is supported by the fact that wool is rarely specified in the inventories (Table 13): on Delos, it is recorded twice (a woollen garment and a woollen *chitōn*), on Samos once (a woollen strip), and in Miletos in connection with two *prosōpidia*, two ribbons, and a belt/girdle. Wool is recorded a dozen times in the Brauron catalogues, both as a dedicated item in its own right, and as part of the description of garments. The rarity of woollen garments recorded as such in the inventories is, however, still puzzling. Perhaps this indicates either a preference for e.g. linen garments, either among the donors or the recipient deities, or perhaps this reflects a specific votive tradition in the choice of offerings. That wool is recorded in the inventory from Miletos is possibly related to the fame of Milesian wool. Milesian sheep and wool is mentioned already in 5th century BC literature, where it is ranked first; in epic poetry, it is described as soft and fine, and the sheep as 'pretty-haired' (*kallitricha*).[249] But as Wagner-Hasel notes, it is unclear whether this refers to actual Milesian wool or wool of equal quality produced by sheep from Attica.[250]

Another animal fibre that needs to be considered is silk. Several Ancient Greek terms have been suggested for silk, but only a few can be securely connected to this particular fibre: *bombykinos* and *sērikos*, of which only the first appears in the material under investigation here. It is recorded in the inventory from Miletos, dated to the 1st century BC, where it denotes four *prosōpidia* and two *epikrēna* (Table 14).

Silk is mainly associated with China, where it was cultivated from the 3rd millennium BC.[251] It has long been debated, though, when the fabric was introduced into the Mediterranean world – this is, indeed, not an easy question to answer. Richter has suggested that it was used in Greece already in the 5th century BC.[252] Archaeological evidence does not, however, provide support for such an early presence of silk in Greece. On the contrary, there are several cases where past

Table 13. Wool

Greek term	Garment	Site	Reference
Erion/ereous	estheta	Delos	IG XI,2, 161B, line 62
ereian (restored)	estheta	Delos	IG XI,2, 162B, lines 49–50
ereoun (restored)	chitōn	Delos	ID 1412A, line 35
ereoun	chitōn	Delos	ID 1417A, II, line 21
erineon	strip/ribbon (spleniskon)	Samos	IG XII 6, 1, 261, line 25
erea	2 prosōpidia	Miletos	Milet VI, 3, 1357, line 13
ereoun	belt	Miletos	Milet VI, 3, 1357, line 20
eria	wool	Brauron	IG II², 1514, line 54
eria malaka	soft wool	Brauron	IG II², 1514, line 57
kateirgas	worked/prepared [wool?]	Brauron	IG II², 1517, line 175
ereoun histos	woollen web	Brauron	IG II², 1518B, line 53
ēmiyphē kai eria kai krokēn	hemihyphes and wool and woof thread	Brauron	IG II², 1518B, line 54
eria malaka en kalathiskoi	soft wool in a basket	Brauron	IG II², 1518B, line 57
eria kateirgasmena malaka en kalathi	soft, worked/prepared wool in a basket	Brauron	IG II², 1518B, line 59
eria	wool	Brauron	IG II², 1518B, line 68
eria glaukea en kalathiskoi	grey wool in a basket	Brauron	IG II², 1518B, line 71
eria	wool	Brauron	IG II², 1521B, line 71
eria malaka kateirgasmena	soft worked/prepared wool	Brauron	IG II², 1522, line 26
eria malaka en kalathiskoi	soft wool in a basket	Brauron	IG II², 1524, line 166
eria kateirgasmena malaka en kalathiskoi	soft worked/prepared wool in a basket	Brauron	IG II², 1524, line 168
eria kateirgasmena	worked/prepared wool	Brauron	IG II², 1528, line 13
eria eirgasmena	worked/prepared wool	Brauron	IG II², 1528, line 17
eriōn gla... katakommenon	grey wool, frayed	Athens	IG II², 1469, line 132
eriōn diktyon katakommenon	wool net, frayed	Athens	IG II², 1469, line 134
eria leuka en kalathiskoi	white wool in a basket	Athens	IG II², 1469, line 136
eriōn katakommenon	wool, frayed	Athens	IG II², 1469, line 137
eriōn katakommenon	wool, frayed	Athens	IG II², 1469, line 138
eriōn	wool	Athens	IG II², 1469, line 140

Table 14. Silk (bombykinos)

Greek term	Garment	Site	Reference
bombykina	4 prosōpidia (masks/veils)	Miletos	Milet VI, 3, 1357, line 12
bombykinon	epikrēnon (headdress)	Miletos	Milet VI, 3, 1357, line 15
bombykinon	epikrēnon (headdress)	Miletos	Milet VI, 3, 1357, line 16

identifications of excavated textile fibres as silk have been recanted or are currently being re-evaluated.[253] One such example is from Greece, where half a dozen textile fragments have been recovered from a tomb (35 HTR73) in Kerameikos, Athens, dated to the second half of the 5th century BC. These fibres were until recently believed to be of pure silk of the species *Bombyx mori*,[254] but this has proved not to be the case.[255] Other examples include the excavated textile finds from a tomb, dated to the 6th century BC, at Hochmichele, Germany, which had previously been identified as silk from *Bombyx mori*, but which have now been shown not to be silk.[256] The textile finds from Eberdingen-Hochdorf, Germany, also from a 6th century BC tomb, previously identified as silk, have also been proven not to be silk.[257] Furthermore, the textile remains recovered from an early La Tène burial from Altrier in Luxembourg are now identified as wool.[258] Bender Jørgensen also rejects the argument that the textile pseudomorph preserved on an Etruscan ceramic urn from Chiusi, dated to the 2nd or 1st centuries BC, can be identified as silk – at least on current evidence.[259] Another example is a silk moth cocoon recovered at Akrotiri, which has been taken as an indication of the presence of wild silk in the Aegean from the mid-2nd millennium BC.[260] It is thought that the cocoon possibly belonged to the species *P. otus. S. pyri*, the largest species of moth in Europe.[261] There is, however, no clear indication that the moths were native to Thera/Santorini, rather than imported as cocoons from elsewhere, and the presence of one cocoon does not necessarily mean that silk garments were either produced or worn.[262] Furthermore, the cocoon is difficult to date since it is calcified and recovered from a disturbed context. Thus, there is no solid archaeological evidence for the presence of silk production in Greece in the Classical Period, and it seems that silk can be attested only from the last centuries of the 1st millennium BC in the Mediterranean. Nevertheless, its presence is attested from at least the 1st century BC, as indicated by the inventory from Miletos and literary texts, especially Roman sources.[263]

The inventories testify to a wealth of terms referring to different fibres and fabrics, yet some are not represented. This applies to garments or clothing items made of skins or leather, for example. Leather had a huge range of clothing uses in antiquity, primarily for footwear, hats, and belts, but also for cloaks and armour.[264] Felt is another material that so far has not been identified in this material. The total absence of these materials from the inventories does not mean that they were not used, but only, as in the case of footwear, that leather items were either not dedicated in these sanctuaries or that they were not recorded (or that it was not stated that they were of leather) in these specific lists.

Plant fibres

The Greek word *linon* and its adjective *linous* could be used to denote anything made of linen (Table 15). It occurs in the inventory of the Sanctuary of Demeter and Kore at Tanagra (a purple *chitōn*, a purple garment for a priestess, and a *chitōn* with a purple border). At Delos linen is often recorded, e.g. in connection with a *chitōn*, a wrap and a cloth for Leto, and a robe for Isis; two statues are also described as clothed in linen. In the inventory from Samos linen is used in connection with a *chitōniskos*, a *proslēmma*, a *periblēma*, two headbands (*sphendonai*), cushions, and strips (*splēniskoi*). Interestingly, one of the strips is described as made of both linen and wool. In the inventory from Miletos, linen is specified for an *epikrēnon*, a belt, 12 pieces of linen, and for an *othonion*. Linen is mentioned only once in the Brauron catalogues, where it is specified in connection with a *kandys*. This is somewhat surprising, since linen was an extremely important textile fibre as far back as the evidence for textiles goes,[265] and one would therefore expect many garments in the inventories to be of linen. This might be explained, though, by the possibility that the fibre was implied by the garment type, or perhaps this information was not deemed important enough to be recorded. But it is also likely that the Brauron scribes used other terms to denote linen: for example, it seems that they employed the term *amorginon* for linen fabrics rather than the term *linon*, and, as we shall see below, several other fibre terms might refer to linen.

Another fibre term that might be related to linen is *othonion*, which occurs several times in the inventory from Miletos and on Delos (Table 16). It is thought to designate fine linen, in general, as well as a garment in this fibre.[266] In the New Testament, for example, the term can mean a linen cloth, strips of linen, or bandages, or more specifically bandages of linen cloth for swathing the dead. As shown above, however, an *othonion* in the inventory from Miletos is described as being of linen, which could argue against the interpretation that this garment term always implied a garment of linen. Perhaps this indicates that *othonion* could be both a fibre term and a garment term. The term is also used in the inventory from Delos for both Leto and Hera. It is worth noting that the term does not occur in the Brauron catalogues. This situation may be explained by the possibility that regional differences determined the term used to express linen/flax. The existence of different preferences for certain terms can also explain the absence (or presence)

Table 15. Linen

Greek term	Garment	Site	Reference
Linon/linous	chitōn	Tanagra	SEG 43: 212 (B), line 9
	lininos	Tanagra	SEG 43: 212 (B), line 39
	chitōn	Tanagra	SEG 43: 212 (B), line 44
	chitōnion	Delos	ID 104(26bis), C, line 7
	pillow	Delos	ID 104(26bis), C, line 12
	chitōn + wrap (emphiesmenon)	Delos	ID 1417A, I, lines 100–103
	wraps (emphiesmena)	Delos	ID 1417A, I, lines 49–52
	emphiesmena linois (in the Heraion)	Delos	ID 1417A, II, line 22
	emphiesmena linois (in the Heraion)	Delos	ID 1426B, II, line 22
	emphiesmena linois (in the Heraion)	Delos	ID 1442B, lines 54–56
	emphiesmena linois	Delos	ID 1424B, lines 3–4
	chitōn, chitōn, wrap (emphiesmenon)	Delos	ID 1425 II, lines 14–17
	robe (stolion)	Delos	ID 1442A, lines 52–54
	chitōniskos	Samos	IG XII 6, 1, 261, line 15
	proslēmma	Samos	IG XII 6, 1, 261, line 26
	periblēma	Samos	IG XII 6, 1, 261, line 18
	2 headbands (sphendonai)	Samos	IG XII 6, 1, 261, line 21
	strip (spleniskon)	Samos	IG XII 6, 1, 261, line 25
	epikrenon	Miletos	Milet VI, 3, 1357, line 14
	belt	Miletos	Milet VI, 3, 1357, line 20
	12 pieces of linen	Miletos	Milet VI, 3, 1357, lines 13–14
	othonion	Miletos	Milet VI, 3, 1357, line 10
	liva penois	Brauron	IG II², 1522, line 22

of certain other garment terms in some inventories and areas. Another possible explanation for the use of a certain term in some contexts, but not in others, is chronological. Since the inventories treated here all belong to the late Classical or Hellenistic Periods, though, this explanation seems unlikely. Alternatively, these differences may be explained by different societies employing or emphasising different garments (and garment terms).

According to Cleland, the term *sindonitēs* refers to a particular garment type, but also directly denotes fibre (Table 16).[267] It is used in the inventory from Miletos, where it denotes an old, useless garment, and in the inventory of the Hekatompedon

Table 16. Other fibre terms: othonion, sindon, karpasos

Greek term	Garment	Site	Reference
othonion	othonion (for Hera)	Delos	IG XI,2, 154A, lines 21–22
	othonion (for Hera)	Delos	IG XI,2, 287A, lines 120–121
	othonion (for Leto)	Delos	ID 1428, II, lines 53–58
	othonion (for Leto)	Delos	ID 1433, line 3
	othonion (for Leto)	Delos	ID 1450A, line 200
	othonai linai	Miletos	Milet VI, 3, 1357, line 10
	2 othonai	Miletos	Milet VI, 3, 1357, line 11
karpasos	karpasos	Miletos	Milet VI, 3, 1357, lines 9–10
sindōn/ sindonitēs	sindonōn	Hekatompedon	IG II², 1478, line 17
	sindonitēs	Miletos	Milet VI, 3, 1357, line 10
	sindona	Thebes	IG VII, 2421, line 7
	sindona	Thebes	IG VII, 2421, line 9
	sindonas	Delos	ID 1426B, I, line 19
	2 sindonas	Delos	ID 1425 II, lines 14–17
	sindōn	Samos	IG XII 6, 1, 261, line 19
	sindoniskē	Samos	IG XII 6, 1, 261, line 24
	sindonitēs	Brauron	IG II², 1524B, lines 131–132

(something made of *sindōn*). The term *sindōn* is recorded twice in the inventory from Thebes, one of which is purple, and twice in the inventory from Samos, where it is recorded once in diminutive (*sindoniskē*), and once just as *sindōn*. It occurs only once in the Brauron catalogues. The *sindonitēs* are usually interpreted as a garment of fine cloth, commonly defined as linen (*sindōn*, fine cloth/linen).[268] Given the similarity to the Babylonian fibre term *sindhu*, though, it is also possible that *sindōn* referred to cotton.[269] Cotton originated in India, and was cultivated in Egypt and Arabia from early on. According to Barber, it was imported from these areas into the northern Mediterranean in the middle of the 1st millennium BC.[270] This is supported by archaeological textile finds, e.g. the textile fragments from a burial at Kerameikos, dating to the second half of the 5th century BC,[271] and in the funerary pyre textile from the the Royal Tomb II at Vergina, dating to the 4th century BC.[272] It thus possible that some of the garments recorded in these inventories could be made from cotton, although they are not identified as such in the terminology.

Other terms possibly denoting linen include a rare term, *karpasos*.[273] This term is recorded in the inventory from Miletos, where it is used as a garment term (an old *karpasos*).

Amorgis/amorginos (Table 17) denotes 'Amorgian cloth' or 'of Amorgos-type'. The term is interpreted as meaning 'fine' on the basis of literary sources, which mention how expensive and luxurious these garments were.[274] Furthermore, this fabric is described as thin or transparent, e.g. by Aristophanes, who writes about women who are 'naked in their Amorgian *chitōns*'.[275] Nonetheless, the term is somewhat confusing and perhaps refers to garments, cloth, or fibres from the island of Amorgos or clothing in the style or quality characteristic of Amorgos.[276] It has been suggested in previous scholarship that the word is derived from *amorgeia/amorgē*, which denoted a red colour obtained from a plant on the island of Amorgos.[277] Another suggestion is that this type of fibre was made of an actual textile plant called *amorgis*, which tentatively has been identified as the *malva silvestris*.[278] Richter also argued that it refers to wild silk,[279] which however seems unlikely, and it seems most plausible to assume that the term refers to linen. Marinatos has proposed a different interpretation based on the etymological connection between *amorgina* and *amorgē* (the watery run-off when olives are pressed) and a comparison with modern Greece, where textiles are treated with olive oil to make them shine.[280] Thus, as Spantidaki suggests, the mentioned transparency of these textiles could be achieved by treating extremely fine linen threads with olive oil.[281] *Amorgis/amorginos* is by far the most common fibre description in the Brauron catalogues, where it is recorded ca. 30 times for different garment types (e.g. *chitōniskoi*, *chitōnes*, a *thryphēma*, a *kandys*). From a statistical perspective, it might therefore be tempting to say that most of the dedicated garments were made of linen, but this is not necessarily the implication. Rather, garments were so described because their design or quality was distinctive (not ordinary linen), and we can thus infer that *amorgis/amorginos* was not a standard linen fibre for garments, although the catalogue indicates that it was rather common among the dedicated garments.[282] Among other inventories, it is only recorded in the inventory from Tanagra, where it is specified in connection with a girl's *chitōn*.

Another fibre term is *stuppinos* (Table 18). It is not entirely clear whether the term relates to fibre or quality. It is defined as meaning 'of, or like, the coarse fibre of flax, hemp, tow', and it thus refers to coarse, woven fabric of either flax or hemp.[283] In the Brauron catalogues, *amorginos* is contrasted with *stuppinos* in a way that suggests that their primary reference is to the quality of cloth, rather than origin or style,[284] and it seems likely that there is a contrast between *amorginos* (fine) and *stuppinos* (coarse).[285] *Stuppinos* is the second most common fabric description in the Brauron catalogues, and it is also recorded in the inventory of the Hekatompedon, where it is used in connection with a *chitōn*, and in the inventory from Samos it is recorded three times in relation to *mitrai* and a *chitōn*. Thus, both *amorginos* and *stuppinos* appear to be plant fibres.

Another fibre or fabric term is *tarantīnon* (Table 19). The term is recorded in the inventory from Tanagra, where it is used for a ragged garment and two girl's *chitōnes* with dark borders. In the inventory from Thebes, it is listed three times, and once

Table 17. Amorgis/amorginos

Greek term	Garment	Site	Reference
amorgis/amorginos	girl's chitōn	Tanagra	SEG 43: 212 (B), line 7
amorginon	chitōnion	Delos	ID 104(26bis), C, line 8
amorginon	chitōnion	Delos	ID 104(26bis), C, line 10
amorginon		Brauron	IG II², 1514, line 2
amorginon	chitōn	Brauron	IG II², 1514, line 10
amorginon	chitōn	Brauron	IG II², 1514, line 22
amorginon	chitōnion aploun	Brauron	IG II², 1514, line 51
amorginon	chitōnion	Brauron	IG II², 1514, line 61
amorginon	chitōnion	Brauron	IG II², 1514, line 63
amorginon	chitōnion diploun	Brauron	IG II², 1514, line 64
amorginon	chitōnion isoptyches	Brauron	IG II², 1514, line 65
amor...		Brauron	IG II², 1517B, line 165
amorg..	chitōniskos	Brauron	IG II², 1524B, line 132
amor...	chitōnion	Brauron	IG II², 1517B, line 120
amor...	chitōnion	Brauron	IG II², 1518B, line 65
amor...	chitōnion	Brauron	IG II², 1518B, lines 65–66
amorgi	chitōnion	Brauron	IG II², 1518B, line 69
amorgi	chitōnion	Brauron	IG II², 1518B, line 70
amorgi	chitōn	Brauron	IG II², 1424B, line 211
amorgi	diptergon	Brauron	IG II², 1424B, line 214
amorgi	chitōn	Brauron	IG II², 1424B, line 216
amorgi	kandys	Brauron	IG II², 1424B, line 217
amorgi		Brauron	IG II², 1424B, line 235
amorginon		Brauron	IG II², 1424B, line 236
amorginon		Brauron	IG II², 1522, line 15
amorginon	chitōn	Brauron	IG II², 1523, lines 20–21
amorginon	tryphēma	Brauron	IG II², 1523, line 21
amorginon	chitōn	Brauron	IG II², 1528, line 19
amor	chitōn	Brauron	IG II², 1529, line 2
amorg	chitōn	Brauron	IG II², 1529, lines 7–8
amorgi	chitōnion	Brauron	IG II², 1529, line 18
amor	chitōn	Brauron	IG II², 1530, line 4

Table 18. Stuppinos

Greek term	Garment	Site	Reference
Stuppinos	chitōniskos	Brauron	IG II², 1517B, line 127
	chitōnion	Brauron	IG II², 1517B, line 125
		Brauron	IG II², 1517B, line 178
	chitōnion	Brauron	IG II², 1518B, line 66
	chitōnion	Brauron	IG II², 1523, line 27
	chitōn	Brauron	IG II², 1523, line 12
	chitōn	Brauron	IG II², 1523, line 13
	chitōn	Brauron	IG II², 1523, line 17
		Brauron	IG II², 1528, line 12
	chitōn	Brauron	IG II², 1529, line 15
	chitōn	Brauron	IG II², 1529, line 16
	chitōn	Brauron	IG II², 1529, line 16
	chitōnion	Brauron	IG II², 1529, line 17
		Brauron	IG II², 1529, line 20
	chitōn	Hekatompedon	IG II², 1414, line 26
	mitrē	Samos	IG XII 6, 1, 261, line 17
	chitōn	Samos	IG XII 6, 1, 261, line 20
	2 mitrai	Samos	IG XII 6, 1, 261, lines 36–37

it is said to be edged with purple. In Brauron, *tarantīnon* is registered several times: as a *tarantīnon* around the old statue,[286] simply as an unspecified garment, and as a half-woven (*hemihyphes*) *tarantīnon*,[287] and the term is preserved without descriptive context in six instances. It has been interpreted as a garment made of a thin, semi-transparent cloth from Taranto ('made of Tarentine cloth').[288] The term is similar in nature to *amorgis/amorginos*, and has similar problems, as noted above. It has also been argued that the garment is determined by its fibre, and it has been suggested that it was woven from sea-silk,[289] which however is very unlikely, since this particular fibre is very rare and not attested until the Late Roman Period. In sum, we do not know if the term *tarantīnon* designates place of origin, quality, fibre, or a special type of garment. Yet, in consideration of Roman sources, the term might also refer to wool. As Wagner-Hasel demonstrates, Milesian and Tarantine sheep represented a form of conspicuous consumption in the literature of the Roman Imperial Era, and Roman writers provide numerous references to differences in the quality of wool: according

Table 19. Tarantinon

Greek term	Garment	Site	Reference
Tarantinon	tarantinon	Tanagra	SEG 43: 212 (B), line 37
	girl's tarantinon	Tanagra	SEG 43: 212 (B), line 46
	girl's tarantine chitōn	Tanagra	SEG 43: 212 (B), line 49
	tarantina (plural)	Thebes	IG VII, 2421, line 3
	tarantinon	Thebes	IG VII, 2421, line 4
	tarantinon	Thebes	IG VII, 2421, line 4
	tarantinon	Delos	ID 104(26bis), C, line 13
	tarantinon	Brauron	IG II^2, 1514, line 37
	tarantinon	Brauron	IG II^2, 1514, line 70
	tarantinon	Brauron	IG II^2, 1514, line 4
	tarantinon × 2	Brauron	IG II^2, 1514, line 68
	tarantinon	Brauron	IG II^2, 1522, line 26
	tarantinon	Brauron	IG II^2, 1517B, line 134
	tarantinon	Brauron	IG II^2, 1518B, line 49
	tarantinon	Brauron	IG II^2, 1524B, line 227
	tarantinon	Brauron	IG II^2, 1523, line 2

to Columella, the Calabrian, Apulian, and Milesian sheep breeds were of outstanding excellence, while the Tarantine was the best of all. He even states that the Tarentines were called Greek sheep.[290] Other authors include Pliny, who ranks Apulian wool first in the hierarchy of wool qualities,[291] and Juvenal ranks Canusian sheep from Apulia among the luxury goods a wife might demand of her husband.[292] Thus, it is a possibility that Taranto already in the Hellenistic Period was renowned for its wool, and thus gave the designation 'Tarantine' to specific textiles.

Several of these different terms (*amorginon, tarantinon, karpasos*) illustrate how the Ancient Greeks often used toponyms in their designations for textiles, a practice which is well-attested already in the 3rd and 2nd centuries BC.[293] As Harlow and Nosch argue, it would be futile to search for special techniques, tools, or textiles in specific places. Instead, these topological textile terms mirror both place of production, origin, and place of sale, but, over time, they come to refer to qualities and types as well as origin.[294] According to Wagner-Hasel, however, 'designations of origin or quality do not refer, as older research suggests, to centres of production where these fabrics were produced, but rather to the places where they were traded: to wool markets.'[295] Similarly, Bücher has argued that designations for articles of clothing, as for example the Milesian *stromata*, do refer not to place of production and consequently a Milesian export industry, but rather to quality labels.[296] This is most likely also the situation in Greece, where these designations do not necessarily refer to imports from these places, but rather specific qualities. These terms were thus originally associated with specific geographical locations, but eventually come to define specific qualities.

Colours and dyes (Tables 20–24)

The inventories describe the colour of garments in relatively few instances, and, if they do so, it usually relates to purple, which thus appears to have been the most important colour to emphasise.

Different terms are used to denote the colour purple in the inventories: *halourgos*, *porphyros*, and *phoinix/phoinikos*.[297] *Halourgos* and its compound terms *parhalourgos* ('with purple borders') and *meshalorgos* ('purple in the middle') is the most commonly used term for purple; second is *porphyros* and its compound terms *periporphyros* ('with purple edges') and *parporphyros* ('with purple borders'); and third is *phoinix*, which is used only twice, once in Brauron and once as a purple garment (*phoinikis*) in an inventory from Delos (Tables 20–21).[298]

Halourgos and its associated terms are recorded 27 times in the Brauron catalogues for different types of garments (*chitōniskoi*, *xenikē*, *chitōnion*, *himatia*, and a *chitōn*), and at Samos eight times (*chitōniskos*, *paralassis*, *proslēmma*, *perizōma*, *kekryphalos*, Lydian *chitōnes* and *chlandia*). The term is used twice in the inventories of the Athena Temple on the Athenian Acropolis (a *mitra* and a *chitōniskos*), twice in the inventories from Delos (two grey *himatia* with purple borders and a *chitōnion*), and twice in the inventory from Miletos.[299]

Porphyros and its compound terms are not recorded at Brauron, but occur five times in the inventory from Tanagra (four *chitōnes* and a linen garment), seven times in the inventory from Thebes (a *tarantinon*, *chitōnes*, *sindōna*, and a *skistos*), eight times in the inventories from Delos, and twice in Miletos (*himatia*). These descriptions include garments dyed completely purple and those with purple borders or edges. For example, white *himatia* with purple borders are relatively common amongst the dedicated garments. These colours seem to have been important features of this garment type, but were not necessarily defining characteristics. Therefore, they were probably important mainly in contrast to more common features possessed by the *himation* in everyday use.[300]

All three terms are considered designations of the so-called 'Royal purple' or 'Tyrian purple', which was the most famous dye in antiquity. The dye is exclusively derived from animal sources, more precisely molluscs. Several species of molluscs were used in antiquity to produce purple dye: two of them are murexes, *Bolinus brandaris* (=*Murex brandaris*) and *Hexaplex trunculus* (=*Murex trunculus*), and one is a type of rock shell, *Stramonita haemastoma* (=*Thais haemastoma*), all three species abundant in the Mediterranean Sea. The dye itself derives from the tiny hypobranchial gland situated diagonally opposite the aperture of the shell.[301] For the dyeing process immense numbers of glands extracted from fresh molluscs are required. The production

Table 20. Halourgos

Greek term	Colour	Garment	Site	Reference
halourgos	sea-coloured/purple	chitōniskos	Brauron	IG II² 1514, line 12
halourgos	sea-coloured/purple	chitōniskos	Brauron	IG II² 1514, line 14
halourgis	sea-coloured/purple	xenikē	Brauron	IG II² 1514, line 49
halourgēs	sea-coloured/purple		Brauron	IG II² 1517B, line 173
halourg	sea-coloured/purple		Brauron	IG II² 1525, line 1
halourg	sea-coloured/purple		Brauron	IG II² 1524B, line 130
halourgē	sea-coloured/purple		Brauron	IG II² 1524B, lines 231–32
halourgoun	sea-coloured/purple		Brauron	IG II² 1522, line 8
halourgēs	sea-coloured/purple		Brauron	IG II² 1522, line 20
halourgou	sea-coloured/purple		Brauron	IG II² 1522, line 23
parhalourgos	sea-coloured/purple	chitōnion	Brauron	IG II² 1522, line 24
halourga	sea-coloured/purple		Brauron	IG II² 1528, line 11
halourgidos	a sea-coloured/purple garment	halourgidos	Brauron	IG II² 1514, line 56
halourgida	a sea-coloured/purple garment	halourgida	Brauron	IG II² 1517B, line 166
halourgida	a sea-coloured/purple garment	halourgida	Brauron	IG II² 1518, lines 50–51
meshalourgē	sea-coloured/purple in the middle	chitōniskos	Brauron	IG II² 1523, lines 16–17
meshalourgos	sea-coloured/purple in the middle	?	Brauron	IG II² 1529, line 9
meshalourgos	sea-coloured/purple in the middle	himatia	Brauron	IG II² 1522, line 234
mesmohalourg	sea-coloured/purple in the middle	chitōn	Brauron	IG II² 1529, line 8
parhalourgē	with sea-coloured/purple border	chitōniskos	Brauron	IG II² 1523, line 18
parhalour	with sea-coloured/purple border	chitōniskos	Brauron	IG II² 1529, line 11
parhalourgēs	with sea-coloured/purple border	himation	Brauron	IG II² 1514, line 27

(Continued)

2. The temple inventories

Table 20. *Halourgos (Continued)*

Greek term	Colour	Garment	Site	Reference
parhalourgēs	with sea-coloured/purple border	white himation	Brauron	*IG* II² 1514, line 69
parhalourg	with sea-coloured/purple border		Brauron	*IG* II² 1517B, line 125
parhalour	with sea-coloured/purple border	himation	Brauron	*IG* II² 1518B, line 60
parhalourgēs	with sea-coloured/purple border	(chitōniskos?)	Brauron	*IG* II² 1517B, line 132
parhalourg	with sea-coloured/purple border	himation	Brauron	*IG* II² 1529, line 16
halourgos	sea-coloured/purple	chitōniskos	Samos	*IG* XII 6, 1, 261, lines 15–16
halourgos	sea-coloured/purple	paralassis with a p. iris	Samos	*IG* XII 6, 1, 261, line 19
parhalourgos	with sea-coloured/purple border	proslēmma	Samos	*IG* XII 6, 1, 261, lines 20–21
halourgos	sea-coloured/purple	perizōma	Samos	*IG* XII 6, 1, 261, line 22
halourgos	sea-coloured/purple	kekryphalos	Samos	*IG* XII 6, 1, 261, line 23
halourgos	sea-coloured/purple	Lydian chitōnes	Samos	*IG* XII 6, 1, 261, lines 27–28
halourgos	sea-coloured/purple	chlandion	Samos	*IG* XII 6, 1, 261, line 30
halourgos	sea-coloured/purple	2 chlandia	Samos	*IG* XII 6, 1, 261, line 36
halourgos	sea-coloured/purple	chitōniskos	Athens	*IG* II², 1475, line 7
halourgos	sea-coloured/purple	mitra	Athens	*IG* II², 1448, line 4
parhalourgos	with sea-coloured/purple border	2 grey himatia	Delos	*ID* 104(26bis), C, lines 3–11
meshalourgos	sea-coloured/purple in the middle	chitōnion amorginon	Delos	*ID* 104(26bis), C, lines 3–11
halourgea	sea-coloured/purple		Miletos	*Milet* VI, 3, 1357, line 7
halourgea	sea-coloured/purple		Miletos	*Milet* VI, 3, 1357, line 23

Table 21. Porphyros, phoinix

Greek term	Colour	Garment	Site	Reference
Porphyros	purple	a boy's chitōn	Tanagra	SEG 43: 212 (B), line 6
porphyros	purple	linen chitōn	Tanagra	SEG 43: 212 (B), line 9
parporphyros	with purple borders	lininos (linen garment?)	Tanagra	SEG 43: 212 (B), line 39
parporphyros	with purple borders	linen chitōn	Tanagra	SEG 43: 212 (B), line 44
parporphyros	with purple borders	chitōn	Tanagra	SEG 43: 212 (B), line 50
parporphyros	with purple borders	tarantinon	Thebes	IG VII, 2421, line 4
parporphyros	with purple borders	chitōn	Thebes	IG VII, 2421, line 5
parporphyros	with purple borders	girl's yellow chitōn	Thebes	IG VII, 2421, line 6
porphyros	purple	sindōna	Thebes	IG VII, 2421, line 7
periporphyros	with purple edges	skistos (open garment)	Thebes	IG VII, 2421, line 8
porphyras	purple	sindōna	Thebes	IG VII, 2421, line 10
porphyras	purple		Thebes	IG VII, 2421, line 11
porphyras	purple dye	for a himation	Delos	IG XI,2, 203A, line 72–73
porphyras	purple dye	for a himation(?)	Delos	IG XI,2, 204, lines 75–76
porphyra	purple	garments (endymata)	Delos	ID 1417A, I, lines 49–52
porphyra	purple	garments (endymata)	Delos	ID 1424B, lines 3–4
periporphyros	with purple edges	kyklos (circle)	Delos	ID 1428, II, lines 53–58
periporphyros	with purple edges	kyklos (circle)	Delos	ID 1429B, lines 1–4
periporphyros	with purple edges	kyklos (circle)	Delos	ID 1433, lines 3–7
porphyran	purple	garment (estheta)	Delos	ID 1442B, lines 54–56
periporphyros	with purple edges	himation	Miletos	Milet VI, 3, 1357, line 6
porphyra	purple	himatia	Miletos	Milet VI, 3, 1357, line 9
phoinikion			Brauron	IG II², 1514, line 41
phoinikis	a purple garment		Delos	IG XI,2, 203A, lines 72–73
euparyphos	with a purple border	chlandion	Miletos	Milet VI, 3, 1357, line 22

of purple dye was thus generally a laborious process, and the final product was therefore extremely valuable.[302] Knowledge of purple dyeing process was lost during Late Antiquity or the early medieval period, but was rediscovered by William Cole in 1684.[303] Since then, many scholars have conducted experiments to determine the exact methods of dye extraction, the dyeing process, and – not least – the range of colours that can be obtained from this type of dye. These experiments have shown that the outcome of purple dyeing is never exactly the same. The colour of the dyed fibre depends on whether it is exposed to sunlight, the amount of glands used, the freshness of the molluscs, the species of mollusc,[304] the dyeing process, and other factors. This means that mollusc purple dye can provide a wide variety of colours, ranging from red to violet to blue.[305]

Scarlet, denoted by the term *kokkinos*, is recorded only once among these inventories. It occurs in the inventory from Miletos, where it describes a ribbon. *Kokkinos* refers to dyer's kermes obtained either from the whole body of the adult female or from the unlaid eggs of a small insect called *Kermococcus vermilio* of the family *Kermesidiae*, which lives exclusively on the Kermes oak (*Quercus coccifera*) native to the Mediterranean region.[306] It produced a brilliant and colourfast dye and is said to be the source of the most highly prized and expensive red dye that ever existed, known in the medieval west as 'scarlet'.[307] It is mentioned by several ancient authors, e.g. Theophrastos, Pausanias, Dioscurides, and Pliny.[308] Direct evidence of this dye has been found very early in a Neolithic cave burial in France,[309] but textiles dyed with kermes have also been recovered in the Mediterranean area, e.g. in the burial towers at Palmyra.[310]

Blue is a rarely recorded colour in the inventories, and it occurs only three times in the inventory from Samos, where it is used to describe the colour of the edges of Lydian *chitōnes* (see Table 24). Two different terms are used: *Isatis* and *hyakinthos*. *Isatis* denoted the colour blue, since the term derives from the Greek name for woad (*isatis tinctoria*) – a plant dye that can be used to obtain a good and very permanent blue.[311] *Hyakinthos* is originally the name of a violet-coloured flower and is commonly thought to describe the colour blue.[312]

White (*leukos*) is a relatively common colour term in the inventories (Table 22). At Brauron, ca. 20 items are described as white (six *chitōniskoi*, a *chitōnion*, seven *himatia*, two *enkykloi*, a *chlaniskion*, a *zōma*, and a *kekryphalos*); at Tanagra white is noted for four garments (*ampechonon, chlanisdiskion*, a headdress (*tegidion*), and 'a white garment'); at Delos white is specified in connection with a woollen garment; and white is recorded twice at Samos for a *himation* and the edges of a Lydian *chitōn*. At Brauron, the *chitōn* is one of the few major garments for which white is not recorded. In fact, of the garment types overall, the *chitōn* has the lowest proportion of described pattern decoration, and is never described in terms of base colour. Cleland therefore argues that this implies that the *chitōn* typically had the neutral colour of its fibres, such as white and beige.[313] Yet, as shown, the inventory from Samos records a Lydian *chitōn* with white edges, which might argue against such an interpretation.

White was not necessarily the natural colour of textiles, as is also the case with sheep's wool. The colour of the coat changes with domestication of sheep: the coats of wild sheep are brown, while a domesticated sheep's coat is either black, grey, spotted, or (the rarest) white.[314] The natural colour of wool could thus range from yellowish

Table 22. White

Greek term	Colour	Garment	Site	Reference
Leukos	white	ampechonon	Tanagra	SEG 43: 212 (B), line 11
	white	a white	Tanagra	SEG 43: 212 (B), line 23
	white	chlanidiskan	Tanagra	SEG 43: 212 (B), line 33
	white	tegidion	Tanagra	SEG 43: 212 (B), line 38
	white	chitōn	Delos	ID 1412A, line 35
	white	chitōn	Delos	ID 1417A, II, line 22
	white	Lydian chitōn	Samos	IG XII 6, 1, 261, line 16
	white	himation	Samos	IG XII 6, 1, 261, line 27
	white	himation	Brauron	IG II², 1514, line 16
	white	himation	Brauron	IG II², 1514, line 20
	white	himation	Brauron	IG II², 1514, line 27
	white	chlaniskion	Brauron	IG II², 1514, line 40
	white	chitōniskos	Brauron	IG II², 1514, line 46
	white	himation	Brauron	IG II², 1514, line 69
	white	kekryphalos	Brauron	IG II², 1525, line 4
	white	chitōnion	Brauron	IG II², 1525B, line 133
	white	chitōniskos	Brauron	IG II², 1517B, lines 123–124
	white	chitōniskos	Brauron	IG II², 1517B, line 136
	white	himation	Brauron	IG II², 1517B, line 138
	white	himation	Brauron	IG II², 1524B, line 205
	white	himation	Brauron	IG II², 1524B, line 210
	white	enkyklon	Brauron	IG II², 1524B, line 206
	white	enkyklon	Brauron	IG II², 1524B, line 223
	white		Brauron	IG II², 1524B, line 238
	white	chitōniskos	Brauron	IG II², 1522, line 21
	white	chitōniskos	Brauron	IG II², 1523, line 14
	white	zōma	Brauron	IG II², 1523, line 16
	white	chitōniskos meshalourgos	Brauron	IG II², 1523, line 16

white to shades of dark brown. Linen fibres contain natural pigments, which in their natural colours fall in the blonde, cream to grey range, and which can be enhanced or modified during fibre processing, since linen bleaches well and its natural shine can be enhanced by 'polishing'.[315] Scarcely anything is known of the art of bleaching as practised in ancient Greece. One of the few ancient authors to mention the practice is Pliny:

> 'There is another kind of wild poppy, known as "heraclion" by some persons, and as "aphron" by others. The leaves of it, when seen from a distance, have all the appearance of sparrows; the root lies on the surface of the ground, and the seed has exactly the colour of foam. This plant is used for the purpose of bleaching linen cloths in summer.'[316]

Theophrastos of Eresos also attributed bleaching qualities to a plant called Herakleia, which he compares to the soap-wort.[317] Soap-wort, *Saponaria officinalis* in Latin or *strouthos* in Greek, is not a dye plant in the strictest sense, but was used for cleaning and for whitening or bleaching linen.[318]

More is known about bleaching in ancient Egypt, where linen cloth was usually bleached. Middle Kingdom paintings depict cloth being wetted beside a river and rubbed with natron or potash,[319] then beaten over a stone or wood base and rinsed in flowing water, before being bleached in the sun. In the New Kingdom linen was sometimes boiled with potash in a vat to bleach it.[320] The same method of bleaching was likely to have been used in ancient Greece. Wool was not bleached in the same manner as linen, since this method would ruin the fibres, but wool fibres are ideal for dying because they take dyes very well. Linen is much more difficult to dye than wool, which makes it unlikely that brightly coloured garments were made from this fibre alone.[321] We do, however, have examples of dyed linen in the archaeological record, so this possibility cannot be wholly excluded.[322] It is most probable, however, that linen was used primarily in its natural colours or bleached, and rarely dyed – and thus as an ideal choice for white garments, whereas wool would be the obvious choice for dyed garments.

Yellow is another quite common colour in the inventories, and is denoted by different terms (Table 23). One term for yellow is *mēlinos*, which can be translated as 'apple-coloured' or 'quince-yellow'.[323] At Brauron the term is used once for a *chitōniskos*, once at Thebes for a girl's *chitōn*, and at Tanagra for two *chitōnes*, an open garment and 'a yellow'. A second term is 'broom-yellow' (*thapsinos*), which is recorded only once in the Brauron catalogues for a *chitōnion*. *Thapsinos* derives from *thapsos* or fustic (*Cotinus coggyria = Rhus cotinus*), a bright yellow dye, imported from the island of Thapsos.[324] A third term for yellow in the inventories is *krokōtos*, which derives from *krokos*, the Greek word for crocus, implying its origin from saffron and thus indicative of saffron-coloured garments. The dye stuff is made from the dried stigmas of the crocus flower (*crocus sativus*), and is obtained by a very labour intensive production process, which must be done by hand.[325] An extract of the dried stigmas

Table 23. Yellow

Greek term	Colour	Garment	Site	Reference
mēlinos/malinos	quince-coloured	chitōn	Tanagra	SEG 43: 212 (B), line 17
	quince-coloured	a yellow	Tanagra	SEG 43: 212 (B), line 23
	quince-coloured	open garment (skistos)	Tanagra	SEG 43: 212 (B), line 30
	quince-coloured	chitōn	Tanagra	SEG 43: 212 (B), line 34
	quince-coloured	girl's chitōn	Thebes	IG VII, 2421, line 6
	quince-coloured	chitōniskos	Brauron	IG II², 1524B, line 132
thapsinos	broom-yellow	chitōnion	Brauron	IG II², 1522, line 24
krokōtos	saffron-coloured	krokōtos	Tanagra	SEG 43: 212 (B), line 8
	saffron-coloured garment	krokōton diploun	Brauron	IG II², 1514, line 61
	saffron-coloured garment	krokōton diploun	Brauron	IG II², 1514, line 62
	saffron-coloured garment	krokōton	Brauron	IG II², 1516, line 52
	saffron-coloured garment	krokōton	Brauron	IG II², 1524B, line 213
	saffron-coloured garment	krokōton	Brauron	IG II², 1524B, line 235
	saffron-coloured garment	krokōton	Brauron	IG II², 1522, line 9
	saffron-coloured garment	krokōton	Brauron	IG II², 1522, line 12
	saffron-coloured garment	krokōton	Brauron	IG II², 1528, line 13
	saffron-coloured garment	krokōton	Brauron	IG II², 1529, line 8
	saffron-coloured garment	krokōtinon	Brauron	IG II², 1529, line 17
	saffron-coloured garment	krokōtion	Brauron	IG II², 1529, line 18
	saffron-coloured	chitōniskos	Brauron	IG II², 1514, line 58
	saffron-coloured	tryphēma	Brauron	IG II², 1517B, line 162
	saffron-coloured	krokōton diploun poikilēn	Brauron	IG II², 1522, line 28
	saffron-coloured	chitōniskos	Brauron	IG II², 1528, lines 22–23

in boiling water will rapidly dye silk, wool, and vegetable fibres yellow or orange even without a mordant, the intensity depending on the amount of colorants used.[326] The term *krokōtos* is recorded 15 times in the Brauron catalogues, but only once at Tanagra. The term is used to describe the colour of garments, in Brauron for two *chitōniskoi* and a *tryphēma*. It is, however, not only an adjective, but also a noun for garments – a *krokōtos*.[327] As a noun, it is recorded 11 times in the Brauron catalogues and once at Tanagra. *Krokōtoi* are known from ancient literary sources. For example, Aristophanes records that little girls at the age of ten wore/shed a saffron robe (*krokōtos*) when performing a specific rite for Artemis Brauronia in her sanctuary on the eastern coast of Attica.[328] During this rite, the young initiates took the name 'bear' in commemoration of a local myth in which a bear was killed after injuring a young girl. The death of the bear aroused Artemis's anger, and she sent a plague that could only be stopped if the Athenians sent their daughters to 'play the bear' in local rituals called the *Arkteia*.[329] Perhaps the saffron-coloured garments recorded in the catalogues are these specific garments worn by the young female initiates. A saffron-coloured garment is also recorded in the inventory from Tanagra, probably for Demeter and Kore, which decreases the possibility that all saffron-coloured garments (*krokōtoi*) are related to Artemis, Brauron, and the *arktoi*. Nevertheless, the saffron colour might have had special ritual connotations, underlined by the fact that there are examples of the use of saffron-coloured garments for divinities in literary sources. For example, in Euripides's tragedy *Hecuba*, the *peplos* offered to Athena at Athens is described as saffron-coloured.[330] The term *krokopeplos* is often used in myth and epic to refer to nymphs and goddesses, e.g. in the *Iliad*, where Eos is wearing a *krokopeplos*.[331] Hesiod uses the epithet for Enyo, the goddess of war, and Telesto, one of the Okeanides.[332] Alcman provides another example when he describes the Muses as saffron-robed.[333] Dionysos is also said to be wearing a *krokōtos*.[334] Pindar depicts the newborn Herakles swaddled in saffron yellow cloth, and Dioskurides calls saffron the 'blood of Herakles'.[335] But *krokōtos* is also used to describe the garments of mortal elite women: at the sacrifice of Iphigenia, Aeschylos writes that she 'shed to earth her saffron robe' (*krokou baphas*).[336] Other examples are present in Euripides, where Antigone's robe is saffron-coloured ('casting from my hair its mantle and letting my delicate saffron robe fly loose, a tearful escort to the dead')[337] and in Aristophanes' *Lysistrata*, where Athenian women wear saffron-coloured garments.[338] The description of the garments as saffron-coloured might be connected with the economic value of the saffron-dye, which was an expensive commodity and, like purple, a marker of social position.[339]

Other colours are quite rare in the inventories (Table 24). For example, grey (*phaios*) only occurs in the inventory of the Athenian Asklepieion (*chlamys*) and in an inventory from Delos (two *himatia*). Bluish-grey (*glaukinos*) is used at Brauron for two *chitōniskoi* and in the inventory from Miletos for a *kalasiris* (*mesoglaukinos*). Green – or in this case 'frog-coloured' (*batracheious/batrachis*) – is another rarely noted colour: it is only recorded at Brauron, where it is used five times, for example in relation to

Table 24. Other colours

Greek term	Colour	Garment	Site	Reference
parorphnidōtos	with dark borders	a girl's chitōn	Tanagra	SEG 43: 212 (B), line 40
parorphnidōtos	with dark borders	a girl's chitōniskos	Tanagra	SEG 43: 212 (B), line 41
parorphnidōtos	with dark borders	a girl's chitōn	Tanagra	SEG 43: 212 (B), line 42
parorphnidōtos	with dark borders	a girl's chitōn	Tanagra	SEG 43: 212 (B), line 45
parorphnidōtos	with dark borders		Tanagra	SEG 43: 212 (B), line 46
parorphnidōtos	with dark borders	a girl's chitōn	Tanagra	SEG 43: 212 (B), line 47
parorphnidōtos	with dark borders	a girl's chitōn	Tanagra	SEG 43: 212 (B), line 48
parorphnidōtos	with dark borders	a girl's chitōn	Tanagra	SEG 43: 212 (B), line 49
phaios	grey	himation	Delos	ID 104(26bis), C, lines 3–4
phaios	grey	himation	Delos	ID 104(26bis), C, lines 5–6
phaios	grey	chlamys	Asklepieion, Athens	IG II², 1533, line 18
glaukeious	bluish-grey	chitōniskos	Brauron	IG II², 1522, line 25
glaukeious	bluish-grey	chitōniskos	Brauron	IG II², 1518B, line 52
mesoglaukinos	bluish-grey in the middle	kalasiris	Miletos	Milet VI, 3, 1357, line 5
isatidos	blue	Lydian chitōn (borders)	Samos	IG XII 6, 1, 261, line 13
hyakinthos	blue/hyacinth-coloured	Lydian chitōn (borders)	Samos	IG XII 6, 1, 261, lines 13–14
hyakinthos	blue/hyacinth-coloured	Lydian chitōn (borders)	Samos	IG XII 6, 1, 261, line 14
kokkinos	scarlet/crimson	spleniskos	Miletos	Milet VI, 3, 1357, line 17
batrachis	frog-coloured		Brauron	IG II², 1514, line 48
	frog-coloured	chitōniskos	Brauron	IG II², 1517B, line 137
	frog-coloured	chitōniskos	Brauron	IG II², 1523, line 14
	frog-coloured	chitōniskos	Brauron	IG II², 1523, line 24
	frog-coloured	kandys?	Brauron	IG II², 1524B, line 220
prasinos	green	2 ribbons	Miletos	Milet VI, 3, 1357, line 17

three *chitōniskoi* and a *kandys*. Another term for green (*prasinos*) is used once in the inventory from Miletos. Finally, the inventory of the Temple of Demeter and Kore at Tanagra records a child's silver garment and six girl's *chitōnes* and a girl's *chitōnion* with dark/black borders (*parorphnidōtos*).

In summary, many different colours are recorded: purple, sea-purple, scarlet, blue, grey, blue-grey, white, black, frog-colour, green, saffron-coloured, yellow, broom-yellow, apple/quince colour. Only some of these colour terms refer to the dye substance used to achieve the colour: purple, *kokkinos*, *isatis*, and *krokōtos*, while the others simply refer to the immediate visual experience. The inventories thus testify to a wealth of colourful garments, with a predominance of purple items. This indicates that white was not predominant for Greek clothing. This is worth emphasising, given the common conception of Greek dress, and it is thus incorrect to assume that when the colour of a garment is not described that it was white. Although white is a commonly specified colour, this very fact precludes it from being either the standard or a neutral colour of dress.[340]

Decoration (Tables 25–29)

All of the Greek preserved textiles produced on the warp-weighted loom are tabbies, and, as Spantidaki argues, the ancient sources indicate that the final appearance of Greek textiles did not depend on weaving technique, but rather on a large variety of decorative techniques.[341] Several ways of decorating textiles were known in antiquity: purely decorative techniques, such as floating wefts, painting, printing, resist dyeing, appliqué, sewing, and embroidery; and certain weaving techniques that are constructional and decorative at the same time, e.g. tablet-weaving, tapestry weave, inwoven patterns, strips and checkers, twill, brocading, taqueté, damask, sprang, and needle-binding.[342] Similarly, many different terms denoting the decoration of the textiles exist in Ancient Greek. Equating the two is, however, a very difficult, and often impossible, task. Nevertheless, some information can be gleaned by compiling the different sources.

Pleated textiles (Table 25)

Pleated textiles are recorded in the inventories from Brauron, and are denoted by the term *isoptychēs*, which is usually translated as 'with equal or similar folds'.[343] It is recorded nine times in the catalogues, primarily for *chitōnia* and two *krokōtoi*. Pleated textiles are common in iconography, especially in the Classical Period. As Spantidaki argues, though, not all the depicted examples represent permanent pleats and could also be achieved by specific draping, for instance. She shows, furthermore, that there are two types of pleated garments depicted: those with pleats evenly distributed over the entire garment, and those with a small number of pleats alternating with stretches of smooth unpleated surface.[344] Permanent pleats were either fixed pleats, like the modern kilt, or ironed pleats. As Cleland argues, the pleated textiles recorded in the

Table 25. Pleated textiles

Greek term	Translation	Garment type	Colour	Decoration/fibre	Site	Reference
isoptychēs	with equal folds (pleated)	chitōnion		amorginos	Brauron	IG II², 1514, line 63
isoptychēs	with equal folds (pleated)	chitōnion		amorginos, anepigraphos	Brauron	IG II², 1514, line 65
isoptychēs	with equal folds (pleated)	chitōnion		diploun, anepigraphos	Brauron	IG II², 1514, line 66
isoptychēs	with equal folds (pleated)	chitōnion			Brauron	IG II², 1524B, line 229
isoptychēs	with equal folds (pleated)			amorginos	Brauron	IG II², 1524B, line 236
isoptychēs	with equal folds (pleated)				Brauron	IG II², 1522, line 4
isoptychēs	with equal folds (pleated)	krokōtos			Brauron	IG II², 1522, line 9
isoptychēs	with equal folds (pleated)	krokōtos			Brauron	IG II², 1522, line 12
isoptychēs	with equal folds (pleated)				Brauron	IG II², 1522, line 16

catalogues were permanent pleats, integral to the garments, rather than a result of draping.[345]

Decorated borders (Table 26)
A common type of decoration is decorated borders, which are denoted by different terms, all recorded only at Brauron. Among such terms are *perikumation*, translated by Cleland as 'wavy border (all around)',[346] which is used only once for a *himation*, and the related *parakumatios* 'wavy border',[347] which is used for a *chitōniskos*. Another term is *parhyphes*, translated as 'with a border',[348] and used three times in relation to a *chitōniskos*, a *lēdion*, and a *tryphēma*. Other terms are: *periēgētos*, meaning 'with a border around',[349] which is recorded 7 times, 6 times for *chitōniskoi* and once as a noun; and *ktenotos*, translated by Cleland as 'spiky-border',[350] recorded 11 times, almost exclusively for *chitōniskoi*. *Thermastis*, translated by Cleland as 'a woven border of tong pattern',[351] is recorded only once. A final term is *purgōtos*, which Cleland translates as 'edged with a border like battlements',[352] and thus the well-known battlemented stitch, made to represent the top of castles. To these examples should of course be added the garments with purple borders described above. The table makes it very evident that these decorated borders primarily, if not almost exclusively, were used for *chitōniskoi*. All of these borders were probably made as inwoven decoration, or as

borders made by tablet weaving and then either applied on the garment or integrated into the weave as a starting or finishing border,[353] since it would be too time consuming and profitless to make them in embroidery.

Patterned textiles (Table 27)
One of the several terms to describe decoration of textiles is *poikillō*, which is translated either as 'work in various colours' or 'work in embroidery',[354] while *poikilos* is translated as 'many-coloured', 'dappled', 'spotted',[355] or simply as 'patterned'.[356] *Poikilos* occurs several times in the Brauron catalogues for different types of garments (e.g. *chitōniskoi, kandys, epiblēma, enkyklon*), and also as a noun. It is also used in the inventory from Samos, for a belt/girdle and two curtains. The possible meanings of *poikillō* and its derivatives vary, and they are used to describe the appearance of animals, the design of textiles (often foreign ones), flowers, music, words etc.[357]

As far as textiles are concerned, there is no clear evidence for the use of embroidery as a technique. On the contrary, Droß-Krüpe and Paetz gen. Schieck argue that most of the precious garments were likely to have been created via tapestry technique or inwoven patterns, and they conclude that *poikillō* is used to describe textiles as colourful and precious, not to denote a specific technique.[358] Indeed, the experiments on weaving the decorative patterns of the *ependytēs* represented on the Peplos Kore that were performed by Harlizius-Klück demonstrated that tapestry weaving on the warp-weighted loom could certainly produce such figurative patterns.[359]

Another technique for producing figurative patterns is resist-dying. Examples of resist-dyed woollen textiles with colourful figurative decoration of Greek mythology, dated from the 5th to the 4th centuries BC, have been recovered in kurgans at Seven Brothers (Figs 16 & 14).[360] Figurative decoration could also be painted on: traces of painting have been identified on a textile fragment from Koropi, which represent a painted diaper pattern with painted decoration in its centre.[361] Associated terms are *parapoikilos* 'bordered with patterns',[362] which is used twice in the Brauron inventories for *chitōniskoi*, and *peripoikilos* 'patterned all over (all around)',[363] 'highly patterned',[364] or simply 'spotted',[365] which is recorded 10 times in the Brauron catalogues for different garments (e.g. *kandys, tryphēma, enkyklon*), but primarily *chitōniskoi*.

Embroidery (Table 28)
Embroidery is often emphasised in connection with Ancient Greek textiles, and many English translations of Ancient Greek texts use this word in relation to textile work. For example, Cleland translates the word *katastiktos* as 'embroidered decoration',[366] whereas Liddell-Scott translates the word as 'spotted', 'speckled', or 'brindled'.[367] It stems from the verb *katastizō* 'cover with marks'.[368] Cleland also translates the term *peristiktos* as 'embroidered all over',[369] whereas Liddell-Scott translates it as 'dappled'.[370] *Peristiktos* is used only once in the Brauron catalogues for a *tryphēma*, while *katastiktos* is registered ca. 20 times, both as an adjective and as a noun. In the inventory from Samos, one *chitōn* is described with the term *katastiktos*.

Table 26. Decorated borders

Greek term	Translation	Garment type	Colour	Decoration/fibre	Site	Reference
perikumation	wavy border (all around)	himation	with purple borders		Brauron	*IG* II², 1514, line 18
parakumation	wavy border	chitōniskos	white with purple borders		Brauron	*IG* II², 1514, line 46
parakumation	wavy border				Brauron	*IG* II², 1521B, line 48
Parhyphes	with a border				Brauron	*IG* II², 1514, line 28
Parhyphes	with a border	tryphēma			Brauron	*IG* II², 1514, line 70
Parhyphes	with a border				Brauron	*IG* II², 1524B, line 221
periēgētos	with a border around	substantive			Brauron	*IG* II², 1514, line 18
periēgētos	with a border around	chitōniskos			Brauron	*IG* II², 1514, line 43
periēgētos	with a border around	chitōniskos			Brauron	*IG* II², 1514, line 52
periēgētos	with a border around	chitōniskos			Brauron	*IG* II², 1524B, line 208
periēgētos	with a border around	chitōniskos			Brauron	*IG* II², 1524B, line 215
periēgētos	with a border around	chitōniskos		cheiridotos	Brauron	*IG* II², 1523, line 23
periēgētos	with a border around	chitōniskos			Brauron	*IG* II², 1529, line 13
ktenōtos	spiky-border	chitōniskos			Brauron	*IG* II², 1514, line 7
ktenōtos	spiky-border	chitōniskos			Brauron	*IG* II², 1514, line 29
ktenōtos	spiky-border	chitōniskos		peripoikilos	Brauron	*IG* II², 1514, line 41
ktenōtos	spiky-border	chitōniskos		periēgētos	Brauron	*IG* II², 1514, line 43
ktenōtos	spiky-border	chitōniskos			Brauron	*IG* II², 1514, line 45
ktenōtos	spiky-border	chitōniskos		amorginon	Brauron	*IG* II², 1514, line 51
ktenōtos	spiky-border	chitōniskos			Brauron	*IG* II², 1514, line 76

(*Continued*)

Table 26. Decorated borders (Continued)

Greek term	Translation	Garment type	Colour	Decoration/fibre	Site	Reference
ktenōtos	spiky-border	chitōniskos			Brauron	*IG* II², 1517B, line 129
ktenōtos	spiky-border	chitōniskos			Brauron	*IG* II², 1529, line 6
ktenōtos	spiky-border	chitōniskos		periēgētos	Brauron	*IG* II², 1529, line 13
ktenōtos	spiky-border	enkyklon		peripoikilos	Brauron	*IG* II², 1529, line 6
thermastis	a woven border of tong pattern	paryphe			Brauron	*IG* II², 1514, line 28
purgōtos	edged with a border like battlements	chitōniskos			Brauron	*IG* II², 1514, line 25
purgōtos	edged with a border like battlements	chitōniskos	white with purple borders	parakumatios, uninscribed	Brauron	*IG* II², 1514, line 46
purgōtos	edged with a border like battlements	chitōniskos	white		Brauron	*IG* II², 1517B, line 136

Table 27. Patterned textiles

Greek term	Translation	Garment type	Colour	Decoration/fibre	Site	Reference
poikilos	patterned				Brauron	*IG* II², 1514, line 1
poikilos	patterned	chitōniskos			Brauron	*IG* II², 1514, line 14
poikilos	patterned	kandys			Brauron	*IG* II², 1514, line 19
poikilos	patterned	epiblēma			Brauron	*IG* II², 1514, line 31
poikilos	patterned	enkyklon			Brauron	*IG* II², 1514, line 48
poikilos	patterned	chitōniskos	purple		Brauron	*IG* II², 1514, line 12
poikilos	patterned	chitōniskos			Brauron	*IG* II², 1517B, line 122
poikilos	patterned	kandys		linen	Brauron	*IG* II², 1524B, line 219
poikilos	patterned			amorginon	Brauron	*IG* II², 1522, line 15
poikilos	patterned				Brauron	*IG* II², 1522, line 28
poikilos	patterned				Brauron	*IG* II², 1523, line 15
poikilos	patterned				Brauron	*IG* II², 1523, line 9
poikilos	patterned	epiblēma			Brauron	*IG* II², 1529, line 19
poikilos	patterned	belt/girdle			Samos	*IG* XII 6, 1, 261, line 22
poikilos	patterned	two curtains			Samos	*IG* XII 6, 1, 261, line 26
parapoikilos	bordered with patterns	chitōniskos	white		Brauron	*IG* II², 1522, line 21
parapoikilos	bordered with patterns	chitōniskos	with purple borders		Brauron	*IG* II², 1523, line 19
peripoikilos	patterned all over (all around)	chitōniskos		ktenōtos	Brauron	*IG* II², 1514, line 7
peripoikilos	patterned all over (all around)	chitōniskos		ktenōtos	Brauron	*IG* II², 1514, line 41
peripoikilos	patterned all over (all around)				Brauron	*IG* II², 1524B, line 217

(*Continued*)

2. The temple inventories

Table 27. Patterned textiles (Continued)

Greek term	Translation	Garment type	Colour	Decoration/fibre	Site	Reference
peripoikilos	patterned all over (all around)	tryphēma		amorginon	Brauron	IG II², 1523, line 21
peripoikilos	patterned all over (all around)	chitōniskos cheiridotos		periēgētos	Brauron	IG II², 1523, line 23
peripoikilos	patterned all over (all around)	chitōniskos	frog-coloured		Brauron	IG II², 1523, line 24
peripoikilos	patterned all over (all around)	chitōnion			Brauron	IG II², 1529, line 5
peripoikilos	patterned all over (all around)	enkyklon			Brauron	IG II², 1529, line 6
peripoikilos	patterned all over (all around)				Brauron	IG II², 1530, line 2

Fig. 16. Resist dyed woollen textile with colourful figurative decoration of Greek mythology, 5th century BC, from kurgan 4 at Seven Brothers (after Stephani 1881, pl. iv).

Truly embroidered textiles, however, are very rare in the Mediterranean area.[371] One example is a fragment of a plain linen tabby preserved in a bronze urn found at Koropi near Athens. It can be dated to the late 5th century BC. The embroidery has been carried out with linen threads wound about with metal foil – probably

Fig. 17. Linen tabby with embroidery from Koropi, Attica, ca. 500–440 BC. © Victoria and Albert Museum, London, inv. no. T.220 to B-1953.

silver – creating a lozenge-shaped pattern with small lions in the centre, although at present only the holes are visible (Fig. 17).[372] In Lefkandi, an early example of embroidery derives from a wealthy burial – perhaps the centre of a hero cult. In a bronze amphora a linen tunic was found with decorated woven bands, one of which embroidered with meander hooks.[373] Other examples of embroidery have been found in the Crimea, e.g. from a female burial near Pantikapaion near Kerch, dated to the 3rd century BC.[374] The textiles recovered from her lead coffin consist of a thin woollen material dyed with murex purple and covered with gold embroidery depicting ivy-leaf garlands.[375] In a burial dated to the 4th century BC in the Pavlovskij Kurgan, also near Kerch, a purplish wool textile was recovered with embroidery in yellowish-white, black, and green in the form of palmettes and leaves, and one fragment even illustrates a horse and rider.[376]

Since the archaeological evidence for embroidered textiles is very scarce, Droß-Krüpe and Paetz gen. Schieck question whether this technique was indeed an omnipresent technique in ancient textile manufacturing.[377] They conclude that the literary and documentary sources that contain phrases usually translated as 'embroidery' provide little evidence for this technique of textile decoration.[378] They are doubtlessly right in their observations, but it should be maintained that embroidery was indeed used in Ancient Greece, and some of the garments listed in the inventories might therefore have been embellished with embroidery. Furthermore, embroidery seems an obvious choice of decorative technique, since it can be made after the textile is finished, and it would be strange to exclude it completely when studying Greek textiles.

Table 28. Embroidery

Greek term	Garment type	Colour	Decoration/fibre	Site	Reference
katastiktos			cheiridotos	Brauron	IG II², 1514, line 6
katastiktos				Brauron	IG II², 1514, line 9
katastiktos			xystidotos	Brauron	IG II², 1514, line 11
katastiktos			dipterygos	Brauron	IG II², 1514, line 38
katastiktos	halourgidos	purple		Brauron	IG II², 1514, line 56
katastiktos				Brauron	IG II², 1514, line 57
katastiktos				Brauron	IG II², 1514, line 67
katastiktos	tryphēma			Brauron	IG II², 1514, line 71
katastiktos				Brauron	IG II², 1518B, line 54
katastiktos	tryphēma			Brauron	IG II², 1518B, line 68
katastiktos			trichaptos	Brauron	IG II², 1524B, line 177
katastiktos				Brauron	IG II², 1524B, line 208
katastiktos				Brauron	IG II², 1521, line 50
katastiktos				Brauron	IG II², 1521B, line 66
katastiktos				Brauron	IG II², 1523, line 9
katastiktos	thōrax			Brauron	IG II², 1523, line 20
katastiktos				Brauron	IG II², 1523, line 25
katastiktos	chitōnion		stuppinos	Brauron	IG II², 1523, line 27
katastiktos				Brauron	IG II², 1528, line 19
katastiktos	chitōn			Brauron	IG II², 1522, line 16
peristiktos				Brauron	IG II², 1525, line 3

Yet it seems that, generally speaking, the inventories describe how the garment looked rather than how it was made. Perhaps the original translation of the terms as meaning 'spotted' and 'dappled' are still valid, and thus straightforwardly describe the visual appearance of the garment. Such 'spots' could be obtained in many other ways than embroidery, for example by inwoven or painted decoration or the use of appliqués (see below).

Gold textiles (Table 29)
The practice of using gold to embellish textiles was not uncommon in the Mediterranean, but was obviously extremely expensive and thus reserved for the highest social strata. It is unclear how early the practice of making gold thread began: already in the Old Testament there are mentions of gold textiles,[379] but they become especially popular after the collapse of the Roman Empire.

Gold could be employed in different ways for textiles: it could be woven alone, as in the three miniscule gold textile fragments found in a South Italian glass urn containing burnt bones (Fig. 18).[380] But, more frequently, it was interwoven with other material, primarily purple wool and silk, which became particularly famous during the Hellenistic Period.[381]

Gold thread could be used for embroidery, as in the textile from Koropi mentioned above, but the vast majority of the archaeological evidence indicates that gold thread was primarily worked on the loom. It could also be used by itself to create hairnets using sprang technique or to make cords or fringes. The gold thread itself was made by twisting gold wire or gold strips around a fibre core (silk, wool, vegetal fibre, or animal gut),[382] or as so-called *lamellae*, made from hammered gold foil cut into fine strips. Such gold strips, dated to the 4th century BC, have been recovered for example at the Mausoleum at Halikarnassos.[383]

The most magnificent example of gold weaving comes from the so-called Tomb of Philip II in Vergina, dated to the 4th century BC. In a gold larnax, together with the cremated remains of a woman, two rectangular pieces of textile woven in gold and purple tapestry were recovered (Fig. 19).[384] Furthermore, many gold woven textiles have been recovered in the Crimea and in southern Russia.[385]

These gold textiles are hard to detect in the inventories, but perhaps they are reflected in garments described by the term *chrysopoikilos*, usually translated as 'gold-embroidered'.[386] I have only found one example of this term, which is in the inventory from Miletos, where it denotes a *strophion* with a thunderbolt motif designated as *chrysopoikilos*. Perhaps it was made in gold-woven textile and not embroidered? The inventory from Samos records a *chitōniskos*, which is described with the term *chrysōi peripoikilmenos*, a variant of *peripoikilos*, and can possibly be translated as 'interwoven with gold' or 'decorated with gold all over', possibly reflecting the technique of weaving gold. The term *diachrysos*, typically translated as 'interwoven with gold',[387] is only recorded on Delos, where it is used to describe a belt/girdle.

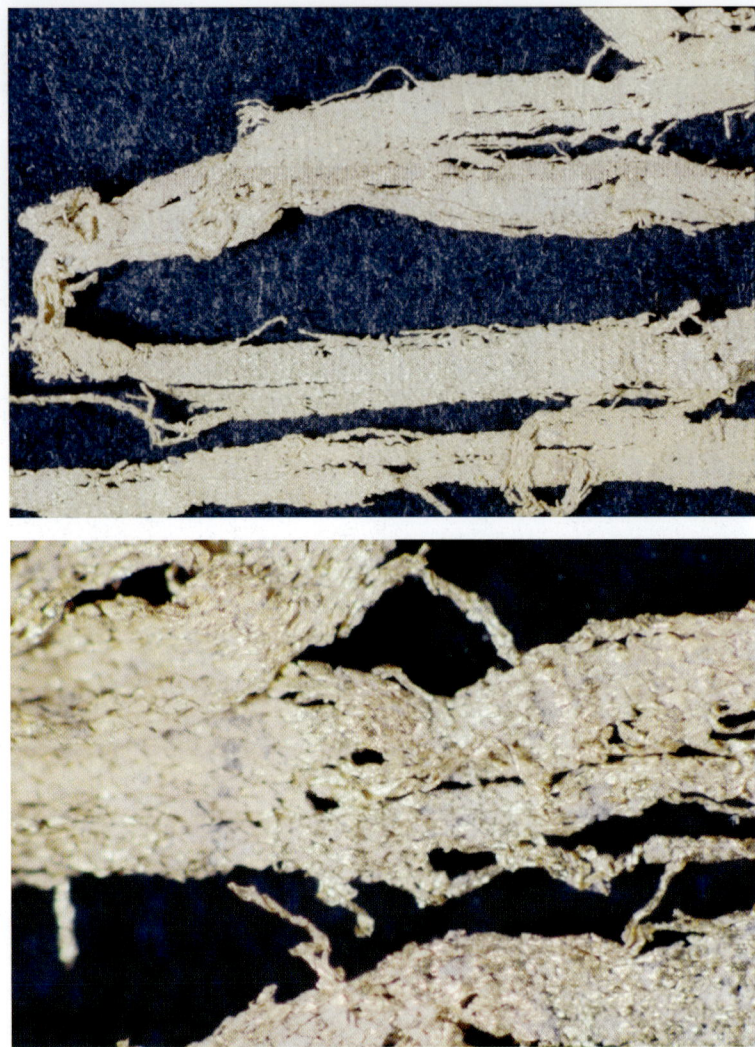

Fig. 18. Miniscule gold textile fragments from Roman urn. © National Museum of Denmark.

Appliqué and gold decoration (Table 29)
Finally, the inventories also testify to the use of attachments to decorate the textiles. One term that provides evidence for this practice are the *pasmatia*, which are interpreted as metal ornaments sewn onto garments.[388] Other possible terms include: *epichrysos*, meaning 'overlaid/plaited with gold', or sometimes indicative of metal decorations on garments;[389] and *epitēktos*, meaning 'overlaid with gold' or 'with gold ornaments'.[390] All three terms are used in the Brauron catalogues in relation to garments (*poikilen, trichapton, kandys*). The inventories from Brauron specify *pasmatia*

Fig. 19. Two rectangular pieces of textile woven in gold and purple tapestry from tomb of Philip II, Vergina. © Hellenic Ministry of Culture and Sports/Archaeological Receipts Fund/Archaeological Ephorate of Hemathia.

overlaid with gold (*epitēkta*)[391] and *pasmatia* of gold (*chrysa*).[392] The former is placed on the garment 'along the border'.[393] The inventory from Miletos records a *kalasiris* with gold border (*perichrysos*), and two *strophoi* and a woollen belt/girdle overlaid/plaited with gold (*epichrysos*); and at Delos, in the Artemision, the goddess is clothed in a purple garment (*esthēta porphyran*) overlaid/plaited with gold (*epichrysos*).[394] These particular terms (*perichrysos* and *epichrysos*) do not occur in the inventories from Brauron, though, or in the other inventories.

According to Miller, the *pasmatia* recorded in the Brauronian catalogues are clothing attachments of gold, known as bracteates, which were a characteristic means

Table 29. Gold decoration

Greek term	Translation	Garment type	Colour	Decoration/fibre	Site	Reference
chrysopoikilos		strophion?			Miletos	Milet VI, 3, 1357, line 19
chrysōi peripoikilmenos		chitōniskos			Samos	IG II², 1522, line 17
diachrysos	interwoven with gold	belt/girdle			Delos	ID 1428, II, line 58
diachrysos	interwoven with gold	belt/girdle			Delos	ID 1429B, line 4
diachrysos	interwoven with gold	belt/girdle			Delos	ID 1443, line 6–7
epichrysos	plaited with gold	poikilen kai pasmatia			Brauron	IG II², 1522, line 15
epichrysos	plaited with gold	2 strophoi			Miletos	Milet VI, 3, 1357, line 18
epichrysos	plaited with gold	belt/girdle			Miletos	Milet VI, 3, 1357, line 20
epichrysos	plaited with gold	garment	purple		Delos	ID 1442B, line 54
perichrysos	with gold border	kalasiris			Miletos	Milet VI, 3, 1357, line 5
epitēktos		pasmatia			Brauron	IG II², 1524B, line 178
pepoikilmenos dia chrysiou					Delos	ID 1428, II, line 58
pepoikilmenos dia chrysiou					Delos	ID 1429B, line 4

of decorating textiles in the Achaemenid Empire.[395] Such Persian gold bracteates, which take the shape of lions and have small rings on the back for attachment, have been recovered from the Sanctuary of Zeus at Dodona and the Sanctuary of the Great Gods on Samothrace.[396] Furthermore, the foundation deposit of the Archaic Artemision at Ephesos revealed several gold and electrum appliques.[397]

The use of gold plaques on garments during the Achaemenid Period is illustrated in Classical texts and is known from archaeological finds, in particular from the assemblages recovered from Scythian burials in the Ukraine and south Russia,[398] as well as in Western Anatolia.[399] These burials provide many examples taking the form of flowers, animals, squares, circles, etc. and possessing small holes for attachment.

2. The temple inventories

Fig. 20. Gold plaques with mythological scenes, from the Great Bliznitsa, ca. 330-300 BC. © State Hermitage Museum, St. Petersburg, inv. nos. BB-44, BB-45. Photo: Leonard Kheifets.

Occasionally, such plaques bear more detailed scenes. One such rare example is the square gold plaques recovered from the 'Tomb of the Priestess of Demeter' in the 'Great Bliznitsa' on the Taman Peninsula. The plaques date to 330–300 BC and have a height and width of ca. 5.8 cm. A total of 42 plaques were recovered from the burial: their subjects are centred on the Eleusinian Mysteries (e.g. heads of Demeter [13], Persephone [15], and Herakles [14]), and they thus stand apart from many other types of more standardised clothing appliqués found in burials (Fig. 20).[400] Other examples are gold appliques in the shape of *nikai* in *bigas* (Fig. 21).[401]

Gold plaques attached to textiles were also used in Greece, where they often take the shape of roundels decorated with rosettes.[402] Such plaques have been recovered in burials from the 6th century BC, e.g. in the tomb of 'the Lady of Archontiko' at Pella (ca. 540–530 BC), where three gold rosettes and four plaques shaped like double triangles were recovered on the upper part of the torso, three rosettes on her abdomen, and silver plaque and two gold plaques with vegetal decoration and two small gold rosettes on her thighs. Finally, two small gold rosettes were found near her feet and probably originally adorned her shoes.[403] Gold rosettes and decorated plaques were also in use in the Hellenistic Period. For example, gold rosettes (width 1.05 cm) and roundels with female heads (diam. 1.05 cm), said to be from Kyme, are dated to ca. 330–300 BC,[404] and gold rosettes from Madytos (diam. ca. 1.9 cm) are dated to ca. 330–300 BC (Fig. 21b).[405]

The tomb of the 'Lady of Aigai' (ca. 500 BC) testifies to the use of gold bands and plaques sewn onto garments. According to the excavators, the deceased wore three garments: a *peplos*, a *chitōn*, and an *epiblēma*.[406] The lower part, or 'apron', of the *peplos* had rich decoration: just below the centre of the waist was a broad rectangular

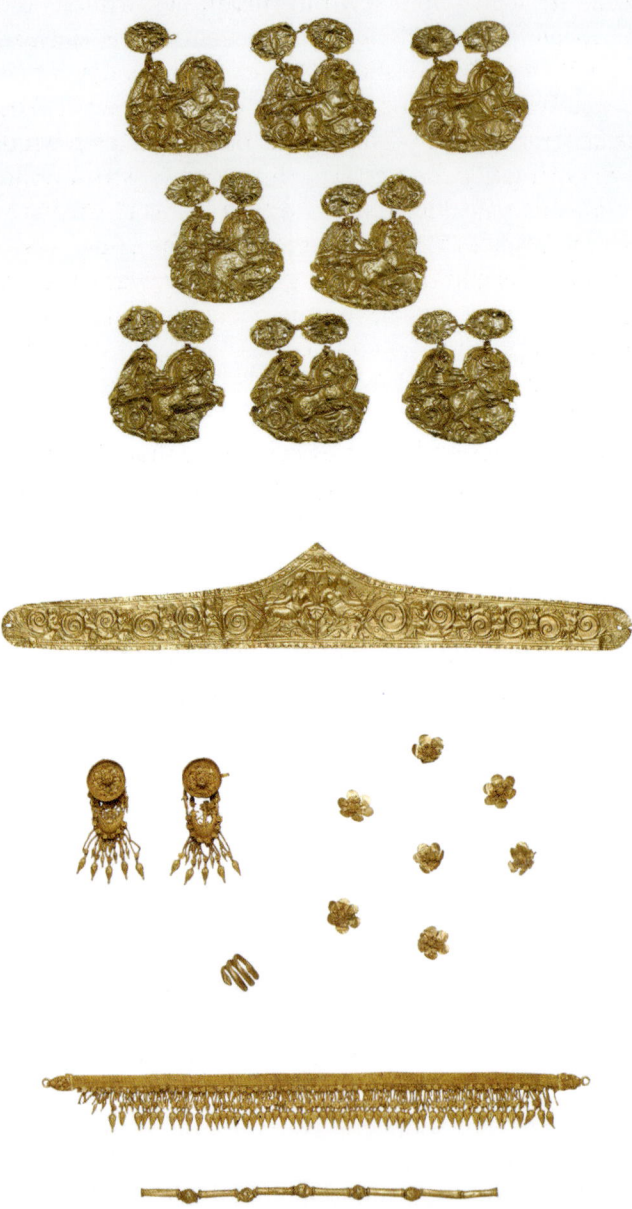

Fig. 21. A. Gold plaques with nikai in bigas, ca. 400–350 BC. © Trustees of the British Museum, inv. nos. GR 1877.9-10.15; GR 1876.5-17.11. B. Set of jewellery including a diadem, earrings, two necklaces, and seven gold rosettes from Madytos on the European side of the Hellespont, ca. 330–300 BC. © Metropolitan Museum of Art, inv. nos. 06.1217.4 -.10.

double strip with repoussé decoration (Fig. 22). Below this, small roundels with rosettes, ca. 2.5 cm. in diam., were sewn in three rows, while two double and four single triangular strips formed a border at the hem.[407] The *epiblēma*, which can be reconstructed from its metal decoration, covered the back down to mid-calf, left the bosom free, and was fastened by a gold double pin at the waist. The entire front of the *epiblēma* was decorated with border bands of gold strips sewn onto the cloth: a long strip (60 cm) with guilloche pattern, three sheets with mythological subjects (27.3, 30.4, 55.7 cm), a damaged long strip (35.7 cm), and 15 smaller strips of varying length with repoussé rosettes.[408] Finally, sewn onto the upper part of the *epiblēma*, in the location of the arms and next to the gold border, was a rectangular strip with three triangular spikes, perhaps marking the openings for the hands.[409]

These examples can possibly be equated with the garments that have *pasmatia* or are described as *epichrysos* or *epitēktos* in the inventories. Especially the use of gold strips can explain why some garments are described as being 'plaited with gold', while the gold plaques and rosettes could be denoted by the term *pasmation*. Although this use of gold was primarily an East Greek (and Persian) custom, it seems that such exotic forms of decoration were also used for garments dedicated in sanctuaries.[410]

As a final note, the inventory from Delos specifically states that the gold-plaited garment is used for dressing the goddess (i.e. her cult statue). This means that it possibly was not a 'regular' garment worn by ordinary people, but something created specifically for this purpose – and is, in this way, similar to funerary garments, as for example those in the burial of the so-called 'Lady of Aigai'.

'Inscribed' textiles (Table 30)

An interesting idiosyncrasy in the inventories from Brauron is the description of garments with inwoven or embroidered letters: one *himation* has 'golden letters' (*chrysa grammata*), another has inwoven letters (*grammata enhyphasmena*),[411] and a white *himation* with purple border has a 'sacred inscription' (*hieron epigeraptai*) of unknown content.[412] Furthermore, an *ampechonon* and child's *chlaniskion* are specifically 'inscribed as sacred to Artemis' (*Artemidos hieron epigeraptai*), and a *chitōniskos* is inscribed with 'to Artemis' (*Artemidi*).[413] In addition, the inventory from Samos records three items (a *chitōn*, a cushion, and a *sindoniske*) with the term *hypogegrammenos*, which is usually translated as 'embroidered', but should possibly be translated as 'inscribed'.

According to Linders, the use of the phrases 'inwoven letters' and 'golden letters' implies that the letters did not convey a meaning to those who wrote the inventory,[414] in contrast to the more specific inscriptions directing the offering to Artemis. She further argues that the sacred inscriptions to Artemis should be interpreted as 'with the inscription sacred to Artemis but no other' or that the garments with these inscriptions did not carry a donor's name because they were not private offerings.[415] Other garments are specified as unwritten (*agraphos*) or uninscribed (*anepigraphos*).

Fig. 22. The 'Lady of Aigai', ca. 500 BC (after Kottaridi 2004, fig. 6).

Such garments are recorded three times in the inventory from Tanagra, once at Delos, and ca. 17 times in the Brauron catalogues. According to Linders, this simply means 'without indication of the donor',[416] or, alternatively, the recipient.

According to Wace, letters on a garment could serve several purposes: they could either record the names of the owners or, if a figural scene was present, explain the subject. He argues that, if the letters were inwoven in the fabric, the inscription would presumably give the name of the owner and would probably be in the nominative or genitive case.⁴¹⁷ If the word instead described the subject represented on the textile, it would be in the nominative, as is the case on several textiles recovered in Egypt. Yet the letters in the dedicated garments may have pertained to a dedicatory inscription rather than to the name of the dedicator. Thus, according to Wace, the inscriptions on the textiles from Brauron were as a rule only dedicatory inscriptions, which he bases on the fact that, in the several cases where the inventory does not state the name of the dedicator the garment is registered as uninscribed. Wace argues, further, that, if the inscriptions on textiles were dedicatory, they would hardly have been inwoven, unless specifically made for the purpose, but would instead have been added in embroidery or appliques.⁴¹⁸ Alternatively, Linders considers it possible that the names of the dedicators were provided on a label attached to the garment, perhaps like a mummy ticket or stuck to it with wax. Such labels would easily come loose, which could explain the relatively high number of garments recorded as uninscribed.⁴¹⁹ Linders also proposes the possibility that the maker of the textile wove her own name into it as a signature.⁴²⁰

Garments provided with the owners name are attested in literature: according to Pliny, the painter Zeuxis from the 4th century BC paraded his wealth by wearing garments with his name in golden letters:

> 'Zeuxis also acquired such a vast amount of wealth, that, in a spirit of ostentation, he went so far as to parade himself at Olympia with his name embroidered on the checked pattern of his garments in letters of gold.'⁴²¹

A further source for this practice is Demosthenes, who describes how a man was accused of temple robbery because he carried the sacred garments marked with the names of the dedicators in golden letters:

> 'After seeing Hierocles carrying sacred garments on which there were letters stitched in gold to denote those who had dedicated them as an offering, Pythangelus and Scaphon accused him before the prytaneis of being a temple-robber, and on the next day the prytaneis took him before the Assembly. Hierocles said that he had been sent by the priestess to get the garments and was supposed to bring them to the Shrine of the Huntress.'⁴²²

There are, however, also examples of dedicated textiles inscribed with the name of the recipient deity. This phenomenon is attested in Apuleius' *Metamorphoses*, where Psyche visits a temple:

> '... she approached the sacred doors. There she saw rich offerings, gold embroidered ribbons, attached to the branches and the doorposts, whose lettering spelled the name of the goddess to whom they were dedicated, with thanks for her aid.'⁴²³

The phenomenon of inscribed textiles is also attested on Mons Claudianus, Egypt. An ostraka bears a small note saying 'To Achillas. I send you your cloak (*pallium*)

Table 30. Inscribed textiles

Greek term/text	Translation	Garment	Site	Reference
grammata enhyphasmena	inwoven letters		Brauron	*IG* II², 1514, lines 8–9
chrysa grammata	golden letters	[himation]	Brauron	*IG* II², 1529, line 14
Artemidos hieron epigegraptai	sacred to Artemis	ampechonon	Brauron	*IG* II², 1514, lines 34–36
Artemidos hieron epigegraptai	sacred to Artemis	ampechonon	Brauron	*IG* II², 1515, lines 26–28
Artemidos hieron epigegraptai	sacred to Artemis	ampechonon	Brauron	*IG* II², 1516, lines 13–14
hieron epigegraptai Artemidos	sacred to Artemis	child's chlaniskion	Brauron	*IG* II², 1514, lines 40–41
hieron epigegraptai	sacred	white himation	Brauron	*IG* II², 1514, line 69
periēgētōn Artemidi	to Artemis	chitōniskos	Brauron	*IG* II², 1514, lines 52–53
anepigraphos	uninscribed	girl's cloth chitōn	Tanagra	*SEG* 43: 212 (B), line 43
anepigraphos	uninscribed	a girl's tarantinon	Tanagra	*SEG* 43: 212 (B), line 46
anepigraphos	uninscribed	a boy's chitōn	Tanagra	*SEG* 43: 212 (B), line 50
anepigraphos	uninscribed	white chitōn	Delos	*ID* 1417A, I, line 7
anepigraphos	uninscribed	chitōniskos	Brauron	*IG* II², 1514, line 28
agraphos	unwritten	chlanis	Brauron	*IG* II², 1514, line 39
anepigraphos	uninscribed	lēdion	Brauron	*IG* II², 1514, line 44
anepigraphos	uninscribed	white chitōniskos	Brauron	*IG* II², 1514, line 46
anepigraphos	uninscribed	xenikē	Brauron	*IG* II², 1514, line 49
agraphos	unwritten	saffron-coloured chitōniskos	Brauron	*IG* II², 1514, line 58
anepigraphos	uninscribed	chitōnion amorginon	Brauron	*IG* II², 1514, line 61
anepigraphos	uninscribed	chitōnion amorginon	Brauron	*IG* II², 1514, line 65
anepigraphos	uninscribed	chitōnion amorginon	Brauron	*IG* II², 1514, line 66
anepigraphos	uninscribed	kekryphalos?	Brauron	*IG* II², 1524B, line 130

(Continued)

Table 30. Inscribed textiles (Continued)

Greek term/text	Translation	Garment	Site	Reference
anepigraphos	uninscribed	sindonitēs	Brauron	*IG* II², 1524B, lines 131–132
anepigraphos	uninscribed	child's chitōnion	Brauron	*IG* II², 1524B, lines 133–134
anepigraphos	uninscribed	chitōnion amorginon	Brauron	*IG* II², 1518B, lines 70–71
anepigraphos	uninscribed	enkyklon	Brauron	*IG* II², 1524B, line 206
anepigraphos	uninscribed	chitōn	Brauron	*IG* II², 1522, line 18
anepigraphos	uninscribed	tarantinon	Brauron	*IG* II², 1522, line 26
agraphos	unwritten	krokōtion	Brauron	*IG* II², 1529, lines 18–19
hypogegrammenos	inscribed?	cushion	Samos	*IG* XII 6, 1, 261, line 23
hypogegrammenos	inscribed?	spleniskos	Samos	*IG* XII 6, 1, 261, line 24
hypogegrammenos	inscribed?	sindoniske	Samos	*IG* XII 6, 1, 261, line 24

inscribed with your name in large letters.'[424] According to Bülow-Jacobsen, the 'flat broad letters' (*platesi grammasi*) might be a kind of embroidery with which the name was written on the cloak, but most probably the word 'flat', in this context, simply means 'spread over a wide space, large' and does therefore not necessarily refer to technique. Whether the name was written directly on the garment or on an attached label remains unknown.[425]

Unfortunately, we do not have any preserved textiles from Greece with inwoven or embroidered letters. Such examples have been recovered from pharaonic Egypt, though, where several textiles from royal tombs have the king's cartouche in tapestry weaving.[426] Most of the woven inscriptions from Late Antique Egypt, however, serve to explain the decoration, especially to label the mythological figures.[427] Since the textiles recorded in the inventories include decorated garments (e.g. denoted by the term *poikilen*), the unspecified inscriptions could have been similar examples of explanations for motifs. This possibility is given support by the textiles from Kurgan 4 and 6 at Seven Brothers, mentioned above. Besides the figurative decoration, the textiles bear Greek letters, which name the mythological figures depicted, including Nike, Athena, Iokaste, the Nereid Eulimene, Phaidra, the Lapith Mopsos, and Hippomedon (Fig. 16).[428]

Another example from Egypt is the so-called weaver's marks, which were used in linen textiles from the Middle and New Kingdom. These weaver's marks consisted of a small woven design or group of hieroglyphs, and represented the logo of the weaver.[429] Similarly, floor mosaics – resembling carpets – occasionally have labels giving the name of the maker. For example, a pebble mosaic from Pella has the inlaid label 'Gnosis made it', and the Hephaistion mosaic from Pergamon has the label 'Hephaistion made it' (*epoisen*).[430] It is thus also possible that the inscribed textiles in the inventories indicate the name of the maker – and donor, if they dedicated their own garments.

State of preservation: Rags or relics? (Table 31)

According to Cleland, although the garments dedicated at Brauron might be denoted as 'expensive' or 'completely finished', they would generally have been used, and often made, by those who dedicated them.[431] Miller stresses that only one of all the known dedicated garments at Brauron is described as new (an *epiblēma* in IG II[1] 1514, lines 30–32), which, she argues, suggests that the majority of the clothing recorded in the catalogue was used.[432]

Other garments are recorded as being either ragged, worn, useless, or old. One such term used in the inventories is *rhakos*, which is commonly found in literature from Homer onwards.[433] As Linders notes, the term is to be regarded as a note on the preservation of the garment, like the numerous remarks on the defective preservation of metal offerings found in inventories.[434] Ragged garments denoted by the term *rhakos* are recorded in several inventories at Brauron, Tanagra, and Samos, while at Miletos many of the recorded garments are described as old (*palaioi*), useless (*achreioi*), and/

or frayed (*katakekommenai*). At Brauron, *rhakos* is used in connection with different garment types (*chitōniskos, chitōnion, chitōn, tryphēma, xenikē, himation, kandys, enkyklos, chlanis, katastikton, tarantinon, lasion, krokotinon,* and *stuppinon*).[435] At Tanagra, *rhakos* is used to describe a Tarantine garment, and at Samos it is used in connection with a linen covering/wrap (*periblēma*), a purple ragged patterned loincloth (*perizōma*), and a table cover. At Miletos, the majority of recorded garments are either old (*palaios*), useless (*achreios*), or frayed (*katekekommenos*). These adjectives are used for different garments: a *kalasīris, himatia*, purple garments, *chlanides, sindonītes, othonai*, ephebic *chlamydes, prosōpidia*, linen pieces, headdresses, ribbons, girdles/belts, children's mantles, children's clothing, and a *karpasos*. There is thus no connection between which garment types were recorded as ragged, old, useless, or frayed, and their state of preservation did not influence their recording in the inventories. Nor does there seem to be any apparent connection between their value and the notation that they were worn, etc., since these garments include large garments as well as ribbons, and are of different fibres and fabrics and different colours – at Brauron, one is even denoted as *stuppinos*, i.e. of coarse fibre.

The worn state of these textiles has been used as an argument that the inventories reveal that textile donations were actual clothes made for and worn by real people, not specifically for goddesses and their statues, or for the heroic and divine characters of tragedies.[436] This may also imply, however, that the garments were not new in the sense that they had been stored in the sanctuary for a long time, and not dedicated recently, but still considered important enough to require that they be recorded in the lists. This hypothesis is possibly supported by the fact that, in the Brauron catalogues, the term *rhakos* is rarely used in the older inventories, but becomes more frequent in the later versions, which could indicate that the garments were deteriorating with time. At present, however, it is impossible to say if they were dedicated as such, or if they were in this state after many years in the sanctuaries, where they could have been used for different purposes that caused them to degrade.

What can be concluded is that the conservation status of each textile appears to have been carefully recorded, which shows that they were inspected. Several of the textiles seem to have been in a very bad state when handed down annually by the magistrates to their successors, and especially the inventory from Miletos can be said to list a large amount of old 'junk' in a very meticulous way.[437] This is, however, most probably the wrong way to perceive it. The recordings of these textiles illustrate their immense importance for the sanctuary – or else they would not have recorded them. These items are thus not just 'junk', but important votive offerings or even ritual items.

Storage and display of the textiles

In several instances the garments are recorded as being stored in some sort of receptacle. In Brauron they are termed *plaisia*, usually interpreted as an oblong box,[438] probably without a lid.[439] Here, they are used only for textiles, not for metal

Table 31. *State of preservation*

Greek term	Translation	Greek term	Translation	Greek term	Translation	Garment	Site	Reference
palaios	old	ēchreiōmenos	useless			kalasiris	Miletos	Milet VI, 3, 1357, line 6
palaios	old	ēchreiōmenos	useless			himation	Miletos	Milet VI, 3, 1357, line 7
palaios	old	achreios	useless	katakekommenos	frayed	8 purple garments	Miletos	Milet VI, 3, 1357, line 7
palaios	old	achreios	useless	katakekommenos	frayed	3 chlanides	Miletos	Milet VI, 3, 1357, line 8
palaios	old	achreios	useless	katakekommenos	frayed	3 purple himatia	Miletos	Milet VI, 3, 1357, line 9
palaios	old	achreios	useless			karpasos	Miletos	Milet VI, 3, 1357, line 10
palaios	old	achreios	useless			sindonitēs	Miletos	Milet VI, 3, 1357, line 10
palaios	old	achreios	useless			3 linen othonai	Miletos	Milet VI, 3, 1357, line 11
palaios	old	achreios	useless	kekommenos	frayed	2 othonai	Miletos	Milet VI, 3, 1357, line 11
palaios	old	achreios	useless			4 ephebic chlamydes	Miletos	Milet VI, 3, 1357, line 12
palaios	old	achreios	useless			4 silken prosōpidia	Miletos	Milet VI, 3, 1357, lines 12–13
palaios	old	achreios	useless			2 prosōpidia of wool	Miletos	Milet VI, 3, 1357, line 13
palaios	old	achreios	useless			12 pieces of linen	Miletos	Milet VI, 3, 1357, line 14
palaios	old	achreios	useless			epikrēnon	Miletos	Milet VI, 3, 1357, line 14
palaios	old					two epikrēna	Miletos	Milet VI, 3, 1357, lines 14–15
				kekommenos	frayed	epikrēnon	Miletos	Milet VI, 3, 1357, line 15
		achreios	useless	kekommenos	frayed	silken epikrēnon	Miletos	Milet VI, 3, 1357, line 16

(*Continued*)

Table 31. State of preservation (Continued)

Greek term	Translation	Greek term	Translation	Greek term	Translation	Garment	Site	Reference
				kekommenos	frayed	silken epikrēnon	Miletos	Milet VI, 3, 1357, line 16
				katakekommenos	frayed	2 woolen ribbons	Miletos	Milet VI, 3, 1357, line 17
palaios	old			katakekommenos	frayed	scarlet ribbon	Miletos	Milet VI, 3, 1357, line 18
palaios	old					2 strophoi	Miletos	Milet VI, 3, 1357, line 18
palaios	old					strophion	Miletos	Milet VI, 3, 1357, line 19
palaios	old			katakekommenos	frayed	woollen perizōma	Miletos	Milet VI, 3, 1357, line 20
palaios	old					2 belts	Miletos	Milet VI, 3, 1357, lines 21–22
palaios	old					2 belts, larger	Miletos	Milet VI, 3, 1357, line 22
				katakekommenos	frayed	two mantles for children	Miletos	Milet VI, 3, 1357, lines 22–23
				katakekommenos	frayed	children's clothing	Miletos	Milet VI, 3, 1357, line 23

offerings.⁴⁴⁰ The inventories from Delos record *plaisia*⁴⁴¹ as well as similar receptacles termed *plintheioi*, also usually translated as oblong boxes,⁴⁴² and thus probably synonymous with *plaisia*. On Delos the *plaisia* are said to be fastened to the wall, but this is not the case in Brauron.⁴⁴³ Instead, Linders suggests that they were placed on the floor or on shelves along the walls.⁴⁴⁴ The inventories from the Athena Temple on the Athenian Acropolis record another type of receptacle, termed *kibōtia*. As an example, two *kibōtia* containing boots (*opisthokrēpides*) are recorded in the inventory of the Hekatompedon.⁴⁴⁵ On Delos, *kibōtia* are recorded in several inventories. They are usually interpreted as boxes or chests, but without any indications of size or shape, and the term does not refer to any specific type of box, in contrast to e.g. a *pyxis*. Prêtre argues, however, that they, in the Delian inventories at least, appear to be have been used primarily for objects of small size, such as jewellery.⁴⁴⁶ They are usually made of wood, but there are also examples of ivory,⁴⁴⁷ bone, stone, or metal,⁴⁴⁸ and others that are covered in leather (*kibōtion eskytōmenon*).⁴⁴⁹ Another form of the term, *kibōtoi*, were also used to store clothing, as is confirmed by Athenaios, who writes that 'up to the time of our grandfathers none ate citron, but as a great treasure they were put away in chests (*kibōtoi*) with clothing.'⁴⁵⁰

In private households, textiles were stored in wooden chests – possibly to be equated with the *kibōtoi* recorded in literary sources. Such chests for storing textiles are depicted in vase-painting, e.g. on an Athenian *lekythos*, which shows a woman placing a folded white textile into a chest.⁴⁵¹ Such large chests could be present in the sanctuaries for storing cultic utensils or votive offerings such as textiles. Support for this possibility can be found in the *pinakes* from the sanctuary of Persephone at Locri, which depict women placing textiles in decorated chests, most probably in a ritual context,⁴⁵² as well as on the eastern pediment of the Parthenon, where two goddesses are sitting on such chests. Boxes of a smaller, more manageable size are also depicted in vase-painting, for example on a red-figure sherd from the Athenian Agora, which depicts a woman holding a sash and a decorated chest, probably of wood (Fig. 23).⁴⁵³ Such chests are preserved from the Classical and Hellenistic Periods, although they are rare. Miniature examples include a wooden box, recovered in the Crimea,⁴⁵⁴ bronze chests from Greece,⁴⁵⁵ and terracotta chests with gabled lids from Smyrna and Myrina.⁴⁵⁶ Larger rectangular wooden chests, some with decoration in wood, terracotta, or stucco, and dating to the 4th century BC, have been recovered in Abusir in Egypt⁴⁵⁷ and Kertch in the Crimea.⁴⁵⁸ The fantastic gold chests/*larnakes* from the tomb of Philip II at Vergina should also be included here as possible receptacles for textiles.

Linders has argued that the dedicated textiles in the sanctuaries were more likely to be displayed than stored away, and she considers the smaller open boxes without lids to be especially 'convenient receptacles' for this purpose.⁴⁵⁹ The preserved examples as well as the ones depicted in vase-painting have lids, though, and are not open as Linders suggests. The chests with flat lids could be placed on top of each other, creating more space for votive offerings, and the textiles would be protected from

2. *The temple inventories* 125

Fig. 23. Red-figure sherd from the Athenian agora, ca. 450–400 BC. Museum of the Athenian Agora, inv. no. P10540. © Hellenic Ministry of Culture, Education and Religious Affairs. Ephorate of Antiquities of Athens.

Fig. 24. Red-figure kylix by the Briseis Painter, ca. 480–470 BC. © Metropolitan Museum of Art, New York, inv. no. 27.74.

pests and vermin such as moths. Perhaps it was only recently dedicated garments that were on display, while older offerings were stored away in chests. Alternatively, the textiles could have been placed on sets of shelves, such as is depicted on a red-figure *kylix* (Fig. 24).[460] Shelves for the storage of small objects are recorded in the Acropolis inventories, specifically in the inventories of the Erechtheion.[461] Cupboards and wardrobes are of course other likely candidates for storing the textiles, but were apparently not used in Greece in the Archaic and Classical Periods, but introduced in the Hellenistic Period.[462] A final possibility is that the textiles were hung on the walls or from the ceiling, e.g. on clotheslines, as depicted in the marble relief from Echinos (see Fig. 2). Of course one possibility did not exclude the others, and several different ways of storing and displaying the textiles could have been used in the same sanctuary.

As far as which buildings housed the textiles, we do not have much information to go on. Due to the records in inventories, which list items from specific temples, we can tell that they were present in the temples. But whether they were assigned to particular areas, rooms, or part of these temples we simply do not know. The inventory of the Treasurers of Athena and the Other Gods,[463] however, refers to a so-called *Peplotheke*, perhaps a closet or a designated area in one of the temples for the storage of the sacred *peploi*, or even a separate building on the Acropolis. Nagy also offers a third interpretation, namely that the *Peplotheke* was a room or large closet in the *Chalkotheke* – a building used for storing votive gifts described in the epigraphic sources, but not yet identified in the architectural remains on the Acropolis.[464]

In the sanctuary at Brauron we have more information to go on: the inventories demonstrate that some of the textiles were displayed on the cult statues in the temple, which also housed other textile offerings. Yet further potential storage locations for the votive offerings have been suggested. Papadimitriou proposed that the cave area also functioned as a storage facility for offerings made at the sanctuary,[465] which appears plausible.[466] Another suggested possibility is the northern section of the stoa, within which was found a row of 37 poros slabs with a central cutting (ca. 80 × 17 cm), possibly for *stelai* (Fig. 25). No *stelai* or slabs have been recovered in these bases or in their vicinity, however, and there are no traces of lead for fastening stone *stelai* in the central cuttings of the bases, in contrast to the stele bases along the northern side of the temple terrace and along the west wall of the western wing of the stoa.[467] Papadimitriou therefore proposed that wooden boards were inserted into the bases and that the clothes of women who had died when giving birth were displayed on these boards.[468] Kontis elaborates on this interpretation and argues that the width of the wooden boards indicates that the only garments that could have been displayed on them were *peploi*.[469] But his conclusion on garment type is nonsense, and, besides, no *peploi* are recorded in the inventories. Moreover, as Ekroth argues, there is no evidence for clothes being displayed on wooden boards in Greek sanctuaries in this manner.[470]

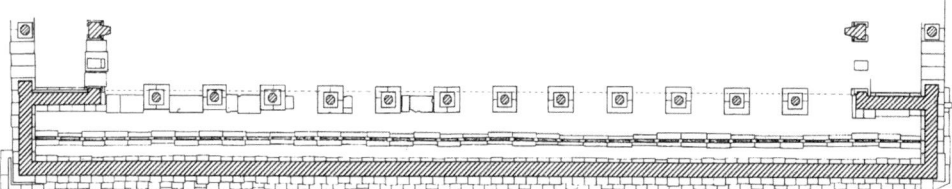

Fig. 25. Ground plan showing the poros slabs with a central cutting from the stoa at the sanctuary at Brauron (after Kontis 1967, pl. 106).

Unworked wool and unfinished textiles

Some of the textiles recorded in the Brauron inventories are described with the term *hemihyphes* or *hemihyphantes*, meaning half-woven.[471] According to Linders, this means that the garment was unfinished, and taken down from the loom before completion.[472] A second interpretation could be that the designation 'half-woven' implied that the textile was made from two types of fibres, understood as half one fibre and half another fibre – perhaps wool and linen, or hemp and linen.[473]

A third interpretation could be that the term represents a warp for a loom, set up for a specific garment type, taken down, and dedicated. As a parallel, such a warp has been recovered from a bog in Tegle, Norway. It was found in a bag made of a coarsely woven fabric, which also contained wool yarn, unspun wool, twisted hair, and textile fragments.[474]

This term is used for different items: garments such as two *chitōniskoi*, a *chitōnion*, a *tarantinon*, a *himation* and a *pteryx*, and also for a web (*histos*). Sometimes in the inventories these items are recorded together with weft and wool, which, according to Linders, could indicate that these items were intended to be used to complete the unfinished textiles.[475] Examples of weft thread (*krokē*) are recorded three times in the inventories at Brauron.[476]

Separate offerings of wool are recorded in the inventories from Brauron and in an inventory of Athena from the Athenian Acropolis. At Brauron these offerings included wool, soft wool (*eria malaka*), and worked/prepared wool (*kateirgasmenon*).[477] The colour of the wool is recorded only once for a dedication of grey wool. At Athens, grey and white wool is recorded, as well as a wool net (*eriōn diktyon*). A further piece of interesting information that the list provides is that four of these wool dedications are frayed (*katakommenos*). Both at Brauron and at the Acropolis, many of these wool dedications are given/stored in a basket (*kalathos*).[478]

In this connection the textile tools recorded in the Brauron catalogues and at the Acropolis should also be mentioned. As an example, at Brauron, the inventories record a distaff (*ēlakatē*)[479] and a weaving-sword (*spathis*).[480] An inventory of the Hekatompedon records spindles (*atraktoi*) and weaving swords in ivory (*spathai elephantinai*),[481] while another records two weaving-swords, one of them of silver.[482] In the archaeological record, textile tools such as spindles, loom weights, as well as fragments of *epinetra* have been recovered from the sanctuary site.[483]

Conclusions drawn from the temple inventories

From the study of these inventories it is evident, first and foremost, that textiles played a far greater part in the votive traditions of ancient Greek sanctuaries than anticipated in many studies. They offer proof that, even though the textiles themselves have disappeared from the archaeological record, this does not mean that they were not dedicated as offerings in the sanctuaries. Furthermore, the custom of dedicating textiles does not appear to have been a feature of a particular geographical area. Study of the inventories demonstrates, further, the wealth of information that we gain about Ancient Greek dress and votive practices when we combine epigraphy, archaeology, and textile research.

With regard to the type of textiles, the inventories display a great diversity of dedicated garments. It is possible, though, to make a few observations: first, the most commonly donated garment is the *chitōn* and its derivatives/diminutives, such as the *chitōniskos* and the *chitōnion*, mantles (especially *himatia*), and veils.

Second, besides the inventory from Oia, there is only one record of an offering of a *peplos* in the temple inventories (although this attestation is highly restored) – a garment type that is claimed to have gone out of use for women in the 6th century BC when it was replaced by the *chitōn*.[484] This supposed change in dress is primarily based on the testimonies of Herodotos, who writes that when the lone survivor from the war with Aegina returned to Athens, he was killed by the wives of those who had fallen in battle: they gathered round him and stabbed him with the pins holding their garments. As punishment, the dress of the women in Athens was changed from the Doric *peplos* to the Ionian fashion, consisting of the linen *chitōn*, stitched on the shoulders, so they should not wear dress pins.[485] The *peplos* does, however, return in the Early Classical Period, ca. 480–450 BC, when it again becomes the predominant garment *represented* in Greek art. But by this time it was not in general use and its appearance on monuments is thought simply to represent traditional female costume and 'an iconographic construct of Hellenic identity.'[486] Therefore, *peploi* have been interpreted in the Classical Period and later as a garment with religious or ceremonial connotations and as an embodiment of traditional female values.[487] This seems supported by the evidence from the inventories, where the *peplos* recorded on Delos is not a 'regular' garment, donated by a mortal woman, but instead a special garment used to dress a cult statue.

Third, the inventories illustrate that the Ancient Greek wardrobe – at least in dedicatory contexts – appear to have been much more varied, decorated, and colourful than often assumed in studies of ancient dress.

Fourth, the inventories show that there are only few foreign 'exotic' garments, and those that are present are specially recorded in the inventories from Brauron. This can indicate several situations: either that these garments were actually worn by the local residents near the sanctuary who then dedicated them, or that people came travelling from afar bringing textiles to be dedicated in the sanctuary. Or perhaps these garments were made especially for the deity, who might have been thought to

2. The temple inventories

prefer such garments. As an example, two of the *kandyes* recorded in Brauron were used to dress the cult image of Artemis.

The choice of votive clothing items reflected in the inventories thus illustrates that the garments given to the gods were often regular clothing items, in fashion and in use at the time and worn by 'ordinary' people, not special garments reserved for deities. Furthermore, there is nothing to indicate that donated garments were made especially for this purpose. This possibility, however cannot be completely excluded, either. For example, the garments from the Brauron catalogues that have inscriptions featuring either the name of the donor or the name of the deity could be an indication that garments were produced specially for dedication or customised for this purpose.

A fifth point that emerges from the inventories is that the majority of the textiles are donated by women, as evidenced by their names recorded on the lists, and only in a very few cases do the lists indicate a male donor. It has therefore been suggested that dedications of textiles were related to the life stages of women, and that such dedications marked the stages of the female ritual cycle,[488] which will be discussed in depth in the following chapter.

Moreover, several ancient textual and iconographic sources testify to the offering of belts or girdles to goddesses on the occasion of marriage. Thus, based on these written sources, and not least the Brauron catalogues, there has been a tendency to argue that the primary recipient of textiles was the goddess Artemis. Therefore, many of the inventory lists for which the name of the deity has been lost have been identified as belonging to Artemis, as exemplified by the previous interpretations of the inventory from Tanagra. This assumption, though, may lead to a circular argument and to an oversimplification of our understanding of ancient votive practices. There is no doubt that Artemis was a primary recipient of textiles, but these offerings were also appropriate gifts for other divinities, as illustrated by the inventory lists. Generally, it appears that primarily goddesses received textiles on a regular basis, but, as has been shown, gods could occasionally also be the recipients of donated garments.

We can conclude that there is rich evidence for the offering of textiles in Greek sanctuaries – a custom which appears to be more widespread than implied by the general archaeological record. Temple inventories can inform us about garment type, fibre, colour, and decoration, and can also reveal information about the donors and divine recipients, and – not least – provide us with knowledge about the ritual practices in the sanctuaries. Hence, it appears that the use of garments as offerings was a prevalent votive tradition, which must be taken more into account in the study of Greek votive practices.

Notes

1. Hom. *Il.* 6.287–304. For this specific textile offering and the interpretation of the priestess Theano, see Nosch 2007.
2. Dignas 2002, 235.
3. Shaya 2005, 425.

4. Blundell & Williamson 1998, 36.
5. Scott 2011, 240.
6. Dignas 2002, 235.
7. The Liber Linteus (Liber Agramensis), dated to the 3rd century BC, is an example of textiles being used to write on. The text is probably a ritual calendar, written in Etruscan. Archaeological Museum, Zagreb.
8. Dignas 2002, 242.
9. Linders 1988a, 38–40.
10. Linders 1988a, 46.
11. See e.g. Lindenlauf 2006.
12. Dignas 2002, 235.
13. Linders 1988b.
14. Scott 2011, 241.
15. Scott 2011, 241, 249.
16. Dignas 2002, 241.
17. Appendix 2 of the temple inventories does not include all of the examples, since the Brauron Catalogues and the Delian and Athenian inventories are very extensive. For these I refer to the respective publications.
18. *IG* II^2 1514–1530. Epigraphical Museum Athens, inv. nos. EM 5294, 7929–7937.
19. For the Greek text and an English translation of the clothing catalogues, see Cleland 2005.
20. Cleland 2005, vii.
21. Dillon 2002, 19. Some of the inscriptions bearing inventories recovered at Brauron were briefly mentioned in Papadimitriou's reports of the excavations, and they have also been commented on by Peppas-Delmousou 1988, but they remain virtually unpublished.
22. For example, inventories said to be identical to those recovered on the Athenian Acropolis (*IG* II^2 1517, 1524, 1529) were recovered in the area of the cave. Ekroth 2003, 80; Papadimitriou 1956, 27–28; 1957, 21–22.
23. Cleland 2005, 6.
24. Cleland 2005, 1.
25. Linders 1972, 68–69, note 18.
26. The inventories of the Treasurers of Athena describe the contents of the Parthenon, the Erechtheion, the Opisthodomos, and the Chalkotheke, but here the focus is on the treasures of the Parthenon.
27. Harris 1995, 1, 10.
28. Harris 1995, 20; Scott 2011, 242; Lapatin 2005, 279.
29. Lapatin 2005, 279–281.
30. Scott 2011, 242. 434–406 BC: separate *stelai* for three separate treasures: Pronaos, Parthenon, Hekatompedon. 405–386 BC: separate *stelai* for three separate treasures: Parthenon, Opisthodomos, Hekatompedon. 385–340s BC: one stele. The Chalkotheke records are added ca. 370 BC. *SEG* 50.70.
31. Harris 1995, 2. In particular, there is considerable debate whether the opisthodomos is the west portico of the Periclean temple, its western chamber, or the remains of the Old Temple of Athena that remained standing on the Acropolis after it was sacked by the Persians. Lapatin 2005, 283.
32. Harris 1995, 5.
33. Lapatin 2005, 283–284.
34. *IG* II^2 1412, 11 (382/1 BC), *IG* II^2 1421, 118 (374/3 BC), *IG* II^2 1424a, 303 (371/0 BC), *IG* II^2 1428, 143 (367/6 BC).
35. *IG* II^2 1414, 26 (385/4 BC).

2. The temple inventories

36. *IG* II² 1475, 7 (after 318/7 BC).
37. *IG* II² 1478, 17 (after 316/5 BC).
38. *IG* II² 1388, 19.
39. *IG* II² 1424a, 55.
40. *IG* II² 1388, 20.
41. *IG* II² 1424a, 336–337; *IG* II² 1425, 269–270; *IG* II² 1428, 222–223.
42. *IG* II² 1448, 4.
43. *IG* II² 1469, 124–140.
44. For the entire Greek text, see Reinach 1899 and Roller 1989, 100–108.
45. Schachter 1997, 279.
46. *LSCGS* 32, 6th century BC; *LSCGS* 33, 3rd century BC.
47. Reinach 1899, 60; Roller 1989, 107; Knoepfler 1977, 68, 86.
48. *IG* VII 2421.
49. According to Roesch, there is no doubt that the inventory list comes from the Kabirion at Thebes, Roesch 1985, 74.
50. Rayet 1881, 264–265.
51. Günther 1988, 232, note 97.
52. The records of the *hieropoioi* and the Athenian temple administrators are kept in the Archaeological Museum of Delos, and are recorded in the museum's inventory catalogue with the numbers Γ.0001-Γ.0770.
53. Hamilton 2000, 1.
54. Linders 1988a, 38.
55. Linders 1988c, 267.
56. The inventories from the Amphictyonic Period are hard to identify with certainty because there is only one complete inventory (*ID* 104) and a few remnants of others, Hamilton 2000, 5, 7.
57. Hamilton 2000, 8.
58. Rutherford 1998, 81.
59. Prêtre forthcoming, 3.
60. For this praxis, see Chapter 6.
61. *IG* XII 6, 1, 261. Archaeological Museum of Vathi, Samos, inv. no. 16.
62. Ohly 1953, 46–48; *SEG* 45:1163, *SEG* 47:314; Shipley 1987, 157.
63. Herrmann et al. 2006, no. 1357; Günther 1988, 221.
64. Kleijwegt 2002, 105.
65. Museum of Miletos, inv. no. 1378.
66. Günther 1988, 219–220.
67. Günther 1988, 223.
68. Blinkenberg 1941, 2. The National Museum of Denmark, inv. no. 7125. The combination of a list of votives and records of epiphanies has led to confusion over whether the inscription should be compared to temple inventories or if it is a work of historiography or a chronicle. Platt 2011, 164.
69. For a detailed study of the stele, see e.g. Shaya 2002; 2005; Higbie 2003. For the epiphanies, see also Platt 2011, 161–169.
70. This information is provided by Hdt. 3.47 and Plin. *Nat.* 19.2.14.
71. Blinkenberg 1941, 2C, 36. Amasis is the Greek name for the Egyptian Pharaoh Ahmose II (570–526 BC), Granger-Taylor 2012, 79; Francis & Vickers 1984, 68.
72. Blinkenberg 1941, 2C, 65–68.
73. Blinkenberg 1941, 2C, 85–89.
74. Aleshire 1989, 6.

75. Aleshire 1989, 39. *IG* II² 1019, 1532, 1533, 1534A, 1534B, 1535, 1536, 1539, *SEG* 28.116.
76. *IG* II² 1533, Epigraphical Museum, Athens, inv. no. 8249.
77. *IG* II² 1533, 8.
78. *IG* II² 1533, 18.
79. *IG* II² 1533, 30.
80. *IG* II² 1533, 85. Aleshire 1989, 162. Finally, it is possible (but somewhat speculative) that line 57 records a garment, Aleshire 1989, 159.
81. *LSJ* s.v. *peirō*.
82. *LSJ* s.v. *porpē*.
83. Prêtre 2012, 198. The *Suda* s.v. *porpē*: defines it as 'the fibula among the Romans'.
84. The term occurs already in Homeric texts, where it is used to fasten clothing, *Il.* 5.425, *Od.* 19.226, 256, *Od.* 18.293.
85. *LSJ* s.v. *peronis* equates the word with *peronē*.
86. *LSJ* s.v. *peirō*.
87. *LSJ* s.v. *peronē*.
88. Prêtre 2012, 193.
89. Deonna 1938, 275.
90. Prêtre 2012, 193.
91. Lorimer 1950, 405. Poll. 7.54.
92. *IG* IV 1588. Blinkenberg 1926, 19–20.
93. Paus. 2.32.2. Figueira 1993, 57.
94. Polinskaya 2013, 272–273, 283.
95. Hdt. 5, 83.2; Dunbabin 1936–37, 85.
96. Hdt. 5.88.2. Translation: Godley 1920.
97. *IG* IV 1588; *IG* IV² 787.
98. Dunbabin 1936–37, 86.
99. Figueira 1991, 35.
100. Polinskaya 2013, 615.
101. Jacobsthal 1956, 98.
102. Dunbabin 1936–37, 86.
103. Possibly several *peploi*. Polinskaya 2013, 280.
104. Jacobsthal 1956, 98.
105. Polinskaya 2013, 281.
106. Jacobsthal 1956, 98–99.
107. Jacobsthal 1956, 98.
108. According to Prêtre, this corresponds to the chronology of the archaeological record of the island. Prêtre 2012, 198.
109. Another possibility is that the differences in terminology reflect different scribes with preferences for certain terms.
110. Prêtre 2012, 198.
111. Prêtre 2012, 198.
112. *IG* II² 1388, 20 (398/7 BC); 1424a, 55 (369/8 BC); 1425 (368/7 BC); 1428 (367/6 BC).
113. *IG* II² 1388, 20; 1425 60–61; 1428 22–23, 33.
114. *IG* II² 1424a, 55; 1425, 54; 1428, 16.
115. IvPerge 11:99,2 = IK Perge 10.
116. Jacobsthal 1956, 104–105; Şahin 1999, 7–12; For Artemis Pergaea, see Pace 1923, 297–314.
117. *IG* VII 2420. Wolters & Bruns 1940, 21, no.2. In Greek mythology, the Kabeiroi were chthonic deities, and, according to genealogical myth, they were sons or grandsons of Hephaistos. Their cult is attested mainly on Lemnos and at Thebes, Burkert 1985, 281.

118. Jacobsthal 1956, 103.
119. Schachter 2003, 133.
120. Jacobsthal 1956, 104.
121. Schachter 2003, 121.
122. As demonstrated in Appendix 4, dress-fasteners were dedicated in several sanctuaries, especially to female deities.
123. *IG* IV 39.
124. *IG* VII 2420.
125. Rehm 1958, nos. 424–478. A possible exception to this omission of garments is no. 447, which records *himatia*.
126. *ThesCra* I, 281.
127. For transvestism in Greek clothing, see e.g. Loraux 1990.
128. Cleland 2005, 73, 79.
129. For the *chitōn*, see Lee 2015, 106–110; for the *chitōniskos*, see Lee 2015, 110–111.
130. Lee 2015, 106.
131. Lee 2015, 110.
132. For the *peplos*, see Lee 2015, 100–106.
133. *ID* 440A, 41. There is no mention of dress-fasteners in connection with this *peplos*.
134. For the *himation*, see Lee 2015, 113–116.
135. Cleland et al. 2007, 92.
136. Sokolowski 1955, 49A, l. 19–27; Gauthier 1985, 149–163.
137. Cleland et al. 2007, 34.
138. Lee 2015, 117.
139. Cleland et al. 2007, 34; *LSJ s.v. chlanis*.
140. *LSJ s.v. chlanidion*.
141. Cleland et al. 2007, 199. The simplicity of the *tribōn* is implied by Thucydides (1.6.4), who writes that the Lacedaemonians were the first to adopt a modest style of dress.
142. Cleland 2005, 108, 113.
143. *LSJ s.v. epiblēma. LSJ* simply translates *periblēma* as 'garment'. Lee identifies the *epiblēma* literally as an 'overgarment', Lee 2015, 113.
144. Llewellyn-Jones 2003, 26–27.
145. *LSJ s.v. ampechō*.
146. Llewellyn-Jones 2003, 26–27.
147. Cleland 2005, 113. Lee suggests that it was a – presumably circular – women's mantle, Lee 2015, 119.
148. *LSJ s.v. proslēmma* 'upper garment'. O'Brien 1993, 30–31 translates it as a 'shoulder wrap'.
149. Dillon 1997, 197; Cleland et al. 2007, 101.
150. Hdt. 2.81.
151. This inscription is examined in chapter 11.
152. *LSCG* 65; *IG* V 1, 1390. For a detailed analysis of the text, see Gawlinski 2011.
153. Günther 1988, 224.
154. Linders 1984, 107. See also Thompson 1965.
155. Poll. 7.58.
156. Xen. *Cyr.* 1.3.2. He also mentions the *kandys* in *An.* 1.5.8.
157. Linders 1984, 109; Miller 1997, 168.
158. Miller 1997, 166.
159. Cleland 2005, 66; Cleland et al. 2007, 102.
160. For Persian objects in Greek inventory lists, see Kosmetatou 2004.
161. Acropolis Museum, Athens, inv. no. Acr. 606. Ca. 520–510 BC. Cleland et al. 2007, 6.

162. Cleland 2005, 129.
163. *LSJ s.v. tryphē.*
164. Cleland 2005, 129.
165. The same garment, donated by a man name Pharnabazos, is recorded in several lists.
166. Lee 2015, 112.
167. The *othonion, sindonitēs, tarantinon*, and the *amorgis* will be discussed in the section on fabrics and fibres, while garments named after colours, such as the *krokōtos*, will be discussed in the section on colours.
168. Cleland 2005, 65.
169. Cleland 2005, 112. The similarity with the term *pteryges* for corselets (see below) could possibly also influence the interpretation.
170. *Gada* is probably a *hapax* and possibly a term used in a local Greek dialect without any other attestations in the written sources. The word is attested in Sumerian, where it is translated as 'flax' or 'linen' (clothing). Sumerian disappeared as a spoken language ca. 2000 BC, however, which makes its connection with this particular inscription unlikely. *Gada* is also used in Akkadian texts of the 2nd and 1st millennia BC as a logogram for the word *kitû*, which means both linen and flax. The term is also attested in Neo-Assyrian tablets. I am grateful to Salvatore Gaspa for this information.
171. *LSJ s.v. summetrios* 'a woman's robe without a train'.
172. Cleland 2005, 128.
173. Gleba 2012, 54.
174. Paus. 1.21.7.
175. Hdt. 2.182.1; 3.47.
176. Gleba 2012, 46.
177. Granger-Taylor 2012, 58–68.
178. Connolly 1981, 58–59.
179. E.g. Tomba del Orco II at Tarquinia, 4th century BC.
180. From the House of the Faun at Pompeii, dated to ca. 100 BC. Granger-Taylor 2012, 64–65.
181. See Prêtre forthcoming, 7–8.
182. Lee 2015, 97.
183. Llewellyn-Jones 2003, 28.
184. Llewellyn-Jones 2003, 28.
185. Llewellyn-Jones 2003, 30, 31.
186. Günther 1988, 227–228.
187. Pfuhl & Möbius 1977, pl. 66, no. 407, pl. 67 no. 410; Günther 1988, 228.
188. Bacch.. 17.37.
189. Aesch. *Ag.* 1178.
190. Eur. *IT* 372.
191. Llewellyn-Jones 2003, 33.
192. Llewellyn-Jones 2003, 31.
193. Jenkins & Williams 1985.
194. Jenkins & Williams 1985. E.g. British Museum, inv. no. 1907,0519.1. 470 BC.
195. British Museum, inv. no. 1848,0804.32. Granger-Taylor 1985.
196. Shamir 2008, 121. Israel Antiquities Authority, inv. no. IAA 577048.
197. Llewellyn-Jones 2003, 62.
198. Metropolitan Museum of Art, New York, inv. no. 1972.118.95.
199. Llewellyn-Jones 2003, 34.
200. Llewellyn-Jones 2003, 64.
201. Günther 1988, 227.

202. Heraclid. Crit. 1.18: 'The covering of their clothes [himation] on their head is such that the whole face seems to be covered by a mask, with only the eyes showing through; the other parts of the face are all covered by the garments [himatia].' Translation: Llewellyn-Jones 2007, 253.
203. *Anth. Pal.* 6.206, 5–7. Llewellyn-Jones 2003, 64.
204. *LSJ s.v. trichaptos*.
205. Cleland 2005, 128; Cleland et al. 2007, 199.
206. In Archaic and Classical literature, the term *zonē* is the most commonly used term to identify what corresponds to the English belt. Other terms are also employed, however, including the *zoster, kestos, mitra, strophion*, and *tainia*, all of which also refer to other articles of dress, from breastbands to headbands. Lee 2015, 134.
207. Papadopoulou forthcoming. See also Stafford 2005, 102–104.
208. Papadopoulou forthcoming.
209. Papadopoulou forthcoming.
210. *LSJ s.v. mitra*. E.g. Ap. Rhod. *Argon.* 1.288; 3.867.
211. E.g. Ar. *Th.* 257; Eur. *Ba.* 833.
212. E.g. Pind. *Ol.* 9.84.
213. E.g. Hdt. 1.195; 7.90.
214. *LSJ s.v. mitra*. E.g. Plut. *Mor.* 2.304c.
215. Hom. *Il.* 4.137, 187, 216, 5.857. An example is the decorated metal *mitra* from Olympia, dated to ca. 650 BC, Archaeological Museum, Olympia, inv. no. B4900. For an in-depth study of the *mitra* and examples of its uses, see Papadopoulou forthcoming.
216. Cleland et al. 2007, 183; Lee 2015, 98. Plut. *Arat.* 53.
217. Stafford 2005, 101. Ar. *Lys.* 931. The visual evidence for such breastbands is sparse. The only instance where it is worn underneath clothing is an Attic hydria from the 4th century BC. This type of flat, undecorated breastband becomes increasingly popular in erotic scenes and images of Aphrodite in the Hellenistic and Roman Periods. Lee 2015, 99. British Museum, inv. no. E230.
218. Dillon 2002, 216; Cleland 2005, 68; Oakley & Sinos 1993, 14–15. Among the literary sources for women loosening their belts are Theoc. 17, 60, Translation: Edmonds 1912: 'the daughter of Antigoné cried aloud to the girdle-looser in the oppression of pain'; Ap. Rhod. *Argon.* I, 288, Translation: Seaton 1912: 'my only son for whom I loosed my virgin zone first and last. For to me beyond others the goddess Eileithyia grudged abundant offspring'; *Orphic hymns* 35, where Artemis is described as 'dissolver of the zone'.
219. Eur. *Heracl.*, 407–415.
220. *Anth. Pal.* 6.201, 6.202, 6. 211, 6.272.
221. *LSJ s.v. anadesmē*.
222. Cleland et al. 2007, 73; Lee 2015, 160–163; Morrow 1985.
223. *Anth. Pal.* 6.21, 6.201, 6.210.
224. *IG* XII 6, 1, 261, 23–26.
225. *ID* 104(26bis), C, 12–13; *IG* XI,2 147B, 13; *IG* XI,2 159A, 5; *IG* XI,2 154B, 4. Prêtre forthcoming, 4.
226. *ID* 1442A, 82.
227. *ID* 403, 8. Prêtre forthcoming, 3.
228. *IG* XI,2 147B, 12.
229. Paus. 5.12.4. Translation: Jones & Ormerod 1918.
230. Eur. *Ion* 1142–1160. See also Vickers 1999, 16–17.
231. Museo archaeologico Nazionale, Palestrina, inv. no. 149000907. Rosenthal-Heginbottom 2009, 166; Schrenk 2009, 152. For a detailed study of the mosaic, see Meyboom 1995.
232. Staatliche Antikensammlungen, Munich, inv. no. 206.

233. A marble relief dated to the 1st century BC depicts Dionysos calling on Ikarios, who is shown reclining, and a large curtain hangs from the buildings behind him. Museo Archaeologico Nazionale, Naples, inv. no. 6713.
234. Similar examples in painting can be recorded in houses in Solunto and Centuripe in Sicily. Moorman 2011, 59.
235. Museo Fori dei Imperiali, inv. nos. FA 2010, 2011, 2012, 2013, 2014, 2015, 2016, 1504 a, b, c, d. 518 fragments of the panel decoration, originally covering 41 m^2, have been recovered. Ungaro & Vitali 2003.
236. Ungaro 2004; 2007; Ungaro & Vitali 2003; Santamaria et al. 2004.
237. Wild 2003, 104; Schrenk 2009, 149.
238. For the use of textiles in late Roman architecture, see Stephenson 2014.
239. Mosaics instead often imitate real carpets, Vickers 1999, 35.
240. Schrenk 2009, 147.
241. Helmecke 2009, 49; De Moor & Fluck 2009.
242. Rosenthal-Heginbottom 2009, 169.
243. See e.g. Rosenthal-Heginbottom 2009.
244. The extensive use of textiles in Christian basilicas is attested in art: for example, a miniature from a 7th century manuscript of the Pentateuch, which shows textiles hung between columns, separating catechumens and clergy from the faithful; and the 6th century mosaic from the Basilica of Sant'Apollinare, Ravenna, depicting Melchizedek, Abraham, and Abel. Stephenson 2014, 7–9.
245. State Hermitage Museum, St. Petersburg, inv. no. UO28. Richter 1966, fig. 179.
246. Vickers 1999, 28.
247. It is important to note, though, that different textile fibres could be mixed together, and that a garment was not necessarily made entirely with one type of fibre. As an example, the Romans wove cotton together with linen to produce a material called *carbasus lina*. Furthermore, silk was also sold as thread that could be interwoven with threads of other fibres (*subserica*). Sebesta 2001, 68.
248. Barber 1991, 20–30.
249. Hom. *Od.* 9.336, 9.426. Wagner-Hasel 2016, 9; forthcoming a, 5–6.
250. The importation of Milesian sheep is recorded in literary sources, e.g. Ath. *Deipn.* 12.540d. For more on Milesian wool, see Wagner-Hasel 2016, 9–11.
251. Barber 1991, 30.
252. Richter 1929, 29, 32.
253. See Bender Jørgensen 2013.
254. Barber 1991, 204; Hundt 1969, 66–70.
255. The Kerameikos textiles were examined using non-destructive methods: stereospectroscopy, Environmental/Scanning Electron Microscopy, (E/SEM), Energy Dispersive X-ray spectroscopy (EDX), Fourier Transform Infra-red (FTIR), Raman spectroscopy, and Cathodoluminescence (CL). The examinations show that the four textiles identified were made from flax and possibly cotton. Margariti et al. 2011; Margariti & Kinti 2014.
256. Margariti et al. 2011, 526.
257. Banck-Burgess, 1999; Margariti et al. 2011, 526.
258. Bender Jørgensen 2013, 582.
259. Bender Jørgensen 2013, 584.
260. Panagiotakopulu et al. 1997, 428.
261. Panagiotakopulu et al. 1997, 422.
262. Panagiotakopulu et al. 1997, 423.
263. Hildebrandt 2009; Hildebrandt 2012a; 2012b, esp. 13 ff.

264. Cleland et al. 2007, 111. Leather and leather clothing items could be denoted by different terms, e.g. *derma, dermatochitōn, diphtheria, nebridopeplos/nebridochitōn* (dressed in fawn skin), *spolas* (a leather garment, jerkin). None of these occur in the inventories investigated here. An inventory from Eleusis, *IG* I³ 386, II, 121; *IG* I³ 387, III, 135 record two leather belts, *himante eskutōmenō*.
265. Barber 1991, 11.
266. Günther 1988, 226, *LSJ s.v. othonion* 'linen cloth'.
267. Cleland 2005, 126.
268. *LSJ s.v. sindōn*.
269. Cleland et al. 2007, 171.
270. Barber 1991, 33.
271. Margariti et al. 2011; Margariti & Kinti 2014.
272. The textile from Vergina is charred and in a fragmented condition. It is only provisionally identified as cotton, and further analysis (e.g. with SEM) is therefore necessary to confirm the identification. Moraitou 2007; Margariti & Kinti 2014, 145.
273. *LSJ s.v. karpasos. Karpasos* is usually interpreted as fine linen, but the word is derived from the Sanskrit word *karpasa*, i.e. cotton. Mayrhofer 1992–2001, 317–318. It is also possible that the term refers to the island of Karpathos or Karpasia, a peninsular of northern Cyprus.
274. Pl. *Ep.* XIII 363a, asks Dionysos to give the daughters of Kebes 'three tunics seven cubits long, not those expensive Amorgian ones, but the more ordinary kind which are made of Sicilian linen.' Translation: Richter 1929. The *Suda* describes *amorginon* as like *byssos* and *polyteles* - very expensive. Ath. *Deipn.* 6.67 (255e) gives a description of an extravagant young man on a silver-footed couch, stretched on a valuable Sardian rug and wrapped in an Amorgian cover. Richter 1929, 27.
275. Ar. *Lys.* l.150 f.; Richter 1929, 27.
276. Cleland et al. 2007, 5.
277. Richter 1929, 27; *Etym. Magn.* 129.
278. Richter 1929, 28.
279. Richter 1929, 29, 32.
280. Marinatos 1967, A4.
281. Spantidaki 2014, 38.
282. Cleland 2005, 93.
283. Cleland 2005, 90; Cleland et al. 2007, 183.
284. Cleland 2005, 107.
285. Cleland 2005, 94.
286. *IG* II² 1514, 37.
287. *IG* II² 1522, 26.
288. Cleland 2005, 127. *LSJ s.v. tarantinos* 'Tarentine'.
289. Cleland 2007, 187. Sea-silk derives from large sea mussels of the species *Pinna nobilis*, which produces long fibres, known as byssal threads, secreted through the mussel's foot to fasten it to underwater surfaces. When the fibres are harvested, they are washed and left to dry, and afterwards combed, carded, and spun into thread, in a reddish to golden brown colour. Burke 2012, 172. For more on sea-silk, see e.g. Maeder 2008; 2009, 2016 or Landenius Enegren & Meo forthcoming.
290. Col. *Rust.* 7.2.3; 7.4.1. Wagner-Hasel 2016; forthcoming, 5.
291. Plin. *Nat.* 8.73.190.
292. Juv. 6.149–152. Wagner-Hasel forthcoming, 7.
293. Michel & Nosch 2010.
294. Harlow & Nosch 2014, 14.

295. Wagner-Hasel forthcoming, 11.
296. Bücher 1922, 47; Wagner-Hasel forthcoming, 3.
297. With regard to the particular nuances, these terms unfortunately do not provide information on the particular purple hue, and they can thus refer to a range of colours ranging from what we would call a deep crimson red, through purple, to violet or even Prussian blue, which indicates that the main characteristic of purple was not necessarily a particular colour nuance. For the ancient terms for purple, see Blum 1998, 20–41.
298. For a detailed study of the purple garments in the inventories, see Brøns forthcoming a.
299. For the term *halourgos*, see Brøns & Droß-Krüpe forthcoming.
300. Cleland 2005, 68.
301. Kanold & Haubrichs 2008, 253.
302. Karali & Megaloudi 2008, 182.
303. Clark et al. 1993, 196; Cole 1685.
304. According to Cardon, *Bolinus brandaris* and *Stramonita haemastoma* produces a violet-red purple, while *Hexaplex trunculus* produces a violet-blue purple. This difference in colour is explained by the different chromogens present in the shells. *B. brandaris* contains only one main chromogen, brominated, which gives 6,6'-dibromindigotin and 6,6'-dibromoindirubin, while *H. trunculus* contains four different chromogens, two of which are brominated. These four chromogens in *H. trunculus* give the dyestuffs indigotin, 6-bromoindigotin, 6,6'-dibromoindigotin (usually in very small quantities). The predominance of indigotin and 6-bromoindigotin gives a dye made from this species of shells a violet to dark blue hue. Cardon 2007, 579–580.
305. Levides 2002, 13.
306. Red dye was also extracted from the bodies of insects belonging to the family of *Coccoidae*. The most famous is cochineal – a New World insect that lives on cacti. Other types are Polish grains or St. John's Blood, which live on the roots of certain northern European plants. Another variant was used for millennia in an area at the foot of Mount Ararat, where an insect like cochineal lives on certain types of grass. Barber 1991, 231; Forbes 1964, 102.
307. Cardon 2007, 614.
308. Theophr. *Hist. pl.* 3.16; Paus. 10.36.2; Dsc. *Mat. Med.* 4.48; Plin. *Nat.* 9.65; 16.12; 21.22; 22.3; 24.4.
309. Barber 1991, 230.
310. Cardon 2007, 616.
311. Pastoureau 2001, 16.
312. Blum 1998, 37–38. It is important to emphasise that the flower was not used for dyeing. The term, when used in relation to textiles, refers only to the optical experience.
313. Cleland 2005, 93.
314. Breniquet 2013, 55.
315. Cleland et al. 2007.
316. Plin. *Nat.* 20.79. Translation: Bostock 1855.
317. Theophr. *Hist. pl.* 9.12.5.
318. Flemestad 2014, 206.
319. Potash is the common name for various mined and manufactured salts that contain potassium in water-soluble form.
320. Allgrove-McDowell 2003, 26.
321. Lee 2015, 95.
322. Balfour-Paul 1997, 6.
323. *LSJ s.v. mēlinos*.
324. Cleland 2005, 127. For a discussion of the colour term, see Pelletier-Michaud 2016, 308–321.
325. Day 2011a, 339.
326. Cardon 2007, 304.

2. The temple inventories

327. Cleland 2005, 97.
328. Ar. *Lys.* 641–647.
329. Connelly 2007, 32; Goette 2006; Nielsen 2009, 84–86.
330. Eur. *Hec.* 466–474.
331. *Iliad* 8.1, 19.1, 23.226, 24.677.
332. Hes. *Th.* 273 (Enyo), 358 (Telesto).
333. Alcm. 85A.
334. Ath. *Deipn.* 5.198c; Ar. *Ran.* 45–46.
335. Pind. *Nem.* 1.38; Dsc. *Mat. Med.* 1.25. In the Roman Period, Chloereus, originally a priest of Cybele and therefore an eunuch, was described by Virgil as wearing a *crocea chlamys*, Verg. *Aen.* 11.775. See also Benda-Weber 2014.
336. Aesch. *Ag.* 239. An alternative translation '*with her robe of saffron hanging down towards the ground*' has been suggested (Sourvinou 1971, 340). The traditional interpretation of the text is that she sheds her robe, a garment dyed with saffron. Other, less convincing, interpretations are that she is shedding blood or tears. Edgeworth 1988, 179; Booth 1979, 85–95.
337. Eur. *Ph.* 1491. Translation: Coleridge 1938.
338. Ar. *Lys.* 215.
339. Llewellyn-Jones 2003, 224; Barber 1999, 117. A further word of caution: the actual evidence for the ritual connotations of saffron-coloured garments in this period is often rather circumstantial, and seems to be influenced by the situation in Minoan Crete, where there is more solid evidence for a connection between saffron-coloured garments and ritual. Day 2011a,b. See also Benda-Weber 2014.
340. Cleland 2005, 96.
341. Spantidaki 2014, 31.
342. Droß-Krüpe & Paetz gen. Schieck 2014, 207.
343. *LSJ s.v. isoptychēs*.
344. Spantidaki 2014, 36.
345. Cleland 2005, 116.
346. Cleland 2005, 123. *LSJ s.v. kumation* 'volute',
347. Cleland 2005, 122.
348. *LSJ s.v. paryphēs*. Cleland 2005, 123, translates it as 'with a woven border'.
349. Cleland 2005, 123.
350. Cleland 2005, 119.
351. Cleland 2005, 128.
352. Cleland 2005, 126. *LSJ s.v. purgōtos* 'made like a tower', 'edged with a pattern like battlements'.
353. This technique has not been identified in the archaeological record, nor has any Greek term been associated with it. Spantidaki 2014, 32.
354. *LSJ s.v. poikillō*.
355. *LSJ s.v. poikilos*.
356. Cleland 2005, 125.
357. Droß-Krüpe & Paetz gen. Schieck 2014, 215.
358. Droß-Krüpe & Paetz gen. Schieck 2014, 215.
359. Ellen Harlizius-Klück, *Gesponnen und Verworben. Textiles zu Zeiten von Römern und Germanen*, Tuchmacher Museum Bramsche.
360. Gleba & Krupa 2012, 415; Barber 1991, 206–207; Gerziger 1975, 51;
361. Spantidaki 2014, 36.
362. Cleland 2005, 122.
363. Cleland 2005, 123.
364. Miller 1997, 166.

365. *LSJ s.v. peripoikilos.*
366. Cleland 2005, 117.
367. *LSJ s.v. katastiktos.*
368. *LSJ s.v. katastizō.*
369. Cleland 2005, 124.
370. *LSJ s.v. peristiktos.*
371. Droß-Krüpe & Paetz gen. Schieck 2014, 219.
372. Droß-Krüpe & Paetz gen. Schieck 2014, 221; Gleba 2008a, 65; Barber 1991, 206.
373. Barber 1991, 197.
374. Gleba 2008a, 66.
375. Droß-Krüpe & Paetz gen. Schieck 2014, 231. Almost no embroidered textiles have been found in Roman Italy. The only exception seems to be a fragment of a woollen textile dyed in murex purple and decorated with embroidery in gold threads. The number of embroidered textiles increases in Egypt in Late Antiquity (4th – 7th centuries AD), but the total amount is still very small compared to the thousands examples in tapestry weave. Droß-Krüpe & Paetz gen. Schieck 2014, 221–222.
376. Barber 1991, 208–209; Gertsiger 1973, 90.
377. Droß-Krüpe & Paetz gen. Schieck 2014, 211.
378. Droß-Krüpe & Paetz gen. Schieck 2014, 219.
379. Gleba 2008a, 61. *Exodus* 39.3.
380. National Museum of Denmark, inv. no. 848. Each thread is ca. 2 mm wide and 17–26 mm long and wound Z-wise around a fibrous core, now disintegrated. The weave has ca. 12–15 threads pr. mm. I am grateful to Bodil Bundgaard-Rasmussen for this information. The skeletal remains indicate a male, ca. 50 years of age. Becker 1993.
381. Gleba 2008a, 61, 62.
382. Gleba 2008a, 68.
383. Jeppesen 2000, 124–140, nos. 40, 48, 61, 66, 88. Bundgaard-Rasmussen 1998.
384. Gleba 2008a, 65. Another example – a gold and purple linen cloth, decorated with vegetal motifs – was supposedly found in a tomb at Derveni. Makaronas 1963.
385. For a list of ancient finds of gold thread from the Mediterranean, Egypt, Ukraine, and Russia, see Gleba 2008a.
386. *LSJ s.v. chrysopoikilos.*
387. *LSJ s.v. diachrysos.*
388. Cleland et al. 2007, 139. Such clothing attachments were used since the Neolithic Period along the shore of the Black Sea and eastwards, where they were made of bone. As metal became more common, they were made from metal or wood covered in metal foil. Barber 1991, 173.
389. Cleland et al. 2007, 59. *LSJ s.v. epichrysos* 'overlaid or plaited with gold'.
390. Cleland et al. 2007, 59. *LSJ s.v. epitēktos* 'overlaid with gold' or 'gilded ornaments laid on'.
391. *IG* II^2 1524B, 178.
392. *IG* II^2 1523, 9 = *IG* II^2 1524, 181.
393. Linders 1972, 26.
394. *ID* 1442B, 54–56.
395. Miller 1997, 42.
396. Miller 1997, 42, 167, fig. 7, 8. Ioannina Museum, inv. no. 4931; Archaeological Museum, Samothrace, inv. no. 65.294. Length: ca. 3.5 cm.
397. Şare-Ağtürk 2014, 263; dated to the 7th century BC. Jacobsthal 1951, 86, pl. 33.
398. Gleba 2008a, 61.
399. Şare-Ağtürk 2014, 262–263.
400. Williams & Ogden 1994, 184, 194–195. State Hermitage Museum, St. Petersburg, inv. nos. BB 44, 45.

401. British Museum, inv. nos. GR 1877.9–10.15.; Gr 1876.5–17.11.
402. According to Tsigarida 1997, 81, the rosette is one of the earliest and most popular decorative motifs on such plaques.
403. Chrysostomou & Chrysostomou 2012, 370, 378. For funerary costumes from the Archaic and Classical Period, see Castor 2008.
404. Williams & Ogden 1994, 104. British Museum, inv. nos. GR 1877.910.15; GR 1876.5–17.11.
405. Metropolitan Museum of Art, New York, inv. nos. 06.1217.4–10.
406. Minimal traces of textiles show that at least one of the garments had a pale purple colour. Kottaridi 2004, 141.
407. Kottaridi 2004, 143; 2012, 417, 431.
408. Kottaridi 2004, 143; 2012, 429–431. Similarly, in tomb 67 at Sindos (510–500 BC), gold strips and sheets decorating the garments were recovered. Ignatiadou 2012, 401.
409. Kottaridi 2004, 143.
410. Gold applications, possibly for textiles, have been recovered in the sanctuary of Brauron. Themelis 1971, 54–55.
411. *IG* II^2 1514, 8–9; *IG* II^2 1529, 14.
412. *IG* II^2 1514, 69.
413. *IG* II^2 1514, 52–53.
414. Linders 1972, 9.
415. Linders 1972, 13.
416. Linders 1972, 12.
417. Wace 1952, 112.
418. Wace 1952, 112–113.
419. Linders 1972, 13
420. Linders 1972, 9.
421. Plin. *Nat.* 35.36. Translation: Bostock 1855.
422. Dem. 25.1; Lib. *Arg.D.* §24 (Dem. 25 & 26). Translation: Gibson 2003.
423. Apul. *Met.* 6, 3. Translation: Kline 2013.
424. O.Claud. Inv. 5360. Translation: Bülow-Jacobsen 2014, 6.
425. Bülow-Jacobsen 2014, 6.
426. Linders 1972, 9.
427. Linders 1972, 9.
428. Gerziger 1975.
429. Examples are e.g. found in linen sheets from the 18th dynasty tombs of Maherpra, Kha and Tutankamon, and range in date from the early 15th to the 14th century BC. Barber 1991, 153.
430. Linders 1972, 9.
431. Cleland 2005, 6.
432. Miller 1989, 323.
433. Linders 1972, 58: LSJ s.v. *rhakos*. In poetic and medical texts, *rhakos* is a bandage used to protect wounds, or an envelope used to protect medicinal mixtures or a sort of tampon. For the literary references and the use of the term in Aristophanes, see Milanezi 2005.
434. Linders 1972, 58. Mommsen has given a whole different interpretation of the term *rhakos*. He considered it unlikely that ragged garments were given as votive offerings, for which reason he argued for a different meaning: since Artemis is a goddess of fertility, he suggested that the textiles termed *rhakos* were sanitary napkins, which young women dedicated to the goddess upon their first menstruation. Mommsen 1899, 344. Yet, as Linders states, such offerings are attested nowhere else in ancient sources, and no other meaning than ragged or tattered, etc. is possible.
435. Cleland 2005, 126.

436. Cleland 2005, 9. Alternatively, Milanezi has suggested that the dedicated garments were deliberately spoiled as complete consecrations. Milanezi 2005, 80.
437. Kleijwegt 2002, 105.
438. *LSJ s.v. plaision*.
439. Linders 1972, 10.
440. E.g. a *chitōniskos* IG II² 1514, 14 and a *kekryphalos* IG II² 1522, 18.
441. E.g. ID 504A, 13, 15.
442. *LSJ s.v. plintheion*.
443. Linders 1972, 11.
444. Linders 1972, 11.
445. IG II² 1424a, 336–337.
446. Prêtre 2012, 80.
447. Prêtre 2012, 140–141.
448. Prêtre 2012, 139.
449. Prêtre 2012, 141. IG XI,2, 205Ab, 20; IG XI,2, 223B, 2.
450. Ath. *Deipn*. 3.26. Translation: Richter 1966, 72. Storing textiles with citrus fruits is also attested in Ar. *Vesp*. 1056: 'Make their ideas your own, keep them in your caskets (*kibōtoi*) like sweet-scented fruit. If you do, your clothing will emit an odour of wisdom the whole year through.' Translation: O'Neil. 1938. Theophr. *Hist. pl*. 4.4.2.
451. Private Collection, Hamburg; Oakley 2004, 25, fig. 6; Beazley archive no. 19743.
452. See Chapter 1.
453. Museum of the ancient agora, Athens, inv. no. P10540.
454. Staatliche Museen zu Berlin, Antikensammlung, inv. no. 11863.
455. Staatliche Museen zu Berlin, Antikensammlung, inv. nos. F8064.185, F8375; National Archaeological Museum, Athens, inv. nos. 8000–8002. Richter 1966, 76.
456. Louvre, inv. no. CA 776, Myrina 404. Richter 1966, 77.
457. E.g. Staatliche Museen zu Berlin, Antikensammlung, inv. no. 16205. See also Watzinger 1905, 1–23.
458. E.g. State Hermitage Museum, St. Petersburg, inv. no. 5155. For Greek wooden sarcophagi, see Watzinger 1905, 25–62 and Vaulina & Wasowicz 1974.
459. Linders 1972, 10.
460. Metropolitan Museum of Art, New York, inv. no. 27.74.
461. IG II² 1489, 1–14; Harris 1995, 1–2; *ThesCra* I, 282.
462. Richter 1966, 79–81.
463. IG II/III² 1462, 11 (ca. 329/8 -322/1 BC).
464. Nagy 1984, 231.
465. Ekroth suggests dual functions as storage facility and dining space or residential quarter. Ekroth 2003, 84.
466. Papadimitriou 1956, 27–28; 1957, 21–22; 1961a, 75–76, 1962, 25–39. Fragments of inventories identical to those from the Sanctuary of Artemis Brauronia on the Athenian Acropolis were recovered in the cave itself and in the area in front of it. Ekroth 2003, 79–80, 82.
467. Ekroth 2003, 87.
468. Papadimitriou 1961a, 29.
469. Kontis 1967, 173–175.
470. Ekroth 2003, 88. Ekroth instead proposes that these wooden boards bore the names of all the girls who had participated in the *arkteia*. Ekroth 2003, 90–91. In contrast, Themelis argues that this building was the stables, and proposes that the boards formed a long wooden manger for horses to eat from. Themelis 2002, 105; 1971, 20.
471. Cleland 2005, 115.

472. Linders 1972, 18, 19.
473. Diocletian's Edict on Maximum Prices records textiles made from two types of fibres, designated by the term *sypseirikos* and its variants.
474. Halvorsen 2012, 284.
475. Linders 1972, 18.
476. *IG* II² 1514, 54, 72; *IG* II² 1518B, 54.
477. From the verb *katergazomai*.
478. On wool baskets (*kalathoi*), see Trinkl 2014.
479. *IG* II² 1517, 209.
480. *IG* II² 1517, 201.
481. *IG* II² 1464, 16–18.
482. *IG* II² 1469, 131; *IG* II² 1469, 163.
483. Linders 1972, 19; Kontis 1967, 189. Loom weights have been recovered from several Greek sanctuaries, e.g. the Heraion on Samos (Buschor 1930, 3); the Sanctuary of Athena at Stymphalos (Surtees 2014); the Heraion at Perachora and the Heraion at Argos (Baumbach 2004, 34–35, 92). For textile tools in sanctuary contexts, see Gleba 2008b, 178–187; Brøns forthcoming b.
484. Lee 2005, 56.
485. Hdt. 5.87.
486. Lee 2005, 55–56, 59.
487. Lee 2005, 55–57; Cleland et al. 2007, 143.
488. Blundell & Williamson 1998, 36.

Chapter 3

Discussion: Textile dedications

The abundant evidence for the use of textiles as votive offerings in Greek sanctuaries raises questions. Who dedicated these textiles? On what occasions were the textiles dedicated? Which deities were recipients of the textiles? What was the function of these gifts? These questions will be addressed in the following chapter.

Greek women and textile production

Archaeological, anthropological, and ethnographical studies have established an association between women and textile work. For millennia, women have been producing textiles, at least in temperate climates where cloth was spun and woven.[1] Among the reasons for this association is that cultures around the world have considered domestic fibre preparation, spinning, and weaving to be ideal women's chores, since they are not dangerous to children and the work can be done at home. Furthermore, the work can, to a large extent, be interrupted and easily resumed, and is thus responsive to the realities of child-rearing.[2] This was also the situation in ancient Greece, where textile production was usually associated with the female sphere, and it is often argued that a Greek woman spent most of her time and energy in producing textiles for her household.[3] A woman would have manufactured the clothing worn by all of the family members in her household, as well as bedding, cushions, curtains, and funerary shrouds. Textiles were thus considered to be synonymous with the domesticity of civilised life.[4] The quintessential skill that Greek men required in a wife – besides the ability to provide offspring – was wool-working (i.e., the preparation of wool for spinning, the spinning itself, and the weaving of cloth), and the acquisition of textile skills appears to have been a prerequisite for marriage.[5] The ability to work wool was so fundamental in a wife that Xenophon, in his treatise on household management, takes for granted that a bride will have mastered all the relevant skills before her marriage.[6]

The association of textile production with women is supported by Greek vase-painting, in which only women are depicted spinning and weaving.[7] Depictions of textile production are more or less confined to the period from the late 6th through the 5th century BC,[8] and include three scenarios, as defined by Bundrick: 1) scenes of actual textile work, 2) scenes in which figures hold or are located near textile implements, and 3) scenes that include textile props such as *kalathoi*.[9] Some of these scenes clearly represent aristocratic mythological women, such as Penelope,[10] or female deities,[11] but the majority of textile scenes are erotic in nature (i.e., they depict women beautifying themselves, bathing, or dressing, as well as receiving male visitors), and they do not have any clear mythological references. Most of these women can be identified as citizen women, but a small number are ambiguous.[12]

It has been suggested, because these women are nude and sometimes accompanied by men holding pouches of money, that the women are rather *hetairai* who are trying to earn additional income by producing textiles either to sell or to entice clientele.[13] Bundrick completely rejects this interpretation and argues that none of these women can be identified as *hetairai*, but only as Athenian citizen women.[14] Ekroth reaches the same conclusion, arguing furthermore that the performance of textile work instead represents the woman's economic contribution to the household, and is the means by which the woman's role in supporting the *oikos* – and, by extension, the *polis* – is expressed.[15] Bundrick likewise argues that these scenes employ a dual metaphor of weaving and marriage to express the *harmonia* of *oikos* and *polis*, a theme that had particular significance under the evolving Athenian democracy.[16] In sum, the images illustrate the important economic and managerial contributions of women to the household and the essential character of these roles for the city as a whole.[17]

This is not to argue that textile production was exclusive to women. Male textile producers are known to have existed, although there were relatively few in the period under investigation here. The dominant view of scholars is that male weavers also existed in Classical Athens.[18] This view is based on literary sources, especially Plato, who attests to the existence of two systems of clothing production at Athens: the familiar production of clothing by women for familial use and a second commercial form of production that involved men.[19] The literary sources indicate a considerable difference in the location and circumstances of production: men worked in small private shops in order to sell cloth and clothing in the market, and thus for profit (or perhaps prestige), whereas women worked in a domestic context, where they made textiles for the home as well as for personal use.[20] Yet this does not means that skill in textile production was linked with male professionalism – women no less so produced complicated, prestigious textiles in the domestic sphere.[21]

The idea that Greek women remained secluded in their homes while producing textiles is widespread in scholarship on ancient Greece. Keuls argues that the women of ancient Greece did not lead lives of leisure, but that their quarters in the house instead resembled a 'sweatshop' where they performed labour for their husbands or fathers.[22] Sebesta likewise argues that an industrious, skilled wife was a commodity herself, a gift of wealth that her husband had received from her father-in-law,[23] since a diligent wife could not only produce sufficient yarn and cloth for household needs, but also cloth of great quality that could increase her husband's prestige.[24]

Studies of Greek house plans and architecture as well as vase-painting have contributed to the impression that Athenian women led lives of repression confined to the women's quarters (*gynaikonitis*). These studies have led scholars such as Barber to conclude that Athenian women in the Classical Period were held in 'haremlike seclusion and scarcely allowed out of the house except for major rituals and festivals. Their duties were to take care of the food and the servants (if any), to spin and weave for clothing and household uses and to care for the children, and to obey their husbands.'[25] Recent scholarship, however, has modified this traditional interpretation of women's roles in the *oikos* and has problematised the idea of seclusion, which was perhaps more of a literary construct.[26] Bundrick, for example, regards the role of women in textile production as positive – nothing like a 'sweatshop'. She argues, instead, that depictions of textile production in the house demonstrate a woman's ability to maintain the *oikos* by providing for it – and thus construct women as sharers in their own *oikos*.[27]

Yet not all female textile production was conducted for household use, and ready-made clothes could be purchased at the market.[28] Evidence for the sale of clothing includes references in literature to the prices of garments: in Aristophanes, an unclad pauper at the Pnyx is said to have announced that he needed 16 drachmas for an outer garment,[29] and in another of Aristophanes' plays, a young man asks an old lady for 20 drachmas to purchase a *himation*.[30]

According to Brock, free women may have turned their textile production skills into profit on an *ad hoc* basis.[31] Textual sources document the existence of women who sold textiles for profit. Examples of women who spin for sale rather than for personal use include a speaker in Aristophanes' *Frogs*, who describes a dream in which she is spinning thread to sell at the market.[32] Homer describes a poor woman who must spin to earn a living.[33] According to Plutarch, the Cynic philosopher Crates of Thebes mentions a husband and wife who comb wool together because of their poverty,[34] and an inscription records that a woman by the name of Thettale sold felt caps for slaves who were engaged in construction in the Sanctuary of Eleusis.[35] These sources indicate that it was primarily poorer women who sold their textile products, although even women of high status could, in exceptional circumstances, be forced to sell their textile products.[36] This is supported by a passage in Xenophon's *Memorabilia* (2.7), which demonstrates that women of high status could also make economic contributions to the household economy through textile production. A certain Aristarkhos had complained to Socrates that he had to house some female relatives and was consequently in financial distress. Socrates therefore advised him to set all the women to producing textiles, which, when sold, provided Aristarkhos with sufficient funds to support both his female relations and himself in comfort.[37]

In van Wees's view, there was a change over time in who was involved in the clothing production: in Homer, high-quality home-produced clothing is a important marker of status, and there are descriptions of women producing textiles needed for the household,[38] but by the end of the Archaic Period the most prestigious forms of clothing were no longer produced by domestic labour, but rather bought and imported from abroad.[39] As evidence for the importation of clothing, van Wees supplies the example of Milesian men's cloaks that came to be banned in the Greek cities of southern Italy before the end of the 6th century BC.[40] He also makes note of allusions to Scythian dye, Scythian shoes, and cloaks made of Scythian wool in the Archaic poets, which he interprets as further evidence of the garment trade.[41] Finally, he cites the adoption of Lydian dress, attested especially in Sappho, who mentions Lydian sandals and compares headbands imported from Sardis and Mytilene.[42] van Wees, however, bases his argument on relatively few sources, almost all poetry, which presents methodological problems. Furthermore, he does not consider the possibility that descriptions of e.g. headbands as 'Lydian' or shoes as 'Scythian' do not necessarily mean that these objects were imported from these places, inasmuch as they could have been produced locally in the style of these locations. He also does not consider the evidence of vase-painting, with the popular depictions of e.g. spinning women.

Altogether, then, there is evidence that textiles could be produced both at home and in workshops. Yet the scarcity of sources that refer to women – at least of higher social status – selling the textiles that they had produced suggests that this scenario was out of the ordinary. This is an important point in connection with the dedication of textiles, since it could mean that women most probably produced the dedicated garments at home, rather than purchasing them at a shop or stall before or while visiting the sanctuary – although this possibility cannot be excluded, either.

A word of caution is in order here, in that these conclusions are based primarily on Athenian sources. Elsewhere in the Classical world, the situation was sometimes rather different from that at Athens. It appears that Spartan citizen women, for example, were not hidden away in the home, but participated in civic life and did not spin or weave.[43] This work was probably done by non-citizen women, or men, among the *perioikoi* or helots. Furthermore, 'women' is obviously a dangerous generalisation, since this gender category included not only free citizens, but also servants, foreigners, and slaves.

The association of women with textile production was so strong that the Greeks used wool-working as a metaphor in conceptualising women. They also viewed raw wool and finished textiles as metaphors: according to Sebesta, 'women's very femininity, in fact, was so co-identified with wool that there was a metaphorical relationship between wool and woman in Greek thought.'[44] Thus, wool and women's bodies were so closely associated that wool on occasion came to serve as a metonym for a woman: a late 6th century source informs us that, when a girl was born, the parents would decorate their doorway with a tuft of wool to announce that a girl – i.e. a new spinner – had come into the world.[45] Wool also functions in myth as a metonym for the womb. In the myth of Hephaistos' attempted rape of Athena, the goddess wipes off his semen from her leg with a tuft of wool, which she subsequently throws on the ground. From this spot Erechtonios is born. In this myth the tuft of wool functions as a substitute for Athena's womb. Athena gave Erechtonios to the daughters of Kekrops and, as a reward, taught them how to weave. Wool is thus associated with conception, and the production of textiles with child-rearing.[46]

The production of textiles was associated not only with mortal women, but also with female deities. In connection with female deities, the production of textiles was employed as a metaphor for the creation of life. The idea that female deities created life by spinning a thread is particularly Greek and runs through Greek mythological thinking on a deep level, and it may have originated in the association of childbirth with female attendants who spun while waiting to serve as midwives.[47] This idea is exemplified in the myth of the Fates, who create a person's life thread out of a pile of amorphous fluff as they spin.[48] Greeks thus conceived of an individual's life span as a thread, formed by the Fates at birth, and the act of weaving the thread symbolised what the individual did with that life, the choices that he or she made.[49] Another myth that illustrates the association between thread and life is that of Ariadne, who saves Theseus by giving him a thread with which he can retrace his steps out of the labyrinth.[50] Baert also points out a ubiquitous association of textile work with the creation of life, observing that 'weaving itself mirrors the potentiality and power of creation from fluid, dark nothingness. The warp yarns are the axes mundi, and the cutting of them is compared to birth itself, with the cutting of the umbilical cord.'[51] Barber likewise argues that 'The analogy of a person's life span to a thread goes beyond length and fragility to the very act of creation. Women create thread, they somehow pull it out of nowhere, just as they produce babies out of nowhere.'[52]

This parallel between the production of new thread and new humans – both activities performed by women – strengthens the association between textile production and the female.[53] This is supported by ethnographic surveys that have demonstrated that cloth can acquire a symbolic force of creation and fertility because women, through weaving, create a new object from disparate elements.[54] In addition, weaving and cloth have been shown to reflect the transition between life and death, as well as the child-bearing, and cloth has been compared to the fertile field, and the warp and weft to the ploughing of the earth.[55] In short, weaving occurs frequently as a metaphor for birth and the generative force of the earth.[56]

Donors of textiles

One should be wary of making the assumption that the producers of textiles also control their distribution.[57] Even so, the evidence of the temple inventories does establish that it was primarily women who donated textiles in sanctuaries. For example, the inventories from Brauron, Tanagra, and Thebes primarily name women as donors of the garments recorded in these texts – on the basis of which it has been assumed that all of the recorded garments were for and from women.[58] The names of the donors are often not recorded, however, so it must remain an open question as to whether men donated textiles. A few examples do exist: e.g. the inventory of the Parthenon records a *xystis* donated by a man named Pharnabasos. Certainly, the donated garments were not exclusively female, and women could donate clothing on behalf of men, women, and children.

Osborne has performed a prosopographical analysis of the 125 names listed in the Brauron catalogues, which provide valuable information about the women who made dedications to Artemis at Brauron. In many cases, only the first name is recorded, without the name of the husband or father, or the name of the deme of origin, which may indicate the extent of women's activity and importance in this sanctuary. We know too little about these female names to be able to reach significant conclusions, but Osborne does note that some of the names appear to be aristocratic (e.g. the names ending in –*ippē*), although this is difficult to prove.[59] Of the 125 names, only 16 include the name of the husband or father and are thus traceable. Seven of these 16 women belong to families otherwise known through inscriptions.[60] All of these women come from the more distant of the demes represented, and the fact that they all lived at a considerable distance from Brauron suggests, as Demand observes, that the garments were not casual offerings made by visitors stopping by the sanctuary,[61] and thus demonstrates Brauron's regional importance. Most importantly, though, if these women were indeed of very high social status,[62] there is further support for the inference that, because many of the offerings were expensive, the donors were most often elite, or at least well-off, individuals.[63]

The dedication of textiles in relation to the life stages of citizen women

The three main occasions for dedicating a votive offering in Greek cult are usually considered to be thank-offerings, preliminary offerings, or propitiatory offerings.[64] Votive offerings have thus been regarded as a part of a more or less transactional relationship between man and god, but can also be viewed as material expression of an individual's gratitude towards a deity for answering a prayer or providing some help or assistance.[65] Votive offerings could also function as a material reminder of the relationship between man and deity, remaining visibly effective as long as they were displayed in a sanctuary. Votive offerings can therefore be conceived of as a means of creating, visualising, and maintaining a relationship between man and god,[66] and be considered a material expression of individual piety.[67] As Gaifman argues, one of several difficulties in studying dedications is the reconstruction of the occasion for which the dedicated objects were produced: that is, whether they were meant to commemorate, supplement, or stand as substitutions for actual rituals – questions that are often impossible to answer.[68] This difficulty is also encountered in connection with textile dedications.

It appears, then, that the majority of textiles were dedicated by women. A question that arises in this connection is the occasion(s) on which women would consider it necessary and appropriate to dedicate these items. Several scholars have emphasised the connection between textile dedications and the stages of a woman's life, more specifically moments of transition such as marriage and childbirth.[69] These transitional moments in the lives of Greek women were surrounded by a variety of rites, including not only votive offerings, but also sanctuary visits, prayers, sacrifices, and dedications.[70] The women's goal in performing these rites and making these offerings was to propitiate the deities and to seek a successful resolution to the life change that they are approaching,[71] as well as to achieve acceptance of newly assigned roles within society for the female and her family.[72] Among the most important life changing transitions were puberty, marriage, and childbirth.[73]

The transition from girlhood to menarche was considered problematic, since it was a dangerous stage of life in which girls were subject to a variety of dramatic and potentially fatal symptoms, including shivering and fever, hallucinations, homicidal and suicidal frenzies, pain, and vomiting.[74] Offerings to a deity might be able to alleviate these dangers.[75] Thus, the Hippocratic essay *Peri Partheniōn*[76] reports that, when girls recover from their first menstruation, they dedicate textiles to Artemis:[77] 'When the female is recovering her senses, the women dedicate to Artemis many other things and especially expensive female clothing (*ta himatia ta poulytelektata tōn gynaikeiōn*) at the orders of the goddess's priests.'[78]

The Hippocratic corpus views this as a misguided practice; rather, the author recommends a quick marriage as the best cure.[79] Because puberty was thus closely related to marriage, it can often be difficult to separate between votive offerings related to the transition to puberty and those connected with marriage. In Athens, for example, the age of marriage for a female is estimated to have been in the mid-teens.[80] In contrast, coming of age rituals are less common for boys. The evidence for boys' maturation rituals is generally late and, according to Lee, likely reflects a Dorian tradition. Even so, there are examples involving textiles. For example, in the Hellenistic Period, youths from Phaistos on Crete participated in a festival called the *Ekdusia* (meaning 'festival of the disrobing'). During this festival the boys removed feminine garments in exchange for masculine dress in order to mark their transition to adulthood.[81]

The roles for women in ancient Greece were almost exclusively those of wife and mother,[82] and the wedding was therefore one of the most significant life events. The wedding essentially marked the transfer of the bride from her father's to her husband's home, but it also marked the principal point of transition from childhood to adulthood, and a fundamental change in status from *parthenos* (marriageable virgin) to *gynē* (wife and ideally also a mother) the two most important categories of the female.[83] The *nymphē* is an intermediate category: a woman who has married, but has not yet had children.[84] Marriage was thus a major rite of passage, and potentially a traumatic experience for the bride.[85] The event was surrounded by pre- and post-nuptial rituals, designed to secure the bride's transition to her new role in society. Before the marriage a special sacrifice was performed, called the *proteleia*.[86] The objective of the *proteleia* was to secure the transition of the young girl from virgin to wife by appeasing the divinity with a substitute for the bride who was leaving its sphere of protection.[87] The morning after the consummation of the marriage was called the *epaulia*. After this event, the bride's incorporation into her house was complete.[88] Eusthatius, in a commentary on the *Iliad* (24.29), details the gifts that a bride might receive at this event, which include clothing (*phoreia*).[89] The groom received a *chlanis*, which the bride had made herself for her husband,[90] which further points up the importance of textiles in such rituals. Another wedding ritual that involved textiles was the *anakalypteria*, meaning 'unveiling'. As Lee has already discussed, the evidence for this ritual is complex and often contradictory, and the written sources are primarily late. It is unclear whether the unveiling took place at the beginning of the wedding ceremony or at the end, in the home of the bride or of the groom.[91] One source describing the *anakalypteria* is the philosopher Pherecydes of Syros from the 6th century BC, who describes the wedding of Zeus and Chthonie: Zeus weaves a garment (*pharos*), which he offers to Chthonie. This is said to have been the first *anakalypteria*, which thereafter became custom among men.[92] Other sources emphasise that the *anakalypteria* involves the bride revealing herself to the groom for the first time, removing the veil of her bridal dress.[93]

Pregnancy was another important rite of passage in ancient Greece. The transitional period of pregnancy would have been especially fraught for a first time pregnancy, when the woman was no longer a *parthenos* but not yet a *gynē*.[94] Pregnancy was generally perceived as a dangerous period, both for the mother and the community,[95] and it was considered a period of crisis because of the high mortality rate during pregnancy and childbirth.[96] Women therefore sought divine goodwill and assistance by making votive offerings,[97] both before and after giving birth.[98] It has been argued that these votive offerings consisted of garments. For example, Lee argues that, 'because of the particular dangers associated with the transition from maidenhood to motherhood, pregnant women supplicated the goddess with one thing over which they did hold control: their dress.'[99] Lee further argues that, after a successful birth, a new mother dedicated garments in part as a gesture of thanks, but in part also as a reflection of her new social role, which required new garments.[100] This practice is related to issues of pollution, since sources confirm that women who had recently given birth were prohibited from entering sanctuaries.[101] Thus the dedication of clothing worn during pregnancy or childbirth could be viewed as a means of re-introducing a woman into society and of purifying her through the renewal of her garments.

Textile dedications and the preservation of *cosmos*

Women dedicated textiles to female goddesses who were most likely to secure their transition through the life events of puberty, marriage, and childbirth. From a more practical perspective, these offerings functioned as a means of ensuring the survival of the woman and the new-born child, and thus of the continuity of the family lineage. Incorrect performance of the dedicatory rites could result in infertility or death. This means that these dedications served to maintain the health and survival not only of the woman herself, but also of the entire *oikos*, on which society was based. By extension, the textile offerings contributed to the maintenance of society, which depended on the birth of new citizens, and thus secured the wellbeing and survival of the polis.[102] In this way, the household as well as the city were dependent on female fertility – a reality that is reflected in the fact that, at Athens, the community provided the sanctuary in which the family placed its offerings.[103]

These rituals were therefore not insignificant or concerned with only one individual, but rather were important on a much higher civic level. That women feature so prominently in these rites demonstrates their importance in Greek cult and ritual. Women were associated with the private sphere, but their religious practices affected the public standing of the *oikos* and its relationship to society as a whole. In this way, women had what Sabetai calls 'a cultic citizenship'.[104] The position of women in Athenian society has thus been considered something of a paradox, inasmuch as women's reproductive capacity places them at the centre of the *oikos* and the polis, because neither can survive without them. But at the same time women are marginalised in a social and political sense, since they are defined by their relationships with men and are inferior in status to men.[105] Yet the role that women played in producing and dedicating textiles does seem to place them in a more important and influential social position.

Recipients of textile dedications

The majority of scholars hold the view that textiles were donated to Artemis and Eileithyia, primarily in connection with childbirth.[106] Besides the evidence of the Brauron Catalogues, this is based on literary sources, especially the *Palatine Anthology*. Nine epigrams testify to the dedication of clothing to these two goddesses, and in seven instances the dedication is said specifically to have been made in connection with childbirth. Yet, as shown in Table 32, the situation is a bit more complex, in that other divinities also received textiles on different occasions. Furthermore, the anthology must be treated with care, since it is based on the lost 10th century collection of Constantine Cephalas. Although the *Palatine Anthology* derives from older anthologies, and is believed to contain material from the 7th century BC through 600 AD, it is a more recent production than the period under investigation here. It should not be assumed, moreover, that these texts reflect reality in a transparent way (i.e., that they are not poetic constructs), nor should it be assumed that they reflect the social realities throughout all of Greece. Simon has argued that the *Palatine Anthology* gives too rigid an impression of a woman's responsibilities over the course of her life, which may not in fact reflect real practice even in the Hellenistic Period.[107] The anthology is thus a deeply problematic source, and the scholarly belief that (with the notable exception of the *peplos* of Athena at Athens) only Artemis and Eileithyia were recipients of textiles is largely based on preconceptions. In fact, other female deities could function as e.g. birth or marriage deities.

There is no doubt that Artemis functioned as a protector of young girls and girls undergoing puberty, as well as women in childbirth and new-born babies.

Table 32. Textile Dedications in the Palatine Anthology

Ref.	Deity	Clothing item	Footwear	Occasion	Donor
6,206	Aphrodite	veil/hairnet, kekryphalos	sandals		F
6,207	Aphrodite	kekryphalos, prokalymma prosōpon	sandals	wedding?	F
6,208	Aphrodite	pharos	shoes		F
6,210	Aphrodite	belt (zōne)	sandals	at her 15th year	F
6,211	Aphrodite	girdle (mēleoukos)			F
6,133	Hera	veil (kalyptre)		wedding	F
6,265	Hera	byssinos (linen garment)			F
6,201	Artemis	mitrē, belt (zōne), hypodyma (undergarment), chitōn, breastband (mastodeton)	sandals	childbirth	F
6,202	Artemis	belt (zōne), garment (kypassis)		childbirth	F
6,271	Artemis	peplos	shoes	childbirth	M + F
6,272	Artemis	belt (zōma), flowery garment (kypassin)		childbirth	F
6,286	Artemis	woven textile			F
6,287	Artemis	textile (border?)			F
6,200	Eileithyia	peplos, hair bands		childbirth	F
6,270	Eileithyia	krēdemna, kalyptra		childbirth	F
6,274	Eileithyia	epiporpida, stefanos		childbirth	F
6,217	Rhea	garment (endyta)		rescue from lion	M (priest)
6,237	Rhea	garment (endyta)		rescue from lion	M (priest)
6,282	Hermes	petasos, double pin, chlamys		adolesence	M
6,21	Priapos	ragged cloak	boots	wealth	M
6,254	Priapos	purple summer dresses	white shoes	before death	M
6,292	Priapos	mitra, purple hypodyma, Laconian peploi, gold for the lōpion?, fawnskin (nebris)		beauty contest	F
6,245	Kabiroi	cloak (lōpion)			M
6,136	Unknown	garment (eima)			F

This functionality is also reflected in her many epithets, which underscore these particular roles: Artemis as Lysizonos, 'She who unties the Girdle',[108] presided over the sexual transition associated with marriage because a woman's belt was a visible sign of the invisible boundary by which she protected her body, and the untying of a female's belt could represent both intercourse and childbirth.[109] Out of gratitude for surviving childbirth and in the hopes that labour pains would be relieved, women prayed to Artemis as Lochia,[110] Eulochia, Eileithyia, and Genetaira,[111] and to Artemis Soodina and Artemis Praiai.[112]

Eileithyia is identified as the foremost deity who presided over women during childbirth.[113] The worship of this goddess dates back to the Bronze Age, and she had several sanctuaries in Greece[114] where women made offerings for a safe and quick delivery.[115] The goddess is sometimes equated with Artemis as Artemis Eileithyia, or even with Hera, who was considered her mother.[116] Several sources attest to her presence and cult in Greece,[117] but only a few – all in the *Palatine Anthology* – attest to the dedication of textiles.[118]

Scholars often describe Hera as 'the goddess of marriage',[119] because of her association in myth and ritual with *gamos*, a term usually translated as 'wedding' or 'marriage'.[120] One of her epithets was Gamelia 'of the wedding/marriage',[121] on which occasion she received offerings. Hera was also involved in the rituals that preceded marriage and the transition between girlhood and womanhood. This is exemplified by the Herai, a festival held at Olympia exclusively for women, which involved the celebration of 'wifely values' and transition rituals for girls as part of the preparation for wifehood.[122] Hera was also goddess of motherhood, although to a lesser extent than of marriage.[123] And, if we can trust the testimony of Apuleius, it seems that Hera was also venerated as a goddess of childbirth. In the *Metamorphoses*, Psyche pleads to Hera for help: 'Free me from the dangers that threaten, for I know you come willingly to the help of pregnant girls in peril.'[124] That Hera received textiles is attested primarily in the inventory form Samos, but also in the *Palatine Anthology*.[125] Furthermore, if dress-fasteners were indeed dedicated with garments, this is further evidence that Hera received such offerings, especially in the Argive Heraion and at Perachora (Appendix 4).

Another goddess who may have been associated with stages in a woman's life is Aphrodite. Not only the goddess of desire, Aphrodite was in some places in Greece also associated with marriage. On Kos, women had to sacrifice, in accordance with their means, to Aphrodite Pandamos within one year of marriage,[126] and, at Hermione, girls before marriage and widows who intended to remarry sacrificed to Aphrodite.[127] None of the inventories examined here testifies to the dedication of textiles to Aphrodite. The *Palatine Anthology*, however, includes four epigrams recording the dedication of clothing items to her.[128]

Finally, Athena could also be associated with marriage, although to a much lesser degree than Artemis and Hera. Among the few sources is Pausanias, who reports that, at Troizen, girls dedicated a girdle to 'Deceitful Athena' before marriage.[129] Literary sources also report that at Athens sacrifices were performed to ancestral deities called the Tritopatores, as well as to the first couple Ouranos and Ge and to Athena.[130] Yet what speaks to an association between Athena and textile dedications is her status as patroness of weaving, as exemplified in the famous story of Athena and Arachne. Arachne boasted that she could weave better than Athena, the patron goddess of weaving. Athena of course wins, as the gods always do, and in wrath she turns Arachne into a spider, doomed to weave for eternity.[131] The inventories from Athens do attest to a few textile offerings to Athena, but no other written sources (except the evidence of the Panathenaic *peplos*, which is examined in Appendix 1), testifies to the dedication of textiles to this particular goddess. Dress-fasteners – which are particularly common in the two Rhodian Athena sanctuaries – could, however, be evidence of such dedications (Appendix 4).

Although not a goddess, but rather a heroine, Iphigenia needs to be considered in this context, because of an important section in Euripides' play *Iphigenia among the Taurians*. At the end of the play, Athena announces that Iphigenia will die and be buried at Brauron, and 'they will dedicate adornment to you, finely woven robes (*agalma peplon*), which women who have died in childbirth leave in their homes.'[132] This has caused scholars to interpret the dedicated textiles at Brauron as the garments of women who had died in childbirth. Neils, however, notes that since there is no mention of *peploi* (the garment term used by Euripides) in the Brauron catalogues, one can assume that the dedicated *peploi* of the dead mothers-to-be were burned for Iphigenia by their relatives and so do not appear in the epigraphic record.[133] This argument is, however, mere speculation, and in a recent article by Ekroth the presence of a cult of Iphigenia at Brauron is rightly refuted.[134]

It is evident, then, that Artemis was a primary recipient of textiles, which were usually donated by women passing through important life transitions. But other goddesses could also receive textiles in connection with important stages in a woman's life. Among these goddesses, Hera especially appears to have been a competitor of Artemis. Concluding that all textiles were dedicated to Artemis, then, disregards the complexity of the situation. The recipient of these votive offerings was instead determined by regional traditions. As Clark argues,

> 'Although worshippers in a given locality must have been familiar with the "general", PanHellenic personality of a deity, there were also local priorities. Different PanHellenic figures might be privileged in different cult systems, and local deities might take on roles elsewhere reserved for others.'[135]

As an example, the cult of Hera does not appear to have been particularly prominent in Laconia, Arcadia, or even Attica. Instead, the cult of Hera is far more prominent in areas outside Attica, especially on Samos, in Argos, and in Boeotia.[136] This regionalism explains the varied recipients reflected in the sources.

Male textile dedications

Female deities appear to have been the primary recipients of textiles, as evidenced in both the temple inventories – which record only few textile dedications to male deities (Asklepios, Hermes, and Dionysos) – as well as in literary sources. A few literary sources do, however, attest to male deities receiving textiles. Thus, Herodotos states that Kroisos burnt expensive clothes for Apollo at Delphi,[137] and Horace refers to the practice of a rescued sailor offering his clothes to Poseidon.[138] It seems that Heroes could also receive clothing, since Herodotos writes that rich garments and precious metals items were offered at the tomb of Protesilaus in the Chersonese.[139]

The *Palatine Anthology* also provides several examples of male gods receiving textiles: Priapos is the recipient of textiles in two epigrams,[140] an epigram attributed to Theodoros records the dedication of a *petasos*, a double pin, and a *chlamys* to Hermes,[141] while an epigram by Diodoros the Younger attests to the dedication of a cloak to the Kabiroi.[142]

In all of the above instances, the dedication is made by a man to a male deity, and, in consideration of the evidence for the female deities, there appears to be a gendered aspect to the dedications: that is, women donate textiles to female deities, and men donate textiles to (primarily, but not exclusively) male deities. Furthermore, the occasion for the dedication of textiles is often different for men and women: women donate at the significant moments of transition in their lives, and men primarily donate as thank-offerings because the god has saved their life. An important exception is the dedication of e.g. a *chlamys* to Hermes, as a thank-offering made for an easy transition into adulthood. One cannot conclude, therefore, that textile dedications were exclusive to female deities, although goddesses receive this type of offering on a far more regular basis than gods.

Which textiles? The relation of garment type and occasion

In a few instances, the temple inventories specify the gender and age group to which a garment belongs, i.e., a woman or a man, a girl or a boy. This is the case for the inventory at Tanagra, which thus provides us with information about the gender associations of garments. For example, the inventory specifies whether that a *chitōn* belongs to a boy or a girl, and it also notes that a specific shawl, *chitōn*, or robe is for a woman. The inventory only twice specifies that a garment is for a man (a cloak, and one unknown garment). When it comes to age, the Tanagra inventory specifies once that a silver-coloured tunic is for a child. The inventory from Miletos likewise records several children's garments, e.g. a purple woollen mantle and garments of unknown type. With regard to the Brauron catalogues, ca. 10% of the descriptions specify the garment's appropriate wearer. Some garments are described as a man's, a woman's, or a child's. The *chitōniskos* and *himation* are the only two garment types that are associated with all three groups, and garments of the *chlanis* type are said to belong to men and children, and one girdle specifically to a woman.[143]

This indicates that the dedicated garments were most often female, or unisex, garments. Since women donated garments in connection with childbirth, a relevant question is whether these garments were of a special type, specifically maternity clothes. Special maternity dress is not, however, attested in antiquity, either in iconography or in the written sources, and there is nothing to indicate such special garments in the temple inventories. With regard to iconography, visual representations of pregnant women are relatively rare in ancient Greek art.¹⁴⁴ Examples include terracotta figurines dating primarily to the Classical and Hellenistic Periods, which wear garment types indistinguishable from those worn by non-pregnant adult women (Fig. 26).¹⁴⁵ According to Lee, however, the way that they were draped and girded may indicate maternity dress, since some of the women simply wear their garments unbelted to accommodate their growing abdomen.¹⁴⁶ She argues that maternity dress was an essential element in the construction of new motherhood, both for the individual woman and for her community.¹⁴⁷ Although pregnant women may have worn their garments in a particular way, there is nothing to indicate that particular garments related to pregnancy or childbirth were dedicated in sanctuaries.

Cole has suggested that the variety of garment sizes indicates the possibility that the dedications pertained to stage of life rituals.¹⁴⁸ Children's clothing is, however, only specified in relatively few instances, and there is nothing to indicate that e.g. *chitōniskoi* are for children especially. By far the majority of the garments appear to be for adults. In general, it often appears that children were dressed as small adults, at least in the same garment types.¹⁴⁹ Kontis has suggested that some of the garments recorded in the Brauron catalogues may have belonged to the girls who served Artemis as bears.¹⁵⁰ Although this possibility cannot be entirely excluded, there is nothing to indicate that

Fig. 26. Terracotta figurine of pregnant woman from Lindos. © National Museum of Denmark, inv. no. 10801.

any of the recorded garments were special ritual garments rather than ordinary garment types. One thing that becomes clear, however, from many of the written sources, especially the epigrams, is that people dedicated their own garments – items of clothing that they had been wearing, and not something bought for the occasion.

The function of the votive offering: Mauss' theories[151] of gift-exchange

Marcel Mauss' 1924 *Essai sur le don* is an important theoretical contribution to the field of gift-exchange and to our understanding of the function of votive offerings. Mauss sets out three obligations in gift exchange: giving, receiving, and reciprocating. He challenges the idea that gifts are given voluntarily and spontaneously. Instead, he establishes that gift-giving was planned, calculating, and obligatory on the part of both giver and receiver. He argues that a gift is never free, but always given as a social obligation, which must be repaid in some way. The gift presupposes reciprocity, since it must be exchanged for another gift in the form of e.g. an object or a favour. Thus, a person gives a gift because he or she is obliged to do so, but, at the same time, there is an expectation that the gift will be repaid at some point. This means that giving a gift is a way to put others in one's debt, since the recipient of a gift bears an obligation.[152] One can therefore feel the urge to decline a gift, if the gift is out of proportion with the relationship that one has with the giver. This means that there must be a balance in the personal relationship between the giver and the recipient and the character of the gift that one receives. The gift received can also be experienced as difficult or impossible to repay, which can result in hurt feelings or shame. Receiving a gift can thus be a way to maintain prestige and honour.[153]

By transferring Mauss' theories to the study of ancient Greek cult, it becomes clear that people gave votive offerings to the gods for one of two reasons: 1) either to ask for a favour, as for example relief of pain in childbirth, or 2) to repay a favour given, for example a successful childbirth. In the first case, if the gods accepted the offering, they were obligated to repay the gift to the donor. In the second case, the person making the offering was in debt to the gods for the favour given. People thus sought to enter into partnerships with the gods, by retaining them in a gift-exchange system.[154]

This relationship, however, was not straightforward. Scholarship has stressed that gifts have a different meaning according to whether the giver is of inferior rank to the receiver or vice versa. According to Godelier, superior gift recipients sometimes are not human beings, and, in all societies, humans make gifts to beings whom they regard as their superiors: divinities, nature spirits, and spirits of the dead. People pray to these beings, make offerings, and sometimes 'sacrifice' possessions.[155] This is the so-called '4th obligation' that constitutes gift-exchange, which Mauss mentioned, but did not elaborate on.[156] Godelier argues that, in analysing a gift, one needs to consider the relationship that existed between the giver and the receiver before the former made a gift to the latter.[157] This is not straightforward, however, when dealing with ancient Greek religion, since the occasion for the offering can be difficult to ascertain. One of the criticisms often levelled against Mauss' work is that he does not take into account the fact that Gods are free to give or not to give.[158] This point is emphasised by Osborne, who observes that to give a gift to the gods is to enter into a relationship in which the return is uncertain, since the donor does not know if, when, and how the deity will react to the gift.[159]

Moreover, people approaching the gods are already in their debt, inasmuch as they have already received the conditions for their existence. Mauss thus fails to take into consideration the fact that gods are already superior to humans, and that the human donors are from the outset inferior to the recipient gods.[160] This is also the case in ancient Greek cult, which means that a person or group could never give 'enough' to gods that they fully enrolled and maintained the deities in a gift-exchange system, as they could do with people.

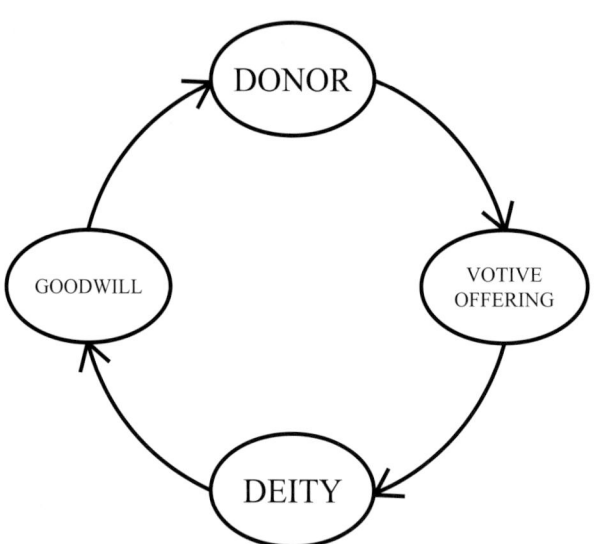

In this context, Mauss' argument that sacrifice is capable of compelling the gods, of making them give back more than they were given is important. He argues that the purpose of destruction by sacrifice is precisely that it is an act of giving that is necessarily reciprocated.[161] This was probably also the case with votive offerings.

With regard to ancient Greek cult, Mauss' theory also fails to consider the 'secondary' recipients of the votive offering: a gift donated in a sanctuary was not only visible to the gods, but also to other sanctuary visitors. Since the offering often bore an inscription with the name of the donor, his or her name was not forgotten (immediately at least), and the value of the gift would reflect the prestige, wealth, and piety of the giver.[162] This indicates that a votive offering served several further purposes: first of all, to engage in a relationship of gift-exchange with the god, either as a sign of gratitude or as a way of seeking a divine favour, but secondarily to advertise the donor's prestige and piety to other sanctuary visitors. The votive offerings thus reflect the phenomenon of 'keeping while giving', or 'inalienable wealth',[163] since it was possible to give the gifts while still keeping them to a certain degree: although the object resided in the sanctuary, it was still connected to the donor.[164]

Mauss further describes how exchanged objects contain part of the spirit of the giver. The object has what the Maori calls *hau*. The *hau* stays in the object given. This means that the gift is somehow imbued with the spirit of the giver, and this spirit stays in the object even if it redistributed.[165] Thus, the obligation attached to a gift is not inert, and even when abandoned by the giver, it still forms part of him. Through the gift, the giver thus has hold over the recipient.[166] As Mauss argues, when giving a gift, 'one gives away what is in reality a part of one's nature and substance, while to receive something is to receive a part of someone's spiritual essence.'[167] To give something is therefore to give part of oneself. This line of argument explains why money in itself is not useful as a gift and is not accepted as gift in many cultures, inasmuch as money does not contain the spirit of the giver in the sense that Mauss describes.[168] This may be the reason why the ancient Greeks preferred to donate objects rather than money to the deities, as is reflected in the inventories and the wealth of votive offerings recovered in the archaeological record. Thus, an individual dedicating a votive offering to a deity, in a sense, gave a part of him- or herself, and thus strengthened his or her gift-giving relationship with the god.

This observation is especially relevant to textiles, since they were probably made by the donors themselves, and often worn by them as well, and therefore were to an even greater degree represented the act of giving 'a part of oneself'. This is exemplified in the *Palatine Anthology*, where women donate textiles that they have woven themselves.[169] As an interesting parallel, in the *Iliad*, Hektor asks his mother Hecabe to give her largest and most beautiful *peplos* to Athena in order to request that the goddess protect Troy.[170] The goddess does not, however, grant their wish.[171] This garment was made by women of Sidon [172] and not produced by the Trojan women themselves, which, according to Karanika, may account for Athena's rejection of this dedication.[173]

In sum, this chapter establishes the close connection between women, textiles, and the use of textiles as votive offerings to female deities, especially in connection with the stages of a woman's life. It further serves to point up the importance and efficiency of textiles, which, as objects often made and worn by the donors themselves, strengthened the relationship of gift-exchange with the recipient deity.

Notes

1. Barber 1994, 33.
2. Barber 1991, 289.
3. Sebesta 2002, 127; Reeder 1995, 200; Håland 2004, 168–170.
4. Reeder 1995, 200.
5. Sebesta 2002, 126–127.
6. Xen. *Oec.* 7.6. Sebesta 2002, 126.
7. The vases especially illustrate spinning women, while weaving scenes are much rarer and other elements of textile production is completely missing from iconography. Larsson Lovén 2013.
8. Bundrick 2008, 285.
9. Bundrick 2008, 286.

10. Museo Archaeologico Nazionale Chiusi, inv. no. 1831.
11. Nereids: Athens, Ephoreia Athinon, inv. no. A1877; Aphrodite: Metropolitan Museum of Art, New York, inv. no. 09.221.40. Bundrick 2008, 295, note 30.
12. Bundrick 2008, 296.
13. Rozensweig 2004, 68–71; Neils 2000, 208–209.
14. Bundrick 2008, 296. See also Ekroth 2011.
15. Ekroth 2011, 11.
16. Bundrick 2008, 283.
17. Bundrick 2008, 286.
18. Barber 1991, 290.
19. According to Plato the most elemental city consists of the farmer, the house-builder, the weaver, and the leatherworker. Pl. *Resp.* 369d-370c. See also Thompson 1982.
20. Barber 1991, 290. There are several references to women performing textile work at home e.g. in Ar. *Lys.* 519–520, 565–586; *Eccl.* 215–218.
21. Wagner-Hasel 2013, 163.
22. Keuls 1993, 124.
23. Sebesta 2002, 134.
24. Sebesta 2002, 133.
25. Barber 1994, 273.
26. Bundrick 2008, 310. Similarly, Wagner-Hasel has proven the importance of female textile production, possibly a positive indication of women's status. She quotes the Law of Gortyn, which contains the regulation that in case of divorce the wife must leave half of what she has produced in textiles during her marriage. Wagner-Hasel 2013, 165.
27. Bundrick 2008, 316.
28. Alden 2003, 4; Geddes 1987, 310–311.
29. Ar. *Eccl.* 413.
30. Ar. *Pl.* 982-983. For the prices of items of clothing recorded on Attic *stelai*, see Kendrick Pritchett 1956, 203–210.
31. Brock 1994, 338.
32. Ar. *Ran.* 1346-1351. Thompson 1982, 219.
33. Hom. *Il.* 12.433-436. Translation: Murray 1919.
34. Plut. *Mor.* 830c.
35. *IG* II² 1672, 70–71.
36. Brock 1994, 338.
37. Sebesta 2002, 133.
38. Helen with the garments she had produced herself: *Od.* 15.104-105; 21.51-52; *Il.* 6.288-290. All the cloth owned by Alcinoos was made by his wife and her maids: *Od.* 7.234-235.
39. Van Wees 2005, 44.
40. Diod. Sic. 12.21.1.
41. Van Wees 2005, 49.
42. Sappho F38 & 39.
43. Barber 1994, 280.
44. Sebesta 2002, 131.
45. Sebesta 2002, 132; Neils 2003, 143; Hsch., *s.v. stephanon ekpherein*.
46. Sebesta 2002, 132.
47. Barber 1994, 236.
48. Barber 1991, 376. This is described in Aeschylos: '*For this is the office that relentless Fate spun for us to hold securely.*' Aesch. *Eum.* 334-335. Translation: Smyth 1926.
49. Barber 1994, 242–243.

50. Aphrodite was also portrayed as a celestial spinner: a female creating something, whether thread or new life out of next to nothing. Barber 1991, 376. Suhr has proposed to reconstruct the famous Aphrodite of Melos as a spinner. Suhr 1958, 56–57.
51. Baert & Sidgwick 2011, 335.
52. Barber 1994, 238.
53. Barber 1994, 236.
54. Barber 1991, 376.
55. Rudy 2007, 237.
56. Håland 2004, 169.
57. Schneider & Weiner 1991, 21.
58. Cleland 2005, 91.
59. Osborne 1985, 158.
60. See Osborne 1985, 159 for the full list of references.
61. Demand 1994, 90.
62. Vikela 2008, 86.
63. Dillon 2002, 22.
64. Jim 2012. However, most surviving dedications give no indication of the circumstances or occasion in which they were made and those that do often do so only in vague and general terms. *ThesCra* I, 279.
65. Van Straten 1981, 72.
66. Van Straten 1981, 74, 80.
67. Gaifman 2008, 86.
68. Gaifman 2008, 87.
69. Cole 1998, 36; Simon 1986, 203; Sebesta 2002, 132.
70. Sabetai 2008, 292.
71. Morgan 2007, 306.
72. Sabetai 2008, 291.
73. According to Cole specifically before puberty, before marriage, between marriage and first pregnancy, during pregnancy, at the time of childbirth, and for mothers at important stages in their own children's developmental cycle, Cole 1998, 32–33.
74. Demand 1994, 103.
75. Sabetai 2008, 291.
76. The *Peri Parthenión* is one in a series of gynaecological treatises in the Hippocratic corpus; it probably dates to the 5th or 4th century BC. King 1993, 113.
77. Cole 1998, 36.
78. Hippoc. *Virg.* lines 36–40. Translation: Flemming & Hanson 1998.
79. Hippoc. *Virg.* lines 36–40. King 1993, 114.
80. Sabetai 2008, 292.
81. Lee 2015, 204; Leitao 1995.
82. Clark 1998, 13.
83. Oakley & Sinos 1993, 14; Clark 1998, 14.
84. Clark 1998, 14; King 1993, 112.
85. Sabetai 2008, 291.
86. Phot. *s.v. proteleia*; Dillon 1999, 72.
87. Sabetai 2008, 292.
88. Oakley & Sinos 1993, 42.
89. Translation: Oakley & Sinos 1993, 38.
90. Poll. 3.39–40: '*The chlanis of the apaulia is sent by the bride to the groom at the apaulia.*' Translation: Oakley & Sinos 1993, 37.

91. Lee 2015, 211. See also Oakley & Sinos 1993, 25-26.
92. Pherecyd. Syr. *Fragm.* 2. Oakley & Sinos 1993, 25.
93. Oakley & Sinos 1993, 25. Ferrari, however, has suggested that the *anakalypteria* does not refer to unveiling, but to the day when the bride emerged from her seclusion. Ferrari 2003, 32-35.
94. Lee 2012a, 33.
95. Lee 2012a, 32.
96. Sabetai 2008, 289. Neils 2003, 143 suggests that 10-20% of women and children died in childbirth.
97. Sabetai 2008, 289; Neils 2003, 143.
98. Cole 1998, 34.
99. Lee 2012a, 33.
100. Lee 2012a, 36.
101. For childbirth and pollution, see Parker 1983, 48-66; Dillon 2002, 252-254; Lee 2012a, 32.
102. Cole 1998, 27.
103. Morgan 2007, 306.
104. Sabetai 2008, 289.
105. Morgan 2007, 309.
106. E.g. Vikela 2008, 85; Sebesta 2002, 132; Lee 2012a, 33.
107. Simon 1986, 205.
108. *Hymn. Orph.* 2.7.
109. Cole 1998, 34; Vikela 2008, 80.
110. Sabetai 2008, 289; Vikela 2008, 80. Artemis Lochia is more common in literary sources, Cole 1998, 42.
111. Cole 1998, 34; Sabetai 2008, 289.
112. Cole 1998, 34; *IG* VII 3407; *IG* VII 3101.
113. Garland 1990b, 66.
114. See Pingiatoglou 1981, 30-49.
115. Neils 2003, 143.
116. Dillon 2012, 3135; Sabetai 2008, 289.
117. For the attestations of Eileithyia in ancient literature, see Pingiatoglou 1981, 144-152; for the attestations of Eileithyia in epigraphy, see Pingiatoglou 1981, 153-171.
118. *Anth. Pal.* 6.200, 6.270, 6.274.
119. Clark 1998, 15; Dillon 2012, 3135.
120. Clark 1998, 13.
121. Plut. *Mor.* 141e-f.
122. Clark 1998, 15, 26.
123. Dillon 2012, 3135.
124. Apul. *Met.* VI, 3.
125. *Anth. Pal.* 6.133, 6.265.
126. Dillon 1999, 63, 66. *ED* 178.
127. Paus. 2.34.12.
128. *Anth. Pal.* 6.206, 6.207, 6.210, 6.211.
129. Paus. 2.33.1.
130. Oakley & Sinos 1993, 12. *Suda* & *Etym. Magn.* s.v. *Ouranis* and *Ge*; *Suda* s.v. *proteleia*.
131. Barber 1994, 239-241.
132. Eur. *IT* 1462-1467. Translation: Neils 2009, 138.
133. Neils 2009, 138.
134. Ekroth 2003.
135. Clark 1998, 15.
136. Clark 1998, 17; Dillon 2012, 3135.

137. Hdt. 1.51.
138. Hor. *Carm.* 1.5. Simon 1986, 203.
139. Hdt. 9.116.2. Simon 1986, 204.
140. *Anth. Pal.* 6.21 & 254.
141. *Anth. Pal.* 6.282.
142. *Anth. Pal.* 6.245.
143. Cleland 2005, 91.
144. Lee 2012a, 24.
145. National Museum of Denmark, inv. no. 10801.
146. Lee 2012a, 27–31; 2015, 213.
147. Lee 2012a, 24.
148. Cole 1998, 38.
149. Brøns 2012.
150. Demand 1994, 90; Kontis 1967, 223.
151. I refer to Mauss' ideas as theories, although they are as argued by Satlow 'rather best appreciated less as a general theory than as providing a set of questions through which we can study antiquity from a distinctive perspective.' Satlow 2013, 6.
152. Mauss 2011, 6–15. It should be noted, however, that Mauss' denial of the existence of a 'free gift' has been challenged, see e.g. Testart 1998.
153. Mauss 2011. Later scholars, however, have criticised that not all gift relations are embedded in obligation and reciprocity, Bille & Sørensen 2012, 126–128.
154. A key artefact contributing to our understanding of ancient Greek votive ideology is a bronze statuette from ca. 700–675 BC, dedicated to Apollo by Mantiklos. The inscription reads: 'Mantiklos donated me as a tithe to the far shooter, the bearer of the Silver Bow. You, Phoebus (Apollo) give something pleasing in return'. Museum of Fine Arts, Boston, inv. no. 03.997. Lapatin 2005, 290, n. 56; Stewart 1990, 43–55, fig. 11.
155. Godelier 1999, 13.
156. Mauss 2011, 14.
157. Godelier 1999, 13.
158. Godelier 1999, 30.
159. Osborne 2004, 2. Gods could also reject a gift, as told in the *Iliad*, where Zeus accepts the sacrifices of the Achaeans, but refuses to accomplish their prayers, i.e. to reciprocate the gift, Hom. *Il.* 2.419–420. Stocking 2010.
160. Godelier 1999, 30.
161. Mauss 2011.
162. Rosenberger 2008, 92, 104. Godelier similarly argues that men who give more than they have been given, or who give so much that they can never be repaid, raise themselves above the other men and are something like gods, or at least they strive to be. Godelier 1999, 30.
163. Weiner 1985, 223; Schneider & Weiner 1991, 26. Mauss is the first to introduce the concept of inalienable wealth.
164. Rosenberger 2008, 92.
165. Mauss 2011, 9. For a discussion of *hau*, see also Weiner 1985.
166. Mauss 2011, 9.
167. Mauss 2011, 10.
168. Bille & Sørensen 2012, 128.
169. *Anth. Pal.* 6.285, 6.286, 6.287.
170. Hom. *Il.* 6.269–310.
171. See Nosch 2007.
172. Hom. *Il.* 6.289–290.
173. Karanika 2001, 85.

Part III

Cult images and dress

Chapter 4

Introduction: Cult statues in ancient Greece

Methodological reflections on how to define a cult image

The custom of dressing statues of gods and goddesses in real clothes has never been examined in order to create an overview of what deities, garments, colours, and fibres were involved, as well as where, when, and why the practice occurred. Epigraphic, literary, and iconographical sources provide information on this subject, and will be treated one by one in the following chapters. This evidence will serve as a point of departure for a discussion on what the use of textiles in dressing cult images can contribute to our understanding of Ancient Greek religion and ritual.

First, it is necessary to make a few remarks about cult images. The *Oxford Encyclopaedia of Ancient Greece and Rome* defines a cult image as 'a sculptural representation of a divinity that served as the earthly substitute for a god or goddess and as a focus of worship and cult activities'.[1] In this study, the term 'cult image' will denote a type of physical object shaped in the round, not necessarily human in form. By the term 'cult statue', scholars in general usually mean the more or less life-size statue of a deity that stood in a temple and received offerings or other forms of worship. Such a statue might have been old or more recently created.[2] It has been argued, however, that the concept of a cult image is a modern construct without validity in the ancient world, and that the Greeks did not classify or conceive of these statues as such that, as Mylonopoulos observes, the term 'cult statue' did not even exist in antiquity.[3] Donohue discusses the semantic problems associated with the term 'cult statue', and she contends that the term as well as the idea should be abandoned, but proposes no alternative to it.[4] Ridgway argues that not only was there no specific term for temple images, but also it was possible for ritual offerings to be made to statues unrelated to a temple and even to statues of mortals thought to possess special powers, such as athletes. Furthermore, she argues that the use of cult images varied from region to region, and that the concept of the temple as a house of the divinity is derived from different ancient cultures and religions, and therefore not necessarily obvious to all Greeks.[5] Nevertheless, these caveats aside, there is no doubt that the Greeks venerated cult statues – whether in a temple or not. As Platt has argued, the absence of a precise termininology does not mean that the category did not exist. The term 'cult image' still has significant heuristic value for modern scholars.[6]

The appearance of cult statues is obviously of great interest in a study on the dressing of these artefacts. Unfortunately, very few identifiable cult statues have survived from antiquity, and it is only seldom possible to determine the garments of the rare surviving examples. Furthermore, it is very difficult – if not impossible – to distinguish these statues from other statues representing divinities, given that not all statues representing a deity were cult statues: they could also be simply dedications or votive offerings. That said, the very opposition of 'cult statue' and 'votive offering' is methodologically problematic. Even those statues placed in the centre of a temple were normally votive offerings to the gods.[7]

Several scholars have tried to establish concrete criteria for the definition and identification of cult statues. One is Mylonopoulos, who identifies three main aspects to the identification of cult statues: 1) cultic use, 2) appearance, and 3) position.[8] His three points will be discussed here:

1: Cultic use: The use of a divine image in cult activities can serve as a basis for the identification of cult statues.[9] Hölscher completely dismisses the importance of a statue's placement in a sanctuary and exclusively defines cult statues on the basis of integration into cult, which had to be repetitive.[10] Mylonopoulos generally agrees, but concludes that attempts to identify cult statues can place too much emphasis on the element of recurring worship, since some statues could be temporarily transformed into cult statues under specific circumstances.[11] Bettinetti, on the other hand, claims that what defined a cult statue was rather the rituals during the act of installation, not the rituals after its erection.[12] In sum, in order to transform a statue into a cult statue, it must be the object of cult, which means that it had to be worshipped in one way or the other.

2: Appearance: First and foremost, it must be stated that there is no way to recognise a divine image used in cult exclusively on the basis of appearance.[13] There are not specific traits that can help us to identify a cult image. According to Mylonopoulos, however, there is an unambiguous criterion for distinguishing between cult images and divine images not used in cult activities: the mythological narrative and, more importantly, the absence thereof can serve as a basis for the semantic and functional separation of cult statues from divine images in general.[14] Thus, if a statue was NOT associated with a mythological narrative or used in any activity, but simply standing or sitting still, it was (more likely to be) a cult statue, while statues used in some sort of activity were not cult statues.

3: Position: Divine images placed centrally in the cella of a temple are usually interpreted as cult statues, while all other divine images in the spatial context of a sanctuary are interpreted as votive offerings.[15] Thus, on rare occasions, statues are recovered *in situ* in a prominent position in a temple, which can contribute to their identification as cult statues. Yet there are problems with this method of identification: first, an image recovered during modern excavations may not be in its original location; second, several temples contained more than one (cult) image; and third, some cult images were not placed in temples at all.[16] This interpretation neglects other religious spaces (such as sacred groves, caves, etc.) as well as venerated statues in sanctuaries dedicated to other deities. Mylonopoulos thus concludes that the exact position of a divine image in a religious spatial context is an important, but not the only, criterion for the identification of a cult statue.[17]

In sum, what defines a cult statue above all is its sacredness and its role in cult activities, but this original function is regularly lost to us today. So, even though there can never be one sure way for us to ascertain clearly what was and what was not a cult statue, the primary approach to identifying a cult statue is therefore consideration of its appearance (even though a statue cannot be classified as cultic through its appearance alone)[18] and, most importantly, its context. Furthermore, many of the reservations listed above, especially those regarding position, may result in the overlooking of possible cult statues, but not the misinterpretation of statues of divinities as cult statues. This means that other cult statues may be present in museums, etc., but that we are unaware of their function as cult images due to the loss of their original context. The following examples therefore consist exclusively of statues that have been recovered in their original context, which resulted in their identification as cult statues.

Archaeological evidence of cult statues

The first example is the three mid-7th century BC hammered bronze statuettes identified as Apollo, Artemis, and Leto found in the rear of the Temple at Dreros in eastern Crete, which are considered to be among the earliest surviving possible cult images (Fig. 27).[19] The male figure is naked and ca. 80 cm high, while the two females each wear a belted *peplos* with *epiblēma* and a low *polos* and are, respectively, 40 and 45 cm high.[20] A second example is the colossal marble statue, dating to ca. 530–520 BC, from Ikaria (southwest of ancient Marathon), which is identified as Dionysos (Fig. 28).[21] According to Romano and Despinis, this statue possibly functioned as a cult image at the site. The statue is seated, and wears a *chitōn*, under a *himation*, and sandals.[22]

Fig. 27. Apollo, Artemis, and Leto from Dreros, eastern Crete, mid-7th century BC. Archaeological Museum of Heraklion, inv. nos. 2445–2447. © Hellenic Ministry of Culture and Sport/Archaeological Receipts Fund.

A third example is the colossal marble sculptural group of Despoina and Demeter, attributed by Pausanias to Damophon of Messene (Fig. 29).²³ The four over life-size figures (ca. 5.6 m high) in the group are likely to have functioned as cult statues.²⁴ Despoina and her mother Demeter are seated on an ornate throne, and are flanked by Artemis and the Titan Anytus. The statues were found on site in the Temple of Despoina at Lykosoura in Arcadia, and date to the beginning of the 2nd century BC. The statues are very fragmented, and, since primarily the heads are preserved, we cannot determine their original garments.²⁵ A large part of the veil of Despoina is, however, intact. It has a tasselled fringe, but is of particular interest due to its complex decoration, which features lines of dancing animal-headed women, eagles, thunderbolts, Nereids, tritons, Nikai, etc.

Fig. 28. Dionysos from Ikaria, ca. 530-520 BC. National Archaeological Museum, Athens, inv. nos. 3072, 3073, 3074, 3897.© Hellenic Ministry of Culture, Education and Religious Affairs/Archaeological Receipts Fund. Photo: Eliades.

Despinis proposes that a colossal marble head recovered from the Athenian Acropolis depicts Artemis, and he dates it to the third quarter of the 4th century BC (Fig. 30). He proposes, furthermore, that this head belonged to an acrolithic image, which he argues is the cult statue of Artemis that Pausanias saw in the Brauronion on the Acropolis.²⁶ Unfortunately, we do not know what the statue was shown wearing.²⁷

A fifth and final example is the colossal group of marble cult images found in pieces inside the Temple of Apollo in Klaros in Asia Minor. The statues can be dated to the 2nd century BC.²⁸ The group was enormous (ca. 7 m high) and consisted of a central seated Apollo flanked by a standing Artemis and Leto.²⁹ Of

Fig. 29. Fragments of the cult images of Despoina, Demeter, Artemis and Anytus from Lykosoura, early 2nd century BC. National Archaeological Museum, Athens, inv. nos. 1734-1737, 2171-2175. © Hellenic Ministry of Culture, Education and Religious Affairs/Archaeological Receipts Fund. Photo: V. Eikstedt and G. Patrikianos.

the Apollo, the only remains are draped legs, which testify to a long *chitōn* or a *himation*. The statue of Leto wears a *chitōn* and a *himation*, while the Artemis is clad in a belted *peplos*.[30]

These few examples suggest that, over time, cult statues changed with regard to size, material, and actual appearance. Furthermore, they indicate that the representation of cult images (if they truly are cult images) follow the same scheme with regard to dress as contemporary statues.

Literary evidence for cult statues

Because so few ancient cult statues have survived, it is necessary to examine other sources in order to determine their appearance. Ancient authors, such as Pausanias, can provide primary evidence for an understanding of ancient cult statues,

Fig. 30. Head of the marble cult statue of Artemis, ca. 350–325 BC. © Acropolis Museum, inv. no. Acr. 1352. Photo: Mavrommatis

even though these descriptions are often limited and open to interpretation. Early cult images are referred to by various terms in the literary sources, but the earliest Greek cult statues that may properly be called statues in the modern sense were known as *xoana*,[31] generally regarded as the most ancient form of the cult image.[32] The word *xoanon*[33] was used consistently by Pausanias when describing a wooden cult image,[34] and has become a common term for an ancient wooden cult statue.[35] The earliest attestation of the word is from the end of the 5th century BC.[36] According to Donohue, however, the word *xoanon* appears to change its meaning between the time of Sophocles and the 4th century, when it is applied broadly to images of gods, a usage that appears only once in Euripides.[37] *Xoanon* in the sense of 'statue' is thus a latecomer to the 5th century.[38] The word literally means '(some)thing scraped' or 'wood carving'. *Xoana* were usually made of wood, although some might have been of stone or a combination of the two.[39] Due to their perishable nature, there are no surviving wooden statues that can be definitely identified as cult images, but representations on coins and vases inform us what they might have looked like: block and plank-like stocks occasionally with a sculpted head, the earliest examples probably lacking hands, feet, and eyes.[40] When exactly these first wooden cult images were made is not known, but they may have existed already in the Bronze Age.[41] From literary evidence it is possible to date the two earliest known *xoana*, the statue of Hera on Samos[42] and the image of Athena Polias at Athens,[43] to the beginning of the 8th century BC.[44]

Hiller has even suggested that Mycenaean cult idols survived into Classical times.[45] Since, however, such early cult images have not survived in the archaeological record (with the possible exception of the female terracotta figurines from Aghia Irini and Phylakopi),[46] we do not have any knowledge of what they looked like.

Judging from the literary as well as archaeological evidence, Archaic cult images ranged in scale from very small to colossal. The statues of Hera at Argos or of Artemis Orthia at Sparta and Messene would plausibly have been of small size, since the rites of bathing and clothing (Hera) or holding them at the altar (Artemis) otherwise could not have been performed.[47] On the other hand, the Apollo of Delos, a *sphyrelaton* by Tektaios and Angelion, located in the first Temple of the Delian sanctuary and dedicated to the god around 530 BC, was a statue of colossal size. There was another even earlier colossal statue (*agalma*) of Apollo at Amyklai, which could be seen from far away.[48] Another example is the colossal wooden cult statue (*xoanon*) of the 8 ft (ca. 2.5 m) high Hermes on Mount Kyllene in Arcadia.[49] Thus, it obviously does not work at all to associate old with small. Cult statues in the Archaic Period cannot be equated with statues that could be treated as human beings. Instead, the size and forms of cult statues depended on their respective functions.[50] According to Romano, however, Archaic cult images (*xoana* in particular) were usually quite small, and the earliest Greek cult images were in general less than life-size.[51] This is indicated by their representation in vase-painting, where both male and female cult images are often of a relatively small size.[52] In terms of their appearance, the *xoana* could be either standing or seated and either aniconic or rudely shaped.[53]

Other words used to describe cult statues are *agalma*,[54] *bretas*,[55] and *hedos*. The word *agalma* has diverse meanings, from 'glory, delight, honour', 'pleasing gift for a god', 'statue', to 'statue in honour of a god',[56] but it could also be used to denote a cult statue. Yet the distinction between a statue of a god and cult image can be difficult to make with regard to this particular word, and it seems to be dependent on context.[57]

The origin of the word *bretas*, which was in use from the 5th century BC, is unclear. It is used mainly in poetry with reference to wooden images of the gods of small size and great age.[58] *Bretades* are usually understood as referring to unshaped planks, which served as cult images, the holy vessels, of particular gods.[59] But it has been suggested that the etymology of *bretas* from *brotos* (mortal), even though incorrect, indicates that a *bretas* would have been anthropomorphic rather than an unshaped plank. According to O'Brien, the word *bretas* thus refers to a rudely-shaped icon.[60]

The word *hedos* also possibly refers to a cult image, at least from the 5th century BC.[61] The word is usually interpreted as meaning 'seat' or 'seated statue of a god' and is applied to many Greek cult images.[62] It is used, however, to describe the statue of Athena Parthenos, which is known to have been standing, and it does not therefore necessarily refer to a seated image.[63] It has been argued, therefore, that the term *hedos* refers to the function of cult images as 'vessels' or 'receptacles' of divine presence, rather than to appearance, and that *hedos* makes concrete the idea that sacred images operate as physical containers for divinity.[64]

The origin and development of cult statues

The Greeks also venerated unworked stones and meteors: for example, 30 square stones at Pharai in Achaia,[65] an unworked stone (*argos lithos*) representing Heracles in his temple at Hyettos,[66] and a stone (*argos lithos*) of Eros at Thespia were all worshipped as gods.[67] Furthermore, at Orchomenos in Boeotia, the Charites were venerated in their temple under three stones thought to have fallen from the sky,[68] and at Megara, a small pyramidal stone was worshipped under the name of Apollo Karinos.[69] These unworked stones were often called *argoi lithoi*, while the meteoric stones were called *baitylia*.[70] They are described by Pausanias, who believes that this rite is older than the cult statues of divinities: 'At a more remote period all the Greeks alike worshipped uncarved stones (*argoi lithoi*) instead of images of the gods (*agalmatōn*).'[71] But whether they were actually completely unworked or rudimentarily shaped and marked is impossible to say. It is tempting to suppose, like Pausanias, that the Greeks began by venerating unworked stones (*argoi lithoi*), then made them into conic shapes, incised them, and finally separated the limbs from the body, and to suppose that this took place before the advent of the great sculptures of the 6th and 5th centuries. This is, however, absolutely not the case.[72] For example, many of the recorded examples of unworked cult objects actually postdate much more anthropomorphic imagery, and there is plenty of evidence that the worship of unworked stones continued throughout antiquity.[73] A well-known example is the *omphalos* (navel stone), which was placed in the *adyton* of the shrine for the oracle at Delphi.[74] The original Delphic *omphalos* cult probably dated back to the Archaic Period. A new representation in marble was set up ca. 330 BC, either in the temple or outside.[75] Recent research has shown that this marble *baitylos* was fitted to a bronze tripod, which stood on the top of a tall marble column.[76] This example is particularly interesting since it has preserved relief decoration, imitating the thick woollen bands or ribbons (called *agrēnon*)[77] that covered the original Archaic *omphalos* (Fig. 31).[78]

Likewise, it is not possible to trace a linear evolution from tree cults to the plank-like cult statues called *bretades*, to *xoana*, to archaic *kouroi* and *korai*, and finally to the more naturalistic Classical and Hellenistic statues, since many of the simpler cult statues existed alongside those by great artists of later times, and they cannot be explained in terms of the persistence of earlier models. The sequence is therefore not linear, evolutionary, or chronological, but typological, although the aesthetic development also played a part.[79] Nevertheless, the appearance of newly produced cult images changes during the first millennium BC, especially in the 5th century BC with the transition from the Archaic to the Classical Period.[80]

The question here is how the development of cult statues played a role in the dressing of cult statues. According to Romano, acts of dressing and adorning early Greek cult images were especially common in the case of wooden images,[81] and it was only cult statues – not votive statues – that wore real clothing.[82] The Classical and Hellenistic cult statues, which were dressed in real garments, were fashioned from a greater range of materials, such as the bronze statue of Apollo at Amyklai and

Fig. 31. Marble omphalos from Delphi, ca. 330 BC. Archaeological Museum of Delphi. © Ministry of Culture and Sports, Ephorate of Antiquities of Phokis.

the stone statue of Artemis at Brauron.[83] This was connected with the changes that became apparent in religious cults, such as the altered attitude towards cult images by the end of the 5th century BC. For example, the old cult images, *xoana*, from the 8th, 7th, and 6th centuries BC were supplemented by new elaborate images of deities in idealised, anthropomorphic forms.[84]

But were the cult statues fashioned as undressed figures? Some early cult images were aniconic and thus lacked any representation of anthropomorphic traits or clothing. Others were iconic, or rudely-shaped images, and could possibly have possessed some vague or cursory representation of clothing. In the case of the later stone cult statues, it is likely, but not an unchallenged fact, that these images had sculpted representations of clothing. Yet this hardly excludes the possibility that the images were clothed. In the case of the statues of Hera on Samos, O'Brien argues that worshippers would more likely clothe a plank than an anthropomorphic statue with clothes already incised.[85] Yet, as we shall see, the rendering of clothing has nothing to do with whether a cult statue could be dressed or not. Similarly, in modern Hinduism, figures of deities already dressed are clad in extra garments, and in the Catholic Church, icons and representations of the Holy Mother can, on certain occasions, be covered with textiles.

Notes

1. Gagarin 2010, *s.v.* cult images. This definition may seem somewhat limited, in that it only includes sculptures, not paintings or equivalents of what we know as icons in the Catholic Church today. We do not know of any such examples, however, from Ancient Greece; therefore, my focus is on cult images in the form of sculpture-like objects.
2. Ridgway 2000, 230.
3. Mylonopoulos 2010, 4.
4. Donohue 1997, 31.
5. Ridgway 2000, 231.
6. Platt 2011, 77.

4. Introduction: Cult statues in ancient Greece   179

7. Mylonopoulos 2010, 4.
8. Mylonopoulos 2010, 6.
9. Mylonopoulos 2010, 7.
10. Hölscher 2005, 52-53.
11. Mylonopoulos 2010, 7.
12. Bettinetti 2001. See also Freedberg 1989, 82-84, 98.
13. Mylonopoulos 2010, 8.
14. Mylonopoulos 2010, 11.
15. For the setting of cult statues in ancient Greece, see e.g. Dawson 2002.
16. Weddle 2010, 17.
17. Mylonopoulos 2010, 6-7.
18. Weddle 2010, 11.
19. Alroth 1989, 19; Romano 1980, 21, 284; Richter 1960, 32, 26. Herakleion Museum, inv. nos. 2445-2447.
20. Romano 1980, 287-291.
21. National Archaeological Museum, Athens, inv. nos. 3072, 3073, 3074, 3897.
22. Romano 1980, 316-327; for detailed studies of the statue, see Romano 1982 and Despinis 2007.
23. Paus. 8.37.
24. National Archaeological Museum, Athens, inv. nos. 1734-1737, 2171-2175.
25. On the sculptures, see Stewart 1990, 94-96; Pollitt 1986, 165-166; Kaltsas 2002, 279-281.
26. Despinis 2010, chapter 6; Despinis 2004; Papalexandrou 2012, 3. Acropolis Museum, inv. no. Acr. 1352.
27. Another example of a possible cult statue is a colossal marble head of Hera from Sparta dated to the 3rd century BC. Again, unfortunately, we have no knowledge of the garments, but she wears a *polos*. Archaeological Museum, Sparta, inv. no. 571. A further example is the limestone head from Olympia, possibly a cult image of Hera. Kardara 1960.
28. For more on the date of the group, see Flashar 1999.
29. Ridgway 2000, 240.
30. Flashar 1999, 56-58.
31. Freedberg 1989, 34.
32. Romano 1982, 8.
33. For the etymology of the word, see Donohue 1988, 9-12.
34. Paus., e.g. 1.32.4; 2.4.5; 3.13.4; 3.14.5; 8.17.2; 9.40.3.
35. This view is criticised by Donohue, who concludes that *xoanon* is a 'historiographic mirage' unsupported by the archaeological evidence or the ancient texts, Donohue 1997, 31.
36. The word *xoanon* appears in Eur. *Tro.* 525-530, 1071-1076 (415 BC), and *IT* 1358-1359 (ca. 412 BC), and *Ion* 1397-1403 (ca. 412 BC). Donohue 1988, 30.
37. Donohue 1988, 30. Eur. *Ion* 1397-1403.
38. Donohue 1988, 30.
39. For example, Paus. (8.31.5-6) records that a *xoanon* of Aphrodite is made of wood, while her hands, face, and feet are of stone.
40. Freedberg 1989, 34.
41. Rutkowski 1973.
42. Paus. 7.4.4.
43. According to Hdt. 5.71, Kylon took refuge by the statue of Athena in the 7th century BC. Furthermore, in Philostr. *V. A.* 3.14, the image of Athena Polias was among the most ancient of the Greek gods.
44. Romano 1982, 8.
45. Hiller 1983, 92.

46. Renfrew 1985, 215–216, 411; Caskey 1981; Rutkowski 1973, 55.
47. Hölscher 2010, 108.
48. Hölscher 2010, 108. Paus. 3.19.2. Pausanias writes that nobody 'has measured the height of the image, but at a guess one would estimate it to be as much as thirty cubits' (13–14 m). Translation: Jones & Ormerod 1918.
49. Paus. 8.17.2. Romano 1982, 9.
50. Hölscher 2010, 108.
51. Romano 1982, 9.
52. See Chapter 5.
53. O'Brien 1993, 18.
54. *Agalma* as a statue of worship is described by e.g. Aesch. *Sept.* 258 (467 BC); *Eum.* 55 (458 BC) and Soph. *OT* 1379 (429 BC).
55. *Bretas* in the meaning of a wooden image of a god is used in Aesch. *Eum.* 80 (458 BC); Eur. *Alc.* 974 (438 BC) and Ar. *Eq.* 31 (424 BC).
56. *LSJ s.v. agalma.*
57. For *agalma*, see also Nick 2002, 12–15.
58. Nick 2002, 20.
59. Freedberg 1989, 33.
60. O'Brien 1993, 19–20.
61. For occurrences of the word, see Nick 2002, 22–23.
62. *LSJ s.v. hedos.* Platt 2011, 104.
63. Isoc. 15.2; Plut. *Per.* 13.9.
64. Platt 2011, 104.
65. Paus. 7.22.4.
66. Paus. 9.24.3.
67. Paus. 9.27.1. Rutkowski suggests that the earliest Bronze Age cult images were rock formations, such as stalagmites and stalactites in Cretan caves. Rutkowski 1973, 54.
68. Paus. 9.38.1.
69. Paus. 1.44.2. On ancient aniconism, see also Gaifman 2005; Gladigow 1988.
70. Freedberg 1989, 66. Pyramidal *baitylia* are depicted on a series of coins (staters) from Caria, dated to the 5th century BC. Robinson 1936, no. 1–5, 266, 270–273.
71. Paus. 7.22.4. Translation: Jones & Ormerod 1918.
72. Freedberg 1989, 68.
73. Freedberg 1989, 66.
74. The Delphic *omphalos* is mentioned by several ancient sources, e.g., Pind. *Pyth.* 4.74, 6.3, 11.9; Aesch. *Eum.* 39, 166; Soph. *OT* 480, 897; Eur. *Ion* 222–225, 461; Paus. 10.16.3. See Herrmann 1959, 13, 14.
75. Herrmann suggests that this *omphalos* was the one Pausanias saw when visiting the sanctuary. Herrmann 1959, 17.
76. Kolonia 2012, 62.
77. *LSJ s.v. agrēnon*: net or net-like woollen robe. Poll. 4.116.
78. Kolonia 2012, 62. There are also representations of *omphaloi* decorated with ribbons or cloth in vase-paintings, e.g. Staatliche Antikensammlungen, Munich, inv. no. 1426; State Hermitage Museum, St Petersburg, inv. nos 1793 and 1792; on reliefs, e.g. Kunsthistorisches Museum, Vienna, inv. no. I 814; and on coins, e.g. silver coins from the reign of Antiochos I, *BMC* 4 pl. 3.4–5; see also Herrmann 1959, pl. 1.1–2.2, pl. 10–11.
79. Freedberg 1989, 68.
80. Tanner 2001, 257; Neer 2010; Stewart 1990.

81. Romano 1982, 5 (primarily since it is assumed that most early cult images (*xoana*) were of wood, and therefore not a conscious choice of dressing wooden statues instead of stone statues, etc.).
82. Romano 1988, 130.
83. Romano 1988, 130. Ridgway, however, has questioned whether temple images were made of bronze, at least in Classical Athens. She suggests that bronze was reserved for votive and athletic sculpture. Ridgway 2005, 115.
84. Romano 1982, 11.
85. O'Brien 1993, 22.

Chapter 5

Iconographic evidence for the dressing of cult statues

Despite the loss of original cult statues, some have survived in replicas produced during the Hellenistic Period. This imitation of famous prototypes and earlier Classical styles is primarily a feature of Hellenistic and Roman art, and happened to such an extent that the Classicising trend is sometimes thought to have originated in the production of cult statues.[1] These replicas date to the Hellenistic or later periods, which raises questions about whether the statues thus reproduced are of ancient origin or a Hellenistic 'archaising' phenomenon.[2] There is virtually no iconographic evidence before the Hellenistic Period for any of these statues, and an Archaic origin can therefore only be hypothesised. Nevertheless, Villing argues that it is difficult to imagine that 'the Anatolian type' of cult statues was a Hellenistic invention, and the late appearance of the type could be explained by antiquarian interest.[3] It therefore seems likely that these replicas reflect older cult statues.

A copy or a replica should not necessarily be understood, however, as an exact duplication of a sculpture in all its details and exact dimensions, but rather as a reproduction of a work to such an extent that its similarity to the original is easily recognisable.[4] An important difference is the change of medium, usually from bronze or perhaps wood to marble. Nonetheless, these reproductions provide important information on the appearance and dress of the original cult statues, assuming that the copyists did not change the clothing. This means that there is some uncertainty as to whether the garments rendered on the statue reflect the 'real' clothing of the original or whether they are artistic constructs with no relation to the original.

Hellenistic and Roman reproductions of Archaic Anatolian cult statues

Identifiable reproductions of cult images from Greece are extremely rare,[5] but several such copies of cult statues are known to us from Anatolia. One of the most well-known and discussed examples is the canonical image of Artemis Ephesia (Fig. 32). Nearly 40 marble replicas exist of this cult statue, but none are earlier than the 1st century AD. The majority belong to the Hadrianic Period or later, and only one was certainly recovered in Greece.[6] Nevertheless, they are thought to represent a much older cult image, now lost to us. The general view of most scholars is, therefore, that we see in these replicas an underlying simple image, possibly an archaic *xoanon*.[7] Thus, Fleischer has shown that the many-breasted cult image is not a Hellenistic invention or replacement, but goes back to an early *xoanon* in wood, probably dating as early as the 7th century BC.[8]

The replicas of the Ephesian Artemis are very distinctive and recognisable due to certain characteristics. The deity is always represented standing rigidly upright almost as if she was a *xoanon*, with her arms bent at the elbows and her hands stretching out towards the viewer; she is adorned with bountiful accessories and a headdress either in the shape of a mural crown or a *kalathos*. But her most crucial attribute is the polymastoid globules on her upper body, usually interpreted as breasts and thus a fertility symbol, although it has also been suggested that they are rather the scrotums of bulls, which were ritually sacrificed to the goddess.[9] With regard to her dress, she wears a long garment, tightly fitted around her lower body so that the garment appears to constrain her. Underneath she wears a long *chitōn*, of which only a small part is visible, opening fanlike to show her feet. The outer garment is heavily decorated, primarily with animal protomes placed in panels, and has long sleeves. In several instances, the decoration is reserved for the front and sides, while the back is undecorated, but with the folds of the garment

Fig. 32. Artemis Ephesia, marble statue. Ephesos archaeological museum, Selçuk, inv. no. 718. © Arachne, DAI. Photo: Hannestad.

rendered, giving the impression that the statue is wearing a sort of apron.[10] This 'apron' seems to follow the seam of the *ependytēs*, which makes it likely that it is part of the same garment and not necessarily an independent item.[11]

A similar cult statue of Aphrodite from the sanctuary of Aphrodisias in Caria has been preserved for us in about 22 marble copies, all belonging to the Hellenistic or Imperial Periods (Fig. 33).[12] The surviving images of the goddess represent her standing upright with her forearms outstretched like the Ephesian Artemis. She is clad in a floor-length *chitōn* with long sleeves, over which she wears a thick, form-disguising *ependytēs* reaching down below her knees as well as a long veil extending to the ground. Her *ependytēs* is always richly decorated with depictions of the Graces, different gods and goddesses, Erotes, and the goddess herself, all placed in horizontal registers. Additionally, she is adorned with jewellery and occasionally a mural crown, a diadem, and a wreath.

Another cult statue known through both replicas and depictions in different media is the *xoanon* of Zeus Stratios from Labraunda at Caria, believed to date back to at least the Archaic Period.[13] It is very close in appearance to the Samian Hera and especially the Ephesian Artemis. The statue is represented in reliefs, bronze statuettes, and on coins.[14] The earliest such representations are on a relief from Tegea dating to the mid-4th century BC and on a coin minted under Mausollos (377–353 BC).[15] On the relief, the god is rendered frontally holding a double axe and a spear. He is bearded and wears a garment decorated with breast-like globules on his upper torso similar to those on the Ephesian Artemis, as well as a *himation* draped over his left shoulder and lower body (Fig. 34).

Fig. 33. Aphrodite of Aphrodisias, marble statue. Aphrodisias Museum. © William Neuheisel 2012. Wikimedia Commons.

All depictions of the god show him with a full beard, and he is always dressed in a long garment, often a *himation*, but sometimes a tight-fitted long skirt with a net-pattern in diamond-shaped sections.[16] Such a garment is represented on a Roman bronze statuette from the 1st century AD (Fig. 35),[17] and on coins (Fig. 36).[18] Carstens interprets this garment as a cloth tied with long strings or some kind of net,[19] possibly like the '*agrēnon*' mentioned by Pollux.[20] This variation in the depiction of the deity indicates either that there were two different cult images, or simply that the statue was dressed in interchangeable garments.[21]

A different – but related – replica of a cult statue is the Late Hellenistic marble statuette of Artemis Kindyas from Bargylia in Caria (Fig. 37), which was recovered at Piraeus.[22] The statuette is 1.05 m high, represented standing upright, and

Fig. 34. Zeus Labraundos, marble relief, ca. 350 BC. © Trustees of the British Museum, inv. no. 1914.0714.1.

dressed in a particular way, with her arms crossed over her chest and held tight to her body by the outermost garment, almost in the manner of a straitjacket. She is clad in a long *chitōn* reaching to the ground and covering her feet, over which she wears a knee-length *ependytēs*, which is fastened by a girdle with fringes at the ends and tied in a Hercules knot at the waist. Finally, she wears a mantle or veil over her head.[23] The statue of Artemis Kindyas is also depicted on Hellenistic and Roman bronze coinage, primarily from the Carian city of Bargylia in south-western Asia Minor. A series of coins even identifies her by name: Artemis Kindyas.[24] The depiction of entire cult statues for goddesses from Asia Minor generally does not become common until Imperial times, but Artemis Kindyas is depicted as such already from the 3rd century BC. In most representations, Apollo stands or sits next to the cult image of Artemis Kindyas, which stands on a base, clad in a *chitōn*, *ependytēs*, and veil, with her arms across her chest.[25] On other examples, the cult statue is shown next to a seated Zeus (Fig. 38). The statue always wears the same garments as the statue from Piraeus, but the veil is occasionally arranged differently.[26]

A similar example to Artemis Kindyas – although not from Anatolia – is the representation of the cult statue of Hera from Samos. Several Samian coins, primarily from the Imperial Period, represent

Fig. 35. Zeus Labraundos, bronze statuette 2nd century AD. Photo: © Museum of Fine Arts, Boston, inv. no. 67.730.

Fig. 36. Zeus Labraundos, Mylasa. Geta. © Trustees of the British Museum, inv. no. HPB,p116.5.A.

Fig. 37. Artemis Kindyas, marble statue, late 4th century BC. Archaeological Museum of Piraeus, inv. no. 3857. © Hellenic Ministry of Culture and Sports.

the cult statue dressed in garments. According to O'Brien, the Imperial coins depict not a Classical, but the Archaic cult image.[27] The arms, though, are not held close to the body like those in the representations of Artemis Kindyas. With regard to the actual garments, the coins are often so worn and sometimes rudimentarily made that it can be difficult to determine exactly what type of garments are represented. It can be ascertained, however, that the statues always wear a long *chitōn*, a sort of mantle, and usually a *polos*. Yet some coins are so well preserved that the garment types can be determined. This is the case for a coin from the reign of Commodus, which displays the cult statue of Hera in a long *chitōn* and an *ependytēs*, across which run two intersecting bands – a similar form of dress of that to Artemis Kindyas (Fig. 39).[28] Furthermore, Hera wears a mantle, several chains or necklaces around her shoulders, and a *polos* on her head.[29] Other coins representing the statue of Hera from Samos depict her wearing similar outfits, including the intersecting bands (girdle), but without the *ependytēs*.[30] In addition to depictions on coins, the cult statue of Hera from Samos is represented in stamps on Samian transport amphorae dating to the 4th century BC,[31] as well as in a badly preserved stamp seal on an Imperial roof tile from the Heraion, which bears a crude image of the goddess, probably representing the *xoanon*.[32]

5. Iconographic evidence for the dressing of cult statues

Fig. 38. Artemis Kindyas, tetradrachm. Alexander III Megas, Macedon (ca. 300–280 BC). © Trustees of the British Museum, inv. no. 1898,0602.70.

A further source for our knowledge of the original cult statue of Hera is a wooden statuette, dated to the 7th century BC, which was discovered in the foundations of the Heraion Temple II (Fig. 40).[33] Although the statuette is a votive statuette rather than a cult statue, it is thought to have been based on or to reproduce the cult statue.[34] The goddess is represented standing upright and somewhat planklike, but with her arms free of her upper body. She wears a very large mural crown with decoration in panels, a long belted garment – over which she appears to wear another garment (possibly an open vest similar to the one worn by the so-called *peplos* kore [see below]) – and a short cape around her shoulders. O'Brien identifies her garment as a typically Cretan 7th century dress.[35]

Fig. 39. Samian Hera. Commodus. © Trustees of the British Museum, inv. no. 1874,1201.10.

Fig. 40. Female wooden statuette from the Heraion, 7th century BC. Museum Vathy, Samos, inv. no. H41. ©D-DAI-ATH-Samos 5476. Photo: Eva-Maria Czakó.

A final artefact of interest for this discussion is a statuette of Artemis of Perga from Cremna in the province of Burdur, south-western Turkey (Fig. 41).[36] It is quite small, with a preserved height of ca. 20 cm, and it is 11 cm broad. The statuette does not resemble a human form, but rather brings to mind a shapeless mass or block. Yet it has human traits such as a small, centrally placed face, and it wears a high *polos* with panel decoration. The lower part of the cult image, separated from the upper part by horizontal garland, consists of two parallel bands featuring primarily female figures in high relief. Furthermore, the statue wears a veil, which is clearly visible on the back and along the sides.[37] The cult image of Artemis of Perga is also represented on coins, dating from around the 2nd century BC until the time of the Emperor Tacitus.[38] Here, the statue is represented similarly as a shapeless mass with a female head wearing a *polos*, and a lower part consisting of two to four figured panels (Fig. 42).[39] Lastly, the statue is often situated in a small temple or *naiskos* with two columns in front. Many scholars have interpreted the cult image as a sacred stone – a *baitylos* – that was in some way decorated with jewellery or metallic finery,[40] but, according to Fleischer, it could just as well have been made of wood.[41] Regardless of its material, it was initially an aniconic image in the shape of a *baitylos* or a wooden column.[42] The panels with figurative decoration could be indications that

Fig. 41. Artemis of Perga from Cremna, marble statue, 1st century AD. Burdur Museum, inv. no. 9300 (after Fleischer 1973, pl. 96–97).

the image was dressed in a garment, perhaps imitating an *ependytēs*, which it seems to resemble. This is supported by the fact that the image clearly wears a veil.

It is evident that the replicas are not always consistent in the various minor details of the specific cult statues. Nevertheless, observation of the general characteristics allows several traits of the original cult statues – such as stance, attributes, garments, etc. – to be discerned. Generally, these Hellenistic and Roman reproductions of the various Archaic cult statues have a number of common traits. First of all, the cult statues are all standing upright and have tapered bodies with – the Artemis Kindyas excepted – their arms stretched forward. Furthermore, it generally appears that these cult statues – or at least their replicas – are all under life-size.

Some conclusions can be drawn with regard to the statues' garments: the overall arrangement is the same, with a long undergarment (*chitōn*) over which another

Fig. 42. Artemis of Perga. A. © National Museum of Denmark, inv. no. KMM 311533. B. © National Museum of Denmark, inv. no. KMM 174557 (Alexander Severus) Photos: Niels J. E. Andersen.

garment is worn. This outer garment is either an *ependytēs*, as in the case of Artemis Kindyas and the Carian Aphrodite, or a heavily decorated long garment, very tightly fitted around the lower body, as is the case for Artemis Ephesia. Furthermore, the Kindyas statue and the Hera of Samos also wear a sort of girdle or cross-band. The wooden statuette of Hera from Samos is the only example of a garment with a frontal, vertical panel.

Generally, the garments of the statues are thus all very long and appear to have been made from thick and non-transparent material, perhaps wool. Usually the outer garments are heavily decorated, almost to the point of seeming cluttered, and often feature panels with figurative decoration in the form of mythological figures, animals, and floral designs. These garments could possibly represent patterned woven panels, perhaps in tapestry technique. Alternatively, this decoration represents metal plaques, like the ones described in Chapter 2. Such gold plaques with hammered decoration in the shape of animals in square panels were used to adorn the dress of a chryselephantine statue of Apollo or Artemis at Delphi, dating to ca. 550 BC (Fig. 43).[43] Furthermore, the female deities all wear head coverings or veils drawn over their heads.

This mode of dress is not attested in artistic renderings of human beings or 'ordinary people' in ancient art, such as marble honorary statues. On the contrary, mortal women from this period are usually dressed in the conventional way with *chitōn* and *himation*, which are never as tightly fitted as the garments of the divinities. Moreover, the garments of these women were undecorated (although some decoration might of course have been painted on, and thus cannot be excluded entirely). This mode of dress, as evidenced by the Anatolian cult statues, therefore appears to have been exclusive to divinities (female as well as male), and thus bears special religious connotations.

Fig. 43. Hammered gold plaques with decoration in the form of animals in square panels for a statue of Apollo or Artemis at Delphi. Archaeological Museum of Delphi, inv. nos. 9796-9797. © Ministry of Culture and Sports, Ephorate of Antiquities of Phokis.

'Fettered' cult images

Some goddesses are depicted with garments binding their arms to their bodies, as in the case of the statuette of Artemis Kindyas. As described above, the goddess wears garments that function almost like a straitjacket, and are fastened by a girdle or rope. The goddesses Artemis Astias of Iasos,[44] Artemis of Sardes,[45] and Eleuthera of Myra in Lycia are similarly depicted on coins without arms – perhaps as an attempt to illustrate that they are hidden under their tight garment.[46]

These bindings are often connected with the fact that the images were mobile – the idea being that the fetters served to prevent the statue from moving, and were intended to secure not the image, but the deity itself.[47] These bindings were probably symbolic rather than functional, since binding a god with ribbons is unlikely to prevent human theft. In support of this idea, the ancient sources do not explain these bindings as a form of practical protection for the image against mortals.[48] Another, simpler explanation is that these representations depict the dressing of cult statues, which due to their shape and design could not be dressed as one would a human being, and the girdle was a method to keep the garments in place.

Numismatics is of paramount importance for a discussion on the possible binding or dressing of cult statues, since a large group of coins from Asia Minor bears depictions of this particular phenomenon. With regard to the extent to which numismatic depictions are reliable for reconstructing the appearance of ancient cult statues, Fleischer's comparative study of imperial coins from western Anatolia (including Samos) is of considerable importance, since he concludes that coins reproduced the true appearance of the image in its official form.[49] This numismatic faithfulness also applies to Imperial coins, which from Hadrianic times onwards reproduced even early Archaic statues.[50]

Several coins from the Imperial Period depict cult statues with a long band attached to each of their wrists and reaching to the ground. Several goddesses are depicted in this way, e.g. Artemis Ephesia,[51] Artemis Leukophryene from Magnesia on the Meander,[52] and Hera of Samos.[53] There are also earlier depictions on coins of goddesses with woollen bands, e.g. on bronze coins featuring Artemis Astyrene from Astyra in Mysia, which are dated to the 4th century BC.[54] This attribute is not exclusive to female deities, though, since there are representations of Zeus Stratios from Labraunda with such ribbons as well.[55]

Only two statues in the round (or fragments thereof) preserve traces of such fillets: one is the statue of Artemis Ephesia from Selcuk. On her lower arms are remnants of twisted fillets or bands, which were most probably attached to the small identical bases (convincingly interpreted as elaborate tassels)[56] placed symmetrically on each side of the statue where her hands (if preserved) would have lingered.[57] A fragment belonging to the statue and preserving part of one of the woollen bands and a falcon has also been recovered.[58] The second example is a fragment of a silver statuette of Artemis Ephesia (Fig. 44).[59] Only the hand is preserved, but it preserves the upper part of a woollen band, similarly to the previous example.[60]

The fillets are interpreted as woollen bands. They are not just any woollen bands, however, but are most likely composed of long bands of unspun, finely carded wool – so-called rovings – either bound with fine yarn or interrupted by

Fig. 44. Artemis Ephesia, fragment of silver statuette. © Münster University Archaeological Museum, inv. no. 2375.

knots. They are therefore not made of spun yarn or woven bands, but unspun bundles of fibres.[61]

It has also been suggested that these bands are supports for the arms of the statue rather than woollen fillets or rovings,[62] although this seems unlikely due to their thinness. The fillets have also been interpreted as chains used to secure the cult image against disappearance or theft.[63] In a few cases, the goddess is clearly shown holding torches or staffs.[64] Still, in the remaining cases, the identification of the bands as woollen fillets appears most plausible, partly due to the widespread use of such fillets for the cultic decoration of different items. For example, coin depictions show Apollo holding a branch decorated with woollen ribbons or rovings, e.g. on a tetradrachm from Kolophon in Ionia, dated to 190–185 BC,[65] and on tetradrachms from Myrina in Aeolis, dated to 155–145 BC (Fig. 45).[66] Furthermore, Pausanias describes how the *baitylos* at Delphi was decorated with woollen bands, currently rendered on the marble *omphalos* from Delphi (see Fig. 31), and how an ancient image of Demeter in the Sanctuary at Stiris was bound with ribbons.[67]

A somewhat similar example is the cult image of Artemis of Anemurium, which is known only from Roman numismatic depictions (Fig. 46).[68] The goddess is rendered with her legs and feet close together; she wears a veil and a *polos*, but, most importantly, she is enclosed from neck to feet in numerous garlands or ropes arranged horizontally, which conceivably were meant to represent ribbons such as the ones attached to the goddesses described above.

Fig. 45. Apollo holding a branch decorated with woollen ribbons or rovings. Tetradrachm, Myrina, Aeolis, mid-2nd century BC. © Trustees of the British Museum, inv. no. 1979,0101.316.

Fig. 46. Artemis Anemurium. Valerian.© Trustees of the British Museum, inv. no. 1872,0709.247.

As previously stated, it has been argued that these ribbons or chains served to keep the cult image in place. Other propositions have been put forward, e.g. by Seiterle, who argues that such woollen bands or ribbons were not decorative, but rather a cultic sign indicating consecration and sacredness.[69] According to Seiterle, the woollen bands of Artemis Ephesia were directly related to the performance of bull sacrifices, although the ribbons were also used to decorate other sacrificial animals, such as boars and rams.[70] Carstens argues, further, that the woollen bands may have been related to the consecration of the sacrificial animal, which could have been replaced – either *in toto* or as a *pars pro toto* – by such a woollen band.⁷¹ At the centre of the Artemis cult was the animal sacrifice on the large altar. During the sacrificial rites, the bull's head was decorated with woollen bands as a sign of its consecration or holiness. According to Carstens, the woollen bands therefore became an iconographical shortcut for the sacrificial bull. This means that animal sacrifices occurred where the cult statue had woollen bands as a decorative element,⁷² and the presence of many bands therefore suggests many sacrifices. Yet, according to Fleischer, it makes little sense to place the ribbons on the cult statue after the sacrifice, since they belong on the bucrania.⁷³ In the case of Artemis statues, the woollen bands may instead have symbolised the asylum institution of the sanctuary, since literary sources inform us that people seeking refuge at the Artemision carried woollen bands.⁷⁴ Thus, Hesychios tells us that the ribbons, which he terms *kleides*, functioned as a symbolic plea for protection at Ephesos.⁷⁵ The *Etymologicum Magnum* likewise provides Artemis Ephesia with the epithet *hikesia*, 'protector of those in flight', and states that people seeking refuge brought flocks of wool (*mallous*) to the sanctuary.⁷⁶ Carstens also suggests that the woollen bands

Fig. 47. Woman at funerary monuments. White ground lekythos by the Vouni Painter, ca. 460–450 BC. © Metropolitan Museum of Art, inv. no. 35.11.5.

may have been related to a stone cult where the binding of woollen bands around the holy stone formed part of the consecration or worship of the stone.[77] An alternative explanation is that this conduct may simply be an expression of the decoration of the cult statue with textile bands, perhaps as a *pars pro toto* effect of dressing the statue. It is impossible to say, however, whether the woollen bands were a permanent decoration or only for special occasions.

White ground *lekythoi* from the 5th to the 4th centuries BC are additional sources on the use of textile fillets or ribbons in cult. The majority feature funerary scenes – visits to the tomb are especially common. These scenes often show a woman carrying a basket with fillets or ribbons in different colours, and a grave monument decorated with them (Fig. 47). This indicates that textile fillets were dedicated and perhaps played a role in specific rituals performed at grave markers, at least in the Classical Period in Athens.

Furthermore, according to Seiterle, the Greeks believed wool to have a purifying, cathartic, healing, and generally apotropaic effect. As examples, he adduces the practice of mystics at Eleusis and other places, who wore white woollen threads around their wrists and ankles, and the custom of placing a tuft of wool on the door when a girl was born in a household.[78] In connection with this, it is relevant to recall the ribbons recorded in the temple inventory of Artemis Kithōnē, which were examined in the previous chapter.

Fig. 48. The 'peplos kore'. © *Acropolis Museum, Athens, inv. no. Acr. 679, ca. 540 BC. Photo: Mavrommatis*

The 'Peplos kore'

Also relevant to a discussion on cult images is the so-called *peplos* kore, recovered in 1886 west of the Erechteion on the Athenian Acropolis (Fig. 48). This marble statue has a height of 1.18 m (not including the plinth) and has been dated to ca. 530 BC.[79] The statue originally held an object in her hand, which is now lost.[80] There has been some disagreement about the type of the statue's dress. During the 1920s, archaeologists considered the outer garment to be a *peplos*, an interpretation that is still reflected in the misleading identification of the statue as the 'peplos kore'. The upper part of the outer garment, however, does not conform to the idea of a *peplos*, since it does not appear to be made of a square piece of cloth folded over, laid around the body, and fastened at the shoulders by pins or fibulas: there are no dress-fasteners and her arms are not as free as they would have been if she had been wearing a *peplos*. There are holes on the shoulders, possibly for the addition of metal dress-fasteners,[81] as attested on the sculptures of the Parthenon, for example.[82] Even so, this does not allow the garment to be identified as a *peplos*, since dress-fasteners were also used for different garment types, such as mantles. Scientific examinations of the polychromy of the statue, moreover, have determined that the statue is not wearing a *peplos*. UV photography has revealed that the vertical band of rosettes is superimposed over, and partly obscures, the wings of what might be a sphinx (Fig. 49). It appears that this band can only be regarded as the vertical seam of the overlying garment, perhaps a long vest.[83] Thus, the statue wears a number of garments: first a long orange under garment, probably a *chitōn*; then an *ependytēs* with horizontal animal friezes on a red background; then a long yellow vest with vertical seams in green and red, fastened by a belt; and finally a short yellow cape covering her upper body.[84] Her dress thus follows, to a certain extent, the dress of the Hellenistic and Roman reproductions of Archaic Anatolian cult statues, which wear long *chitōnes* and *ependytai* featuring figural decoration in friezes. It especially resembles the dress of the wooden statuette of Hera at Samos, which also wears a long under-garment, possibly an *ependytēs* and long vest (although this is difficult to determine with certainty), and a short cape. This could support the theory that the statue represents a cult image of a goddess, and it has been suggested that the statue might be a marble version of a *xoanon*.[85] It is difficult to determine which goddess is represented, especially since her attributes have disappeared, but likely candidates are Artemis or Athena.

Depictions of cult images in vase-painting

Cult images are also depicted in vase-painting. The relationship between cult image and depiction is, however, complex, in part because it is influenced by Greek religious

Fig. 49. The 'peplos kore', UV reflectography and contour drawing. © Ulrike Koch-Brinkmann

and philosophical thought.[86] Furthermore, although vase-painting can serve as an important source for us to detect iconographic traditions in sculpture, it does not provide accurate and faithful information about the details of a given statue.[87] In each case, the figurative context, the creativity of the artist, and the production standard play a central role in the ways that patterns, types, and fixed formulas are reworked and deployed. Two points are important to keep in mind in this regard: 1) the existence of encoded figurative repertoires, circulating in the workshops of sculptors and painters; 2) the creativity of painters and the development of formulas that figuratively translate a 'mythical memory' shared by the craftsman and his customers.[88] Cesare also emphasises the importance of considering the context in which the image is inserted. She lists three contexts of importance: 1) a mythological context or narrative, 2) a cultic context, in which the statue is the object of worship,[89] 3) the context of the craftsman and the processes of artistic production. In this respect it is important to note that, although many of the vases originated in Attica, they were purchased and used by a Western market.[90] It is also important to keep in mind that the statues can play different roles in the vase-paintings: the statue can be the focal point, as in many depictions of supplication; it can play a marginal role, as a simple element of scene setting; it can serve as a help with interpretation of the scene, suggesting the background or the outcome of the story; or, finally, the statue can intensify the *pathos* of the scene.[91]

Another challenge to understanding cult images depicted in vase-painting is the concept of the 'split' or 'double present' deity – that is, scenes in which a deity is present as itself and in the form of its cult statue. This scenario appears to be restricted to Athena, Aphrodite, Apollo, Artemis, and Dionysos.[92] The deities are usually depicted in a different manner than the images, which are not only smaller, but also wear different garments and hairstyles.[93] The depiction of a deity next to its image indicates that there are several levels of reality reflected in the vase-painting. Cesare argues that a complete fusion between the human and divine world takes place in the 6th century BC, which is exemplified by the emergence of 'living and moving statues' at the end of the 6th century BC. The separation of the divine from the human world, introduced with the rise of rationalistic thinking in the 5th century BC, manifests in vase-paintings through the representation of the deity as *xoanon* and/or through the splitting of the image (mid-5th – 4th century BC). This phenomenon begins with depictions of Apollonian myths, and continues with cases in which deities appear alongside their *xoana*, as well as 'split' figures of the deceased on Apulian funerary vases.[94] The interpretation and analysis of cult images depicted in vase-painting is thus not a straightforward task, but it is possible provided that one keeps such considerations in mind.

Vase-painters had several visual means of showing that a statue of a god – not the god itself – was meant: one was the statue base (which was not used in all representations to indicate 'statue', but was one of the options).[95] Another was representing the god as smaller than the acting persons or depicting him in an eye-catchingly archaistic or stiff posture in order to show the viewer that the god is not involved as a 'person' and is not an acting figure.[96] A frontal position was a further means of indicating a statue. Finally, the centrality of the cult statue could be accentuated by framing it between two columns, which mark sacred architecture or a sanctuary.[97] All of these visual devices allowed the artists to paraphrase a statue, and made it possible for the vase-painters to represent 'living' gods next to their statues as well. This formula for representing a cult statue became a pictorial convention from around the end of the 6th century BC.[98] Even during the 5th century BC, however, after the invention of a formula for 'statue', there are several cases where modern scholars are not sure if a statue or a living god is represented.[99]

Vase-painting is still a rich source of information on the dressing of cult statues, even though they can be difficult to identify and to separate from ordinary statues of gods and goddesses. One method of recognising cult statues is only to consider statues that stand before altars or inside temples, that are attended by worshippers, or that appear in scenes which can be identified from literary sources. Unfortunately, early Greek cult images are only very rarely identified in vase-painting.[100] Generally, cult statues are depicted in Greek vase-painting only in the late Archaic and Classical Periods, and in south Italian vase-painting from the mid-5th century BC through the late 4th century BC.[101]

One of the most interesting examples of an early cult image is represented on an Attic black-figure plate, dated to ca. 600–550 BC (Fig. 50).[102] In the central scene there is a depiction of a *xoanon*, which is tightly wrapped in a patterned textile decorated with horizontal friezes.[103] In my view, this vase illustrates the dressing of a *xoanon* in real textiles, similar to the garments worn by the women surrounding the cult image.

The red-figure examples can be divided into two main groups that depict Greek cult images: Attic vases and South Italian vases. As Alroth emphasises, there are both similarities and differences in the representations of the cult images in the two groups of vases. In both groups, the image is placed on a base, often in front of an altar, whereas on the South Italian vases the temple building or *naiskos* is far more common.[104]

Several scholars have treated the subject of cult statues in Archaic, Late Classical and Hellenistic Greek vase-painting: Schefold and Bielefeld address the depiction of cult images in Attic vase-painting,[105] and Schauenburg and Schneider-Herrmann discuss statues of gods on South Italian vases.[106] Romano has produced an appendix of around 30 Greek vases with depictions of what may be Greek cult images – mythical and real.[107] De Cesare and Oenbrink have also studied depictions of statues and cult statues in Greek vase-painting in depth.[108] None of these scholars, however, has discussed the dress of these cult images.

There are identifiable representations of cult images of Athena, Artemis, Aphrodite, Apollo, Dionysos, Zeus, and herms. The goddesses are always dressed, while the

5. Iconographic evidence for the dressing of cult statues

Fig. 50. Cult image. Attic black figure plate by the Polos Painter, ca. 600–550 BC. National Archaeological Museum, Athens, inv. no. 18717. © Hellenic Ministry of Culture, Education and Religious Affairs/ Archaeological Receipts Fund. Photo: K. Xenikakis.

gods – primarily Apollo, but also Zeus – are often rendered in the nude, at least in red-figure vase-painting. The cult statues of goddesses are usually depicted in a long garment reaching to their feet. Often, the garment is wound tightly around the lower body in a way that appears to be unprecedented in depictions of human beings, mythological figures, as well as the deities themselves in vase-painting. This way of arranging the garment very tightly around the lower body is also depicted on replicas of cult statues from Anatolia. Moreover, the cult statues' garments are often decorated with different colourful patterns and ornaments that probably represent inwoven decoration. With regard to the specific garment types, the majority of the female divinities are often interpreted as wearing *peploi*, while vase-paintings from the

4th century also provide examples of female cult statues in *chitōnes*. But it can often be difficult to distinguish between the two types of garments in vase-paintings. A few cult statues also appear to wear mantles over their long garments, and several from the 4th century BC wear headgear in the form of a high or low *polos*. A *pelike* from the late 5th century BC depicting a female cult statue on a column is of particular interest, inasmuch as it depicts the image dressed in a long garment over which is draped what appears to be a decorated *ependytēs* (Fig. 51).[109]

There are only a few known representations of male cult images wearing garments. An example is the cult statue of Dionysos on an Apulian volute krater from the 4th century BC, where the god wears a short, heavily decorated tunic and a *polos* (Fig. 52).[110] Other examples include a krater dating to the 4th century BC, which depicts an ithyphallic figure carrying a caduceus and dressed in a long patterned garment (Fig. 53).[111] Furthermore, an Attic black-figure vase, dated to ca. 560 BC, depicts a god, perhaps Apollo,[112] standing in his temple and dressed in a long garment underneath a mantle (Fig. 54).[113] Several of these vases can be grouped according to their representation of specific deities in particular settings, which often appear to be rendered on certain vase shapes. These groups will be treated separately in the following sections.

Fig. 51. Female cult image on a column. Pelike, late 5th century BC. © The State Hermitage Museum, St. Petersburg, inv. no. BAK7. Photo: Vladimir Terebenin, Leonard Kheifets, Yuri Molodkovets, Aleksey Pakhomov.

The Palladion of Troy

A famous scene from the *Iliad* involving a cult statue is often depicted in vase-painting: during the sack of Troy, Kassandra took refuge in the Temple of Athena, hiding by the cult statue of the goddess, the so-called Palladion, in an attempt to

5. Iconographic evidence for the dressing of cult statues 205

Fig. 52. Cult image of Dionysos. Apulian volute krater. Museo Archaeologico Nazionale, Naples, inv. no. 2411 (after Pfuhl 1923, pl. 359, fig. 801). © Ministero dei Beni e delle Attività Culturali e del Turismo.

escape from Ajax.[114] Davreux has compiled a comprehensive catalogue of ancient depictions of this particular scene,[115] to which Romano adds seven examples.[116] The iconographic depictions of the story first emerge at the beginning of the 6th century BC in the Peloponnesian workshops,[117] but by far the majority belong to the 4th century BC, and only a few examples date to the 5th or 3rd centuries BC.[118] The cult statue is usually depicted as smaller than life-size, helmeted, and holding spear and shield.[119] With regard to dress, the vases depict the Palladion differently: some show it in a long white dress without decoration (Fig. 55),[120] but the bulk depict it in a long garment with decoration. The decoration often takes the form of a border down the front of the garment and along the hem, in some instances supplemented by singular 'dots', perhaps imitating embroidery, inwoven decoration, or stitched on objects

Fig. 53. Ithyphallic figure with a caduceus. Red-figure krater from Gela, 4th century BC. Museo Archeologico Regionale Paolo Orsi, Syracuse, inv. no. 22934. © Assessorato Beni Culturali e dell'Identità Siciliana della Regione Sicilia.

(Fig. 56).[121] Other examples clearly represent the Palladion in a patterned garment (Fig. 57).[122] Furthermore, one particular depiction stands out, since the Palladion is clad differently: it wears a very tightly-fitted garment on the lower body, over which it wears a belted, decorated garment with long sleeves (Fig. 58).[123]

The Palladion also appears in scenes where it is stolen from Troy. The theft of the Palladion is depicted in red-figured vase-painting. Oenbrink lists 12 examples: six vases

5. Iconographic evidence for the dressing of cult statues

Fig. 54. Cult image of Apollo. Attic amphora, ca. 560 BC. © Trustees of the British Museum, inv. no. B49.

from the 5th century BC,[124] five Apulian vases from the first half of the 4th century BC,[125] and one Campanian vase dated to ca. 340 BC.[126] In these scenes the Palladion is smaller than life-size, and carries a shield and spear. The statue is dressed, for the most part, in white, undecorated garments, but there is an example where the Palladion is dressed in a decorated garment with a vertical border down its front and long sleeves.[127] In one example, the Palladion is dressed in a long, spotted garment.[128]

Artemis as represented in Euripides' play Iphigenia in Tauris

A number of Attic vases depict scenes from Euripides' play *Iphigenia in Tauris*, produced in Athens ca. 413 BC. In this play Iphigenia is to be sacrificed to Artemis by her father

Fig. 55. Palladion. Red-figure hydria, ca. 340–320 BC. © Trustees of the British Museum, inv. no. F209.

5. Iconographic evidence for the dressing of cult statues

Fig. 56. Palladion. Red-figure volute krater, ca. 460–450 BC. © Trustees of the British Museum, inv. no. E470.

Fig. 57. Palladion. Red-figure volute krater from Basilicata, ca. 370–350 BC. © Trustees of the British Museum, inv. no. F160.

5. Iconographic evidence for the dressing of cult statues 211

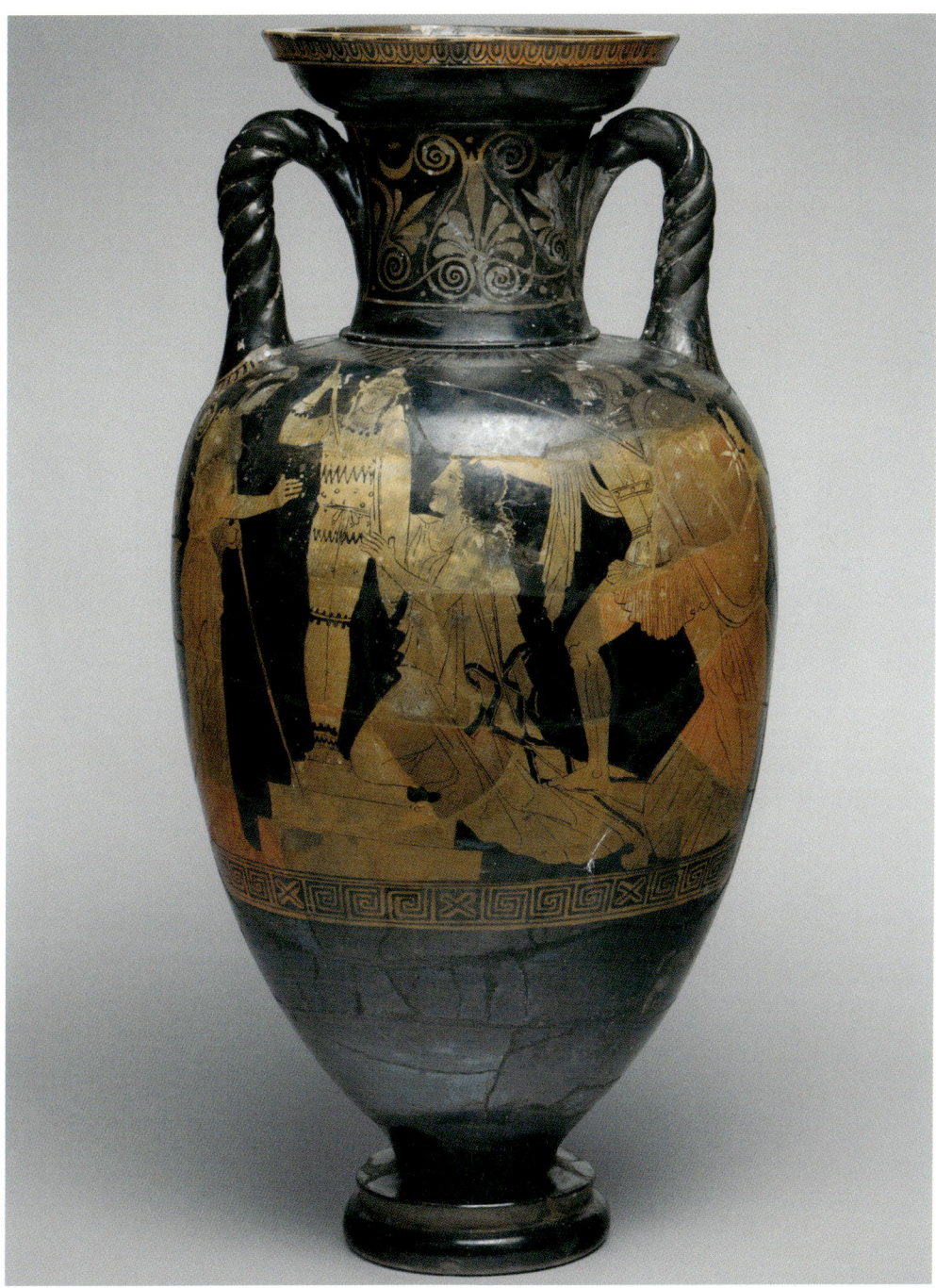

Fig. 58. Palladion. Red-figure Attic amphora, 5th century BC. © The Fitzwilliam Museum, Cambridge. Loan Ant.103.22, Lewis Collection, reproduced by courtesy of the Master and Fellows of Corpus Christi College, Cambridge.

Fig. 59. Scene from the play Iphigenia among the Taurians. *Attic calyx krater by the Iphigenia Painter from Spina, ca. 400–380 BC. Museo Archaeologico Nazionale, Ferrara, inv. no. 303.*

Agamemnon, but the goddess spares her life at the last minute, and takes her to Tauris on the Black Sea, where she serves as priestess of the goddess. A small number of vases depict a scene in the play where Iphigenia's brother Pylades comes to Tauris and finds her at the Temple of Artemis.[129] One such rare example is an Attic vase recovered at Spina, which depicts Iphigenia as priestess leaning against the temple of the goddess (Fig. 59).[130] The cult statue of Artemis stands in the temple behind an altar table. The standing cult statue is painted white, and is dressed in a long, belted garment, possibly with an *ependytēs* and a girdle. The garment is tightly fitted around her lower body, and tapered towards her feet.

Chryse

Another scene from ancient literature involving a female cult image that appears in vase-painting is the myth of Philoctetes being bitten by a snake. This scene is sometimes accompanied by the cult image of Chryse. Chryse was a nymph or a minor goddess of Lemnos who lured Philoctetes away from his companions, which resulted in the misfortune of the snake bite.[131] Oenbrink lists six vases on which the cult image of Chryse is depicted. Only two of these depictions are identifiable by inscriptions that provide the deity's name (Fig. 60).[132] Oenbrink identifies the remaining four depictions on the basis of the typological conformity of the statues and the consistency of the environment in which the cult images are depicted.[133] The statues do indeed have several traits in common: they are female, they are placed on a podium or pillar, they are of small size, and they are clad in garments that are tightly wrapped around their lower body and tapered towards the feet. One should be cautious about making an identification, though, since these traits also occur in depictions of other female cult images. The dress of these images is either white with a vertical border down the front or heavily decorated.[134]

Aphrodite Ourania en Kēpois

Several red-figure squat *lekythoi* depict a female *xoanon* standing in an open-air sanctuary.[135] The vases are all from Athens, and belong to the Late Classical Period, from ca. 400 through the early 4th century BC. The female *xoana* depicted on these vases have been identified as renderings of Aphrodite because they often show the image together with Eros, and because this specific vase shape contained perfume and was a common gift offering to the goddess. These vase-paintings have been interpreted as representing an ancient cult image of Aphrodite Ourania ('heavenly') in her various shrines at Athens.[136] Langlotz specifically identifies the depictions as Aphrodite Ourania *en Kēpois* ('in the gardens').[137]

The cult images on the vases show Aphrodite standing on a base in a stiff frontal position with her feet close together. Her lower arms and hands are often stretched forward, either empty or holding *phialai*; in other depictions, her arms are raised or are hidden under her garment. Several of these squat *lekythoi* depict the cult image

Fig. 60. Cult image of Chryse. Red-figure krater, 5th century BC. © Kunsthistorische Museum, Vienna, Antikensammlung, inv. no. 114.

as small, ca. half life-size (Fig. 61).[138] On these vases, the cult image is shown frontally and rendered quite sketchily without recognisable details. These images, however, all have certain traits in common: they are small, their hands are visible, and they wear a very tightly fitted garment around their bodies. Another squat *lekythos* depicts a similar type of dress for the cult image. It is depicted in right profile, is smaller than life-size, and is clad in a tightly fitted red textile around the entire body, which conceals the goddess' arms and hands; it also wears a *polos* decorated with three globules (Fig. 62).[139]

Other vases in the group depict the cult image as larger, even life-size, as is indicated by the size of the surrounding female worshippers. This is the case on a *lekythos* where the cult image is almost as large as the females next to it (Fig. 63).[140] The statue is depicted frontally with feet close together on a base. Her arms are raised, and she is clad in a light garment, tightly fitted around her lower body, and she wears a decorated *polos* on her head. Finally, a *lekythos* dating to ca. 400 BC bears an exceptional depiction of this cult image (Fig. 64). The statue is rendered life-size, stands frontally with feet close together, and holds a *phiale* in each hand.[141] The image wears a *polos* and a tightly fitted, belted, heavily decorated garment with an overfold, which features a frontal vertical decorative panel and a star-shaped pattern. The statue is very similar to the Anatolian cult images of e.g. Artemis Ephesia. These cult images of Aphrodite Ourania thus render the goddess similarly in a tightly fitted garment. Perhaps the range in size and the different posture of the arms and hands indicate that different *xoana* are reflected in the vase-paintings, or alternatively this is simply an expression of artistic freedom.

Aphrodite had several sanctuaries in and around Athens: one sanctuary

Fig. 61. Cult image of Aphrodite. Lekythos, early 4th century BC. © Trustees of the British Museum, inv. no. E714.

is known from literary sources to have been located at the Ilissos River, and was dedicated to Aphrodite *en Kēpois* ('in the Gardens'), but it still remains to be found; a second sanctuary for Aphrodite and Eros was located on the north slopes of the Acropolis;[142] and a third was located at Daphni, northwest of Athens along the Sacred Way to Eleusis.[143] Karousou has suggested that the three sanctuaries belonged to the Oriental Aphrodite, more properly named Aphrodite Ourania *en Kēpois* ('Heavenly in the Gardens').[144] Moreover, Pausanias writes:

Fig. 62. Cult image of Aphrodite. Lekythos, ca. 380–370 BC. National Archaeological Museum, Athens, inv. no. 1283. © Hellenic Ministry of Culture, Education and Religious Affairs/Archaeological Receipts Fund. Photo: Fafalis.

5. *Iconographic evidence for the dressing of cult statues* 217

Fig. 63. Cult image of Aphrodite. Lekythos, 4th century BC. © Staatliche Antikensammlungen und Glyptothek München, inv. no. 2264. Photo: Renate Kühling

Fig. 64. Cult image of Aphrodite. Lekythos, ca. 400 BC. © Ashmolean Museum, University of Oxford, inv. no. 1966.714 (MM 80).

> 'Concerning the district called The Gardens, and the temple of Aphrodite, there is no story that is told by them, nor yet about the Aphrodite which stands near the temple. Now the shape of it is square, like that of the Hermae, and the inscription declares that the Aphrodite Ourania is the oldest of those called Fates (Moirai).'[145]

This short description indicates that the image of Aphrodite was squareish, similar to a herm, and that her cult was very old. This suggests that the vase-paintings possibly depict an early, real, small cult statue in the shape of a standing *xoanon* rendered in an Oriental manner with garments like those of the Anatolian cult images. Pausanias, moreover, states that the statue of Aphrodite was made by the sculptor Alcamenes, who worked in the second half of the 4th century BC. Perhaps there was more than one cult image in the sanctuary, as in the Sanctuary of Artemis at Brauron, the Heraion on Samos, and the Argive Heraion (Pausanias 2.17.3–5).[146]

Hekate

The goddess Hekate is occasionally depicted on Attic vases, where she is recognisable due to her two torches. It is often not possible, however, to draw a clear distinction between depictions of the goddess and depictions of her priestesses. There are a few clear depictions of her cult image on red-figure vases, dating from the second half of the 5th century BC to the late 4th century BC. Oenbrink records five examples.[147] The cult image is depicted in different garments, often white.[148] An example is a relief oinochoe from the late 4th century BC depicting the cult image of Hekate on a pillar in front of an altar with a sacrificial fire. The cult image is dressed in a white garment concealing her arms and a purple *polos* (Fig. 65).[149] A *skyphos* fragment from the Kerameikos depicts the cult image of Hekate quite differently: the statue is triple-bodied – a so-called Hekataion.[150] Of special interest is the textile arranged around the chests of the cult image. This might be an *agrēnon*, the woollen bands or rovings also used to decorate the omphalos at Delphi.[151]

Dionysos Lēnaios

As these examples demonstrate, when it comes to depictions of cult statues in vase-painting, it is generally very difficult to tell whether the garments of the cult images are painted on and/or sculpted or whether they were meant to represent actual garments used to dress the images. A group of vases – usually termed Lēnaia vases – can help in this regard, since they represent without a doubt a cult statue dressed in real garments. The *Lēnaia* was an annual winter festival in Athens in honour of Dionysos Lēnaios. '*Lēnaia*' probably comes from '*lēnai*', which is another name for the maenads.[152] The Lēnaia vases consist of a group of ca. 100 Athens-produced vases that depict maenads perhaps at the *Lēnaia* or at the *Anthesteria*, a festival of the new wine held in the spring.[153] Some of these vases bear a representation of a cult statue of Dionysos Lēnaios as their central motif. This particular cult statue took the form

Fig. 65. Hekate. Oinochoe, late 4th century BC. © Trustees of the British Museum, inv. no. G17/ 1871,0722.1.

of a column or pillar draped with garments, and had a bearded mask occasionally adorned with twigs.[154] The earliest example of such a Dionysos cult column dressed in a garment is a black-figure *skyphos* from Athens dating to the beginning of the 5th century BC (Fig. 66).[155] There are only a few black-figure examples, but the motive becomes more common in red-figure vase-painting.

A famous representation of a cult statue of Dionysos Lēnaios is a red-figure *kylix* dating to ca. 480 BC (Fig. 67).[156] On one side of the cup, the cult statue is depicted standing in front of an altar and surrounded by maenads dancing and playing music instruments. The statue is clad in two garments: a long inner garment (perhaps a *chitōn*) and a patterned mantle with depictions of dolphins draped loosely around his 'shoulders'. On other red-figure Lēnaia vases the cult statue is unmistakably a column and is even surmounted by a capital. These examples are dressed in one garment only, which takes the form of a piece of fabric wrapped around the column. Often the textile is patterned (Fig. 68),[157] but in some cases it is undecorated.[158]

A group of *stamnoi* from ca. 470–460 BC depict the Dionysos column statue behind a table with two large *stamnoi* and occasionally also offerings (Fig. 69).[159] On these vessels, the statues are clad in two undecorated garments: an inner *chitōn* and a piece of cloth that runs around it without any visible opening. According to Frickenhaus, the earlier examples have shorter garments, the later ones longer.[160] In all of these examples, the statues have two circular ornaments, one at each shoulder. Frickenhaus interprets them as *plakountes*, small, flat cakes.[161] Two very similar *stamnoi* dating to the mid-5th century BC depict the Dionysos statue somewhat differently (Fig. 70).[162] The statues do not look like columns and are depicted in profile, standing on a podium or base and surmounted by branches and leaves. The statues are clad in two garments – a long inner *chitōn* and a *himation* with a coloured stripe along its border – and wear headbands. Next to each statue there is a table with textiles. A final example is a *stamnos* from Naples dating to the end of the 5th century BC.[163] The cult statue is depicted frontally in the middle of the scene. The lower part of the column is visible and an ornamented capital is depicted above the mask. The cult statue is clad in a short undergarment, over which it wears a belted *ependytēs* decorated with fringes at the neckline and lower hem. Above each shoulder is a white circular ornament/disc (*plakous*).

The depictions are quite uniform, suggesting a specific ritual occasion in the same cult, and Frickenhaus consequently argues that it is the same cult statue which is depicted on all the vases. The column itself is always rendered without fluting, with a Doric capital, but on a base, which could indicate that they were made of wood like the early *xoana*, reinforcing the interpretation that they were cult images.[164] The cult statue is, as shown in the vase-paintings, clad in different garments, either one or two, with or without decoration. This could possibly indicate – if it truly is the same cult image – that it received new garments on special occasions at periodic intervals. Alternatively, the different garments simply represent the artistic freedom of the vase-painters, who perhaps rendered the cult image in dress popular or considered appropriate at the time.[165]

222 Gods and Garments

Fig. 66. Dionysos Lenaios. Black-figure skyphos, ca. 500–525 BC. National Archaeological Museum, Athens, inv. no. 498. © Hellenic Ministry of Culture, Education and Religious Affairs/Archaeological Receipts Fund. Photo: Sp. Pistas

Fig. 67. Dionysos Lenaios. Red-figure kylix by Makron, ca. 480 BC. Staatliche Museen, Antikensammlung, Inv. no. 2290. Photo: Johannes Laurentius.

Fig. 68. Dionysos Lenaios. Red-figure pelike, ca. 500–450 BC. © Musée du Louvre, inv. no. G227. Photo: H. Lewandowski.

Herms

Herms were not typical cult statues, but they were venerated with offerings and were considered to have an apotropaic function.[166] There are examples of herms being venerated as cult images in vase-painting.[167] Some herms are depicted 'nude', while others are rendered as covered with textile: a *pelike* from ca. 480 BC depicts an ithyphallic herm dressed in a garment or mantle with a coloured stripe at the edge and standing in front of an altar (Fig. 71),[168] while a *lekythos* with a similar image, dated to 480–440 BC, depicts a herm, also in front of an altar, with what appears to be a square, patterned piece of textile placed on its 'chest'.[169] A *lekythos* from ca. 500 BC provides another example (Fig. 72).[170] According to Boardman, the scene shows three men doing obeisance before a mummy. He admits that the mummy looks like a Greek herm, but emphasises that it lacks arm stumps or a

Fig. 69. Dionysos Lenaios. Red-figure stamnos by the Villa Giulia Painter, ca. 450 BC. Photo: © Museum of Fine Arts, Boston, inv. no. 90.155a–b.

phallus, and that its criss-cross pattern resembles the bandaging of a mummy.[171] In my opinion, however, this scene is a clear illustration of a cult image dressed in real textile, and worshipped by three men. The body of the herm is thus covered from neck to feet with a patterned piece of cloth, wrapped tightly around its body. This interpretation is given further support by the two cloths hanging in the background above the adorants. The act of dressing a herm in textile is depicted on a Lucanian bell *krater*, dated to ca. 440 BC (Fig. 73).[172] In the middle

Fig. 70. Dionysos Lenaios. Red-figure stamnos by the Eupolis Painter, ca. 450–440 BC. © Trustees of the British Museum, inv. no. E452.

of the scene is a herm with a *caduceus* rendered on its side and wearing a type of headgear. In front of the herm there is a woman, who is about to place a large textile ribbon on the herm.

Dressed herms are also depicted in statuary. Female examples include a marble herm, possibly depicting Artemis, dressed in what appears to be a *peplos*. It was recovered from the Agora of Athens and is dated to the 2nd century BC (Fig. 74).[173] Another example is a Roman headless marble herm from the temple

Fig. 71. Herm. Red-figure pelike, ca. 480 BC. © State Hermitage Museum, St. Petersburg, inv. no. B 4515 (Gr-10480). Photo: Vladimir Terebenin, Leonard Kheifets, Yuri Molodkovets, Aleksey Pakhomov.

of Aphrodite at Cyrene, which is dressed in a mantle that covers her upper body and arms (Fig. 75).[174] The example from Athens is more evidently a herm, while the herm from Cyrene is rendered with a more naturally draped mantle, which conceals her arms and upper body. The obvious objection is that these herms only have their garments rendered in stone, and that this is no way connected with the 'real' practice of dressing cult statues in textiles. I believe, however, that these depictions reflect a custom of dressing herms in garments, perhaps in connection with specific rituals or as a way to signal or transmit certain characteristics. In support of this interpretation, Goldman has argued that the side brackets of the herm do not represent rudimentary or stylised arms, but that they were functional in origin and developed during a time when the primitive shaft or herm was habitually draped, in order to support the garment and keep it from slipping off.[175]

The dress of cult statues depicted in vase-painting

It is hard to draw firm conclusions on the dress of cult statues from the vase-paintings, since the depictions are rather diverse. Some observations, however, can be made on the basis of these depictions: as demonstrated, it is primarily female cult statues that are dressed, and they are never rendered naked, in contrast to male cult statues, which, as noted above, were usually depicted naked. We might therefore imagine, based on this evidence, that female cult statues were more likely to be dressed in textiles than male cult statues (with the exception of Dionysos).

Fig. 72. Herm. Athenian lekythos, ca. 500 BC. © Staatliche Antikensammlungen und Glyptothek München, inv. no. 1871. Photo: Renate Kühling

Fig. 73. Herm. Red-figure krater, ca. 440 BC. © Royal Museum of Art and History, Brussels, inv. no. A724.

The dress of the female images can, furthermore, be divided into three types of costume:

1. First, they are dressed in 'normal' contemporary garments, consisting of a white *chitōn* with a coloured hem (occasionally an entirely white *peplos*), rendered as if worn by a human being (and not tightly fitted).
2. Second, they are clad in a more special costume similar to that of the Hellenistic cult statues from Asia Minor. This costume is often heavily decorated, occasionally with a frontal, vertical panel, and is tightly fitted around the lower body of the statue. It is possible that these statues may represent an attempt to depict a cult statue of Anatolian type, such as those preserved in Hellenistic marble copies. Yet

Fig. 74. Female marble herm, 2nd century BC. Museum of the Athenian Agora, inv no. S1086. © Hellenic Ministry of Culture, Education and Religious Affairs. Ephorate of Antiquities of Athens.

Fig. 75. Marble herm, Roman. © Trustees of the British Museum, inv. no. 1861.1127.161.

    we cannot tell whether the South Italian vases depict actual local or foreign cult statues, or whether they are simply generic renderings of 'a cult statue'.[176]
3. Third, some statues are wrapped in textiles that conceal their entire 'body', including arms and hands. This type of dress is also employed for some of the male cult statues, e.g. Dionysos Lēnaios and male herms. Furthermore, the female

statues (as well as some male statues, most commonly Dionysos) wear headgear, often in the form of a *polos*, or alternatively what appears to be a crown.

Based on the case studies of the Trojan Palladion, Artemis in Tauris, Chryse, Aphrodite Ourania, Hekate, and Dionysos Lēnaios it appears that these particular deities were often dressed in real garments. Such a conclusion, however, has several pitfalls: first, we are often not entirely certain what deities the cult images on other vases depict, and the fact that cult images of other deities are not identified on the vases does not mean that those deities were less likely to be dressed. Second, the scenes on the vases might simply have reflected the popularity of certain mythological scenes: for example, the scene of Kassandra taking refuge at the Palladion was extremely popular in vase-painting primarily in the 4th century BC, which explains the frequency of its occurrence (and identification). Perhaps scenes featuring a dressed cult image of Hera or Artemis were not as common because scenes involving such images did not occur in popular literature, which was often the inspiration for Late Classical and Hellenistic vase-paintings. Nevertheless, these vase-paintings confirm that cult images were dressed in real garments, at least in the case of Athena (the Palladion), Aphrodite, and Dionysos, as well as several other deities.

The vase-paintings illustrate, furthermore, that different types of cult images were dressed in garments: *xoana* (e.g. the Palladion and Aphrodite Ourania *en Kepois*), column images (Dionysos Lēnaios), and herms. The mode of dress, however, differs, perhaps due to the statues' different properties: the *xoana* often appear to be tightly wrapped, the garments appear to be 'hung' on Dionysos Lēnaios, and the herms could either be swaddled or draped with a piece of cloth or have cloth tightly wrapped around the body. Furthermore, without exception, all of these recognisable cult images are standing.

Another significant observation arises on the basis of these representations: the cult image often possesses Archaising features, very much as we imagine early cult images. Hiller states that the models of cult images on vases from the 5th century BC date to the Geometric Period.[177] It has been suggested that this is because of nostalgia and a renewed interest in the old religion at the end of the 5th century BC.[178] Alroth, however, raises the question as to whether these depictions are simply standard versions of a cult image,[179] and Hölscher firmly believes that the archaising features must be attributed to iconographic conventions:

> 'it is not at all a question of reality, but a mode to put a god out of action, as an additional way to tell the viewer that he was not looking at Athena but at her statue.'[180]

Oenbrink also considers the depictions of cult images on vase-paintings to be deliberately archaising.[181] These depictions, however, might just as well represent an attempt to depict the cult image in question in as realistic a fashion as possible, and thus illustrate old wooden *xoana*. This leaves us with the question of cult statues, which are contemporary to the vase-paintings. Perhaps they were not depicted on the vases - or, more probably, the deities were rendered 'in person' instead of as statues, thus often making them unrecognisable to a modern audience.[182] This is an important point as far as the analysis of the dressing of cult images is concerned, since the conclusions drawn here therefore refer primarily to the dressing of archaising, old cult images.

Dressing technique

A question arises as to how exactly garments were attached to the statues. This depends, of course, on the individual statue, since cult images, as discussed, had different appearances, with regard to size, material, stance, etc. In the case of standing statues with arms free of their bodies, it seems rather straightforward to dress it in, for example, a *peplos* or a mantle. Many statues, however, were either seated or had arms positioned alongside the body, as for example the wooden *xoana*. In such cases, it must have been difficult or even impossible to dress the statue like one would dress human beings. For the seated statues, the garments could instead have been placed (folded) on their labs or merely on tables in front of them, as depicted on reliefs from Lokris, Italy (see Fig. 6). Yet, in these instances, it can be difficult to ascertain whether this specific act should be characterised as dressing or merely as offering textiles. In other instances, the garments must have been placed 'around' the statue and thus 'enclosed' or 'wrapped' the statue, concealing its arms. Various goddesses are represented in this way, with garments binding their arms to their bodies,[183] – an excellent example of which is the Late Hellenistic marble statuette of Artemis Kindyas from Bargylia in Caria (see Fig. 37).

As demonstrated above, iconography gives clear examples of the dressing of cult statues and can thus often provide information on how this dressing was done. For instance, the Hellenistic replicas of Anatolian cult statues, such as Artemis Ephesia, tend to depict the deities in a very tightly-fitted undergarment, a type of dress which also appears on several cult images in Attic and South Italian vase-painting. This particular type of undergarment is never employed in the depictions of 'living' deities, heroes, or human beings, and therefore seems to be exclusive to the rendering of statues, whether in vase-painting, numismatics, or marble statuettes. It seems likely that this tightly-fitted garment indicates a mode of dress reserved for cult statues – especially the standing *xoanon*-like examples. Other cult images rendered in vase-paintings show the statue wrapped in a piece of textile, which conceals its entire body, including arms and hands. Such examples are somewhat reminiscent of the wrapping of infants or shrouding of the dead: both were tightly wrapped in cloth, occasionally with additional ribbons to hold the textile in place.[184] Another example showing the actual mode of dressing a cult statue is an Attic vase-painting that depicts the cult image of Dionysos Lenaios. This column-shaped cult image was draped in garments, which seem simply to hang on the pillar.

Finally, there are several examples of cult images being decorated or bound with ribbons, possibly instead of dressing the image in real garments, as a *pars pro toto* effect. The reason may have been that a textile ribbon was much easier to attach to a statue, and was indisputably much cheaper than an entire garment. That this practice served as *pars pro toto* is especially evident in case of *baityloi*, which could not be dressed in garments due to their shape. Yet shape alone is not necessarily a determining factor when it comes to the choice of type of dress, since cult images with a less human form also were dressed in real garments.

These examples show that there were many different ways of dressing or adorning a cult image with textiles, and that the practice appears to have been widespread. In a few cases, however, we can determine exactly how the garments were attached to the statue, but for the most part we cannot tell. It is reasonable to suppose that something was needed to attach the garments, either dress pins or fibulas. Alternatively, the garments could have been stitched on. Nevertheless, fibulas and pins are only rarely depicted (e.g. on the girdle of Artemis Kindyas).[185] But does this mean that they were not present? Perhaps they were placed discreetly, hidden in the garments or on the back of the statue. Or perhaps they were simply not considered necessary in a reproduction of a cult image. In this connection, the records of dress-fasteners in the temple inventories are of great interest, since they might have been used for the dress of the cult images.

Notes

1. Ridgway 2000, 233.
2. Villing 1998, 148.
3. Villing 1998, 151–152.
4. Ridgway 1984, 6.
5. Romano 1985, 348. For the reproductions of the cult images of Athena from the Athenian Acropolis, see Appendix 1.
6. Jucker 1967, 143.
7. Alroth 1989, 25; LiDonnici 1992, 391.
8. Fleischer 1973; Fleischer 1999, 605.
9. LiDonnici 1992, 393; Seiterle, 1979. This interpretation is rejected by e.g. Fleischer 1983, 81–93.
10. E.g. Ephesos archaeological museum, Selçuk, inv. no. 717. See Fleischer 1973, pl. 24–27.
11. For epigraphical evidence mentioning *chitōnes* for Artemis at Ephesos, see Wankel 1987; Sokolowski 1965.
12. Jucker 1967, 143. Fleischer lists ca. 28 marble replicas or fragments of the statue. Fleischer 1973, 146–155.
13. Romano 1980, 468.
14. See Fleischer 1973, 310–314.
15. Romano 1980, 465. Relief: British Museum, inv. no. 1914.0714.1. *BMC* 13 181 no. 6.
16. Romano 1980, 466–467.
17. Museum of Fine Arts, Boston, inv. no. 67.730.
18. E.g. on a coin from Mylasa, *BMC* 13 133, no. 37.
19. Carstens 2012, 143.
20. Poll. 4.116.

21. Romano 1980, 467.
22. Archaeological Museum of Piraeus, inv. no. 3857. Artemis Kindyas is a local goddess from the city of Kindye in Caria *LIMC s.v.* Artemis Kindyas, 763–764. Jucker dates the statue to the late 4th century BC or ca. 300 BC. Jucker 1967, 134–135.
23. It has been suggested that the girdle with fringed edges originated in Syria, but came into fashion in Greece in the first half of the 6th century BC. Fleischer 1973, 226; Riis 1948–49, 78, 82, 87.
24. Weddle 2010, 142; Jucker 1967, 135. The earliest depictions of the goddess on coins are not, however, from Bargylia like the statue itself, but on Seleucid coins of Antiochos III from the 3rd century BC. Fleischer 1973, 223.
25. Jucker 1967, 137; *BMC* 4, 25, no. 11.
26. Jucker 1967, 138.
27. O'Brien 1993, 28.
28. *BMC* 16, pl. 37.2.
29. Fleischer 1973, pl. 86 b; *BMC* 16, 374, no. 242, pl. 37.2.
30. Fleischer 1973, 204–205, pl. 87a, b; *BMC* 16, pl. 37.5 & 37.6.
31. Romano 1980, 29.
32. Ohly 1953, pl. 4.1; O'Brien 1993, 29. Kardara, however, dates it to the 5th century BC. Kardara 1960, 355.
33. Archaeological Museum of Vathi, Samos, inv. no. H41.
34. O'Brien 1993, 26.
35. O'Brien 1993, 28.
36. Burdur Museum, inv. no. 9300. The site of the sanctuary is unknown.
37. For further description of the statue, see Fleischer 1973, 234–235.
38. Fleischer 1973, 236; Merkelbach 1978, 2.
39. *BMC* 19, 125, no. 31, pl. 24.5.
40. E.g. Akurgal 1970, 329; Bean 1968, 47.
41. Fleischer 1973, 249.
42. Fleischer 1973, 251.
43. Delphi archaeological museum, inv. nos. 9796 and 9797. Kolonia 2012, 25. The plaques were recovered from the 'Halos' in the Sanctuary of Apollo at Delphi. The deposit included numerous chryselephantine statues and statuettes. Lapatin 2001, 147.
44. This goddess is only known from one coin depiction, dating to the reign of Commodus, Fleischer 1973, 228, pl. 92a; Jucker 1967, 137, pl. 49.4; Imhoof-Blumer 1913, 7, no. 22.
45. Fleischer 1973, pls 80–83.
46. Jucker 1967, 137, pl. 49.5.
47. Merkelbach 1978, 1. An example of such a perception is in Ath. *Deipn.* 15.11–15.
48. Weddle 2010, 142.
49. Fleischer 1973. Alroth 1989, 16, also finds that coins are a fairly reliable and good source of information.
50. Fleischer 1973, 215–216; Ridgway 1984, 98.
51. Fleischer 1973, 102–111. pl. 53–57.
52. Fleischer 1973, 141–144. pl. 62–63.
53. Fleischer 1973, 214. pl. 85–88.
54. Cahn 1985, 587–588.
55. Romano 1980, 468; Fleischer 1973, 313–314. pl. 142b, 143a.
56. Carstens 2012, 141.
57. Others have interpreted the bases as urns or tripods, *BMC* 16, 43, no. 53.
58. Ephesos Museum, Selcuk, inv. nos. 718 and 1637; Carstens 2012, 141; Fleischer 2002, 211.
59. Münster University Archaeological Museum, inv. no. 2375.

60. Fleischer 2000–2001.
61. Seiterle 1999, 251–252; Seiterle also suggests that the astragal-moulding used in Greek architecture is a stylised version of such woollen bands. Seiterle 1999, 252.
62. Lange 1881b, 70.
63. E.g. Merkelbach 1971, 554, 557; Merkelbach 1978, 1.
64. E.g. on a relief from Rijksmuseum von Oudheiden, Leiden, inv. no. E72. Fleischer 1973, pl. 40, 43a.
65. Fleischer 2002, 213, fig. 14.
66. *BMC* 17, 136, no. 19, pl. 27.6.
67. Paus. 10.24.6, 10.35.10.
68. Fleischer 1973, 258–259, pl. 109.
69. Seiterle 1999, 251.
70. Seiterle 1999; Fleischer 2002, 209.
71. Carstens 2012, 146.
72. Carstens 2012, 145–146.
73. Fleischer 2002, 209.
74. Carstens 2012, 146; Fleischer 2002, 210.
75. Hsch. *s.v. klēides*. *LSJ s.v. kleis*: VI 'sacred chaplets'.
76. *Etym. magn. s.v.* Ephesos. *LSJ s.v. mallos*: 'flock of wool'. It is important to note, however, that the function of asylum was not reserved for sanctuaries of Artemis. Zeus, for example, was also accorded the epithet *hikesios*. *LSJ s.v. hikesios*. Sinn 1993, 97.
77. Carstens 2012, 146.
78. Seiterle 1999, 251.
79. Karakasi 2001, 168; Richter 1968, cat. 113.
80. Karakasi 2001, 124 suggests a branch. Brinkmann 2009, 75, 78 suggests bow and arrow.
81. Richter 1968, 72. cat. 113.
82. Brøns 2014.
83. Brinkmann 2014, 127.
84. Brinkmann 2014, 126–127. For the colours and their reconstruction, see also Brinkmann 2009.
85. Brinkmann 2014, 127; 2009, 75.
86. Cesare 1997, 220.
87. For a discussion of the models for the depictions of the cult images in vase-painting, see Oenbrink 1997, 227–292, who concludes that it is usually impossible to determine an exact cult image as the model for the depictions in vase-painting. Notable exeptions are the Palladion of Troy and possibly the Athena Promachos rendered on the Panathenaic amphoras, if this latter example can indeed be considered a cult statue.
88. Cesare 1997, 220.
89. In scenes of worship before a statue, it is, according to Cesare, not the representation of the divine (i.e. the cult statue), but the ritual act itself, in its repetitiveness, which is the focal point of the representation. Examples of this phenomenon include the representations of the sacrifice offered by Heracles and Chryse or scenes of sacrifice performed by Oenomaos before the chariot race (4th century BC). Cesare 1997, 158.
90. Cesare 1997, 221.
91. Cesare 1997, 221.
92. According to Oenbrink, this phenomenon does not occur in black figure vase-painting, but it does appear in Attic red-figure vase-painting in 17 out of 93 representations. It is more popular in South-Italian vase-painting, where the deity is represented next to its image in 20 out of 55 cases. Oenbrink 1997, 203.
93. Oenbrink 1997, 206.

5. Iconographic evidence for the dressing of cult statues 235

94. Cesare 1997, 220–221.
95. The statue could also be portrayed on a column or pillar, which raises the question whether these statues are cult images or votive images. We do not know of any examples of cult statues placed on pillars or columns. Such a placement appears to have been reserved for votive statues. The fact that cult images are only rarely portrayed thus in scenes of worship is taken by Cesare as a confirmation that the statues placed on columns are *anathemata*. Cesare 1997, 160.
96. Hölscher 2010, 112.
97. Cesare 1997, 160.
98. Hölscher 2010, 113–116.
99. For statues represented in vase-painting, see Cesare 1997; Oenbrink 1997.
100. Romano 1980, 23.
101. Oenbrink 1997, 363.
102. National Archaeological Museum, Athens, inv. no. 18717.
103. Another black-figure example is British Museum, inv. no. B80.
104. Alroth 1992, 26.
105. Schefold 1937; Bielefeld 1954/55.
106. Schauenburg 1977; Schneider-Herrmann 1972.
107. Romano 1980, 455–464.
108. Cesare 1997; Oenbrink 1997.
109. State Hermitage Museum, St. Petersburg, inv. no. BAK7.
110. Museo Archeologico Nazionale, Naples, inv. no. 82922 (H2411).
111. Museo Archeologico Regionale Paolo Orsi, Syracuse, inv. no. 22934. Romano 1980, 462, no. 28.
112. Nick 2002, 33.
113. British Museum, inv. no. B49.
114. Eur. *Tro.* 70, which states that Kassandra was dragged away by Ajax. See also Ov. *Met.* 13, 408ff.
115. Davreux 1942, 139–209.
116. Romano 1980, 463–464. On the Archaic and Classical representations of the Kassandra myth, see Schefold 1937, 41; Daveux 1942, 140–141; Shapiro 1989, 36–37; on the epiphanic aspect of the myth, see Platt 2011, 92–100.
117. Nick 2002, 30; Connelly 1993a.
118. For a black-figure example, see Hölscher 2010, 114, fig. 35; Künze-Götte 1992, pl. 33.
119. Romano 1980, 463.
120. E.g. Davreux 1942 no. 87 (British Museum, inv. no. F209), 92 (Museo Civico Archeologico, Bologna, inv. no. 18108), 111 (Museo Archeologico Nazionale, Naples, inv. no. 8166.9, H2422).
121. E.g. Davreux 1942, no. 91 (Louvre, inv. no. G458), 98 (British Museum, inv. no. E470); Romano 1980, 464 (Museo Nazionale Archaeologico, Taranto, inv. no. 52.665).
122. Davreux 1942, no. 112. National Archaeological Museum, Athens, inv. no. LG212; British Museum, inv. no. F160.
123. Davreux 1942, no. 83. Fitzwilliam Museum, Cambridge, loan ant. 103.22 (from Corpus Christi College).
124. Medelshavmuseet, Stockholm, inv. no. 1963.1; State Hermitage Museum, St. Petersburg, inv. no. 1543; Archaeological Museum of Brauron, inv. no. BE 431; Museo Archeologico Nazionale, Naples, inv. no. H 3235; Ashmolean Museum, inv. no. 1931.39.
125. Museo Archeologico Nazionale, Taranto, inv. no. unknown; British Museum, inv. no. F366; Louvre, inv. no. K36; Museo Archeologico Nazionale, Naples, inv. no. H3231; Archäologische Sammlung der Universität Heidelberg, inv. no. 26/75.
126. Oenbrink 1997, cat. no. A62 (Basel).
127. Museo Archeologico Nazionale, Naples, inv. no. H3235.

128. State Hermitage Museum, St. Petersburg, inv. no. 1543.
129. Connelly 2008, 198.
130. Museo Archaeologico Nazionale, Ferrara, inv. no. 3032.
131. Soph. *Phil.* 1327; Hyg. *Fab.* 102 (this source does not mention Chryse).
132. Louvre, inv. no. G413; Kunsthistorisches Museum, Vienna, Antikensammlung, inv. no. 114. Without inscriptions: Louvre, inv. no. G342; British Museum, inv. no. E494; Museo Nazionale Archeologico, Taranto, inv. no. 52.399; State Hermitage Museum, St. Petersburg, inv. no. 43F.
133. Oenbrink 1997, 143–144. See also Hooker 1950.
134. Heavily decorated: State Hermitage Museum, St. Petersburg, inv. no. 43F and Kunsthistorische Museum, Vienna, Antikensammlung, inv. no. 114.
135. Connelly 2008, 215.
136. Alroth 1992, 29; Bielefeld 1954–55, 382, 383, 388.
137. Langlotz 1954, 29–30.
138. British Museum, inv. no. E714, National archaeological Museum, Athens, inv. no. 1538, State Hermitage Museum, St. Petersburg, inv. no. 1863a.
139. National Archaeological Museum, Athens, inv. no. 1283.
140. Staatlichen Antikensammlung, Munich, inv. no. 2264.
141. Ashmolean Museum, inv. no. 1966.714 (MM 80).
142. Furthermore, Beschi identified a sanctuary for Aphrodite Pandemos and Peitho on the southwest side of the Acropolis. Delivorrias 2008, 111.
143. See Rosenzweig 2004.
144. Karousou 1956, 179.
145. Paus. 1.19.2. Translation: Jones & Ormerod 1918.
146. For the Classical cult images of the three sanctuaries of Aphrodite, see Weber 2006.
147. Oenbrink 1997, 153.
148. Martin von Wagner Museum, Würzburg, inv. no. H4307; National Archaeological Museum, Athens, inv. no. 12543.
149. British Museum, inv. no. G17, 1871,0722.1.
150. Kerameikos museum, Athens, inv. no. 4961.
151. Oenbrink 1997, 155.
152. The main sanctuary of the *Lēnaia* has not been identified with certainty, but it has been suggested that it was situated in the Athenian Agora. Simon 1983, 100.
153. Neils 2008, 247.
154. Simon 1983, 100.
155. National Archaeological Museum, Athens, inv. no. CC1001.
156. Staatliche Museen zu Berlin, Antikensammlung, inv. no. 2290.
157. E.g. *Pelike*, Louvre, inv. no. G227; *Kelebe*, Museo Caputi, Ruvo, inv. no. 328.
158. Frickenhaus 1912, 7, pl. 2. no. 14 and National Museum, Warsaw, inv. no. 142351.
159. Museum of Fine Arts, Boston, inv. no. 418; Louvre, inv. no. G408; British Museum, inv. no. E451; Museo Nazionale Etrusco di Villa Giulia, inv. no. 983; Martin von Wagner Museum, Würzburg, inv. no. HA126; Museo Archeologico Etrusco, Florence, inv. no. 4005.
160. Frickenhaus 1912, 9.
161. Frickenhaus 1912, 9.
162. British Museum, inv. no. E452; Louvre, inv. no. G407.
163. Museo Archeologico Nazionale, Naples, inv. no. 2419.
164. Frickenhaus 1912, 17–19.
165. See also Frickenhaus 1912, 18.
166. *OCD* s.v. herms.
167. For herms in the vase-painting of the 6th–4th centuries BC, see Cesare 1997, 161–163.

168. State Hermitage Museum, St. Petersburg, inv. no. 4515.
169. British Museum, inv. no. E585. Goldman 1942, 64 interprets the textile as a stylised *nebris* (fawnskin).
170. Antikensammlungen, Munich, inv. no. 1871.
171. Boardman 1999, 151.
172. Royal Museum of Art and History, Brussels, inv. no. A724.
173. Museum of the ancient Agora, Athens, inv. no. S1086, *LIMC s.v.* Artemis, pl.447, 76.
174. British Museum, inv. no. 1861.1127.161.
175. Goldman 1942, 61.
176. For a discussion on the identification of specific cult images in vase-painting, see Oenbrink 1997, 207–292.
177. Hiller 1983, 92.
178. Bielefeld, 1955/54, 399ff.
179. Alroth 1992, 35.
180. Hölscher 2010, 112.
181. Oenbrink 1997, 301.
182. Alroth 1992, 39.
183. The goddesses Astias from Iasos and Eleuthera of Myra in Lycia are both depicted on coins without arms - perhaps concealed under their garments, Jucker 1967, 137. Furthermore, Pausanias (10.35.10) records that the statue of Demeter has ribbons, but it is not clear whether it is bound or decorated with them.
184. Some of the images may even have been wrapped not in real garments, but in linen strips arranged in elaborate patterns, similarly to Egyptian mummy wrappings. I thank Ellen Harlizius Klück for this suggestion. For the swaddling of Greek babies, see Lee 2015, 97.
185. For depictions of dress-fasteners in Greek art, see Brøns 2014.

Chapter 6

Written evidence for the dressing of cult statues

Epigraphic and literary evidence for the dressing of cult statues

When dealing with the epigraphic sources, it is important to take into consideration that in the written evidence it is very difficult to distinguish between textiles dedicated as offerings to a deity and a garment used to dress a cult image. The sources often use the Greek verb 'ekhei', which can be translated as either 'has' or 'holds'.[1] But it is quite clear in this particular epigraphic context, as Culham notes, that the Greek *ekhō* conveys the idea of 'wearing' and not 'possessing'.[2]

The Sanctuary of Artemis at Brauron

The Brauron catalogues describe different cult statues present in the sanctuary. The catalogues mention an 'old (seated?) statue' (*tōi hedei tōi archaiōi*),[3] a '(seated?) statue' (*tōi hedei*),[4] and the 'stone (seated?) statue' (*to lithinon hedos*),[5] which possibly, but not necessarily, are one and the same. The catalogue also mentions an 'upright statue' (*tōi agalmati tōi orthōi*).[6] These terms occur in the same inscription, which could indicate that several different statues were meant. Other inscriptions record a 'statue' (*agalma*),[7] a 'standing statue' (*tōi agalmati tōi hestēkoti*),[8] and an 'upright statue' (*agalma to orthon*).[9]

Based on a passage in Pausanias, recording a statue of the Brauronian Artemis by Praxiteles as well as a *xoanon* called the Tauric Artemis,[10] and the catalogues, Mansfield argues that there were three cult statues in the sanctuary. The first and earliest statue was the ancient wooden statue (termed *hedos* in the catalogues, but *xoanon* by Pausanias), and the second was the stone statue (also termed *hedos*) from the mid-5th century BC, which Mansfield argues was probably a replacement for the Archaic statue carried off by the Persians in 480 BC. The third statue was the upright or standing statue (termed *agalma*) from the 4th century BC – the work of Praxiteles later removed to the Athenian Acropolis (see Fig. 30).[11] Linders, however, contends that none of the statues mentioned in the catalogues was the Artemis of Praxiteles, since this statue, according to Pausanias (1.23.7), stood on the Acropolis.[12] With regard to *to lithinon hedos* and *to hedos to archaion*, Linders argues that two different statues are meant, since these two terms occur in consecutive lines. She suggests, furthermore, that *to hedos* is a shorthand reference to one of the other statues – probably the old statue, which was of wood.[13] Linders also doubts that these statues were actually seated: although the word *hedos* is usually understood to imply that a statue is seated, not all literary sources use the term in this way. Linders therefore suggests that *hedos* simply signifies a holy, venerable statue.[14]

The questions about the number and appearance of the statues in the sanctuary at Brauron are thus not easy to answer. It may be assumed, however, that the sources indicate the presence of at least two different statues termed *hedos* (one of stone and one old – probably of wood). Three statues are denoted by the term *agalma* – of which at least two out of the three are standing or upright, but perhaps these are in fact one and the same. The catalogue thus mentions at least three different cult statues, at most five.

The statues termed *hedoi* (possibly seated) are described in connection with various garments: there is a *tarantīnon*,[15] two wraps (*ampechonoi*),[16] and an embroidered garment (*katastikton dipterygon*)[17] around the old (seated) statue (*peri tōi hedei tōi archaiōi*), a *chitōn amorginon*[18] around the (seated) statue (*peri tōi hedei*), and a white *himation* with purple border is placed around the (seated) stone statue (*to lithinon hedos*).[19] Another inscription also records something around the statue (*peri tōi agalmati*)[20] and around the old (seated) statue (*peri tōi hedei tōi archaiōi*),[21] but the text specifying the type of garment is not preserved.

There is a patterned (*peripoikilos*) *chitōniskos* around the upright statue (*tōi agalmati tōi orthōi*),[22] and another inscription records that the upright statue has (*to agalma to orthon ekhei*) something – perhaps the double-layered *krokōtos* recorded immediately before.[23] Finally, there is also a record of the statue (*to agalma*, and thus unknown exactly which one) having (*ekhei*) a *kandys*.[24]

There is thus evidence of at least two different cult statues being dressed, one seated and one upright.[25] Different terms are used to describe this phenomenon: either the garments are placed 'around' (*peri*) the statues or the statue 'has' (*ekhei*) garments. A rather wide variety of garments is used to dress the statues: wraps, *tarantīnon*, *chitōn*, *himation*, *chitōniskos*, and *kandys*. Only a few of these garments are described as having colour: the white *himation* with purple border and the *krokōtos*, which indicates a yellow colour. But two garments also had some sort of colour decoration, since one is embroidered (*katastikton dipterygon*) and another is patterned all over (*peripoikilos*).

The Heraion on Samos (Appendix 2.3)

The inventory of the Heraion on Samos describes 'the goddess' (Hera) as having (*ekhei*) a *mitrē* (a belted girdle or snood to tie up women's hair).[26] The inventory also records a *proslēmma* with purple edges 'belonging to the goddess', possibly indicating that a statue of Hera wore this garment and that a *sindōn* was spread before the goddess, i.e. her statue. Later in the inventory, it is recorded that, under Thrasyanax, the goddess had *chitōnes*; under Hippodamas, the goddess had two *chitōnes*; and, finally, when Demetrios was archon, the goddess had two *chitōnes*.

According to the inventory, other statues in the Heraion also wore garments: 'the Euangelis'[27] had one of the seven veils (*krēdemna*) and the so-called 'Goddess at the back' had a white *himation*. According to O'Brien, 'the goddess' is the earlier wooden cult image, while the 'Goddess at the back' is a more recent 6th century statue.[28] The inventory also records that the statue of Hermes had one of the 38 *chitōnes* and one of the 48 *himatia*, and that the statue of Hermes in the Temple of Aphrodite had two of these *himatia*. The inventory thus establishes that both the statue of Hera and that of Hermes were likely to have been dressed in real garments.

Another later source on the dressing of Hera's cult statue at Samos is the early Christian apologist Lactantius (240–320 AD), who quotes a now lost text by the Roman author Varro (1st century AD) stating that the image of the Samian Hera was dressed as a bride (*habitu nubentis*), and that an annual festival with wedding rites was celebrated in her honour: 'Varro writes that the island of Samos was early called Parthenia because there Juno grew up and there she also married Jupiter. Therefore her most noble ancient temple is at Samos, where her statue, dressed in the garb of a bride, and her annual worship are celebrated in a nuptial rite.'[29] This worship was probably the *Tonaia*, or 'Roping': a ritual performed at the Samian Heraion during which the old wooden statue of Hera (*axoos sanis* or 'unwrought plank') was bound to a tree. This was accompanied by an annual festival with musical and athletic competitions, communal feasts, and, at least in the later periods, the presentation of bridal dresses to Hera.[30] Unfortunately, there is no information on whether these bridal dresses were made especially for the occasion or whether they were used by women – perhaps in a wedding ceremony – and then offered to Hera.

The sanctuary on Delos
On Delos, the statue of Artemis is described, in 146/5 BC, as being clothed in[31] a purple garment (*esthēta porphyran*) interwoven/plaited with gold.[32] Earlier, in 279 BC, the statue is reported to have had (*ekhontos*) simply a woollen garment (*ereian esthēta*).[33]

The statue of Leto seems to have been wearing a purple mantle in 269/8 BC, since purple dye is twice reported to have been bought specifically for her *himation*.[34] In 156/5 BC, the wooden statue of the goddess is described as wearing a linen *chitōn*, a linen wrap (*ēmphiesmenon linōi*),[35] and a pair of shoes or half-boots,[36] and an inscription from 140/39 BC records the presence of Leto's *chitōn*, gold diadem, and *othonion*, along with a belt with inwoven gold.[37] The inventory of 155 BC records the *chitōniskos* of Leto, which could indicate that her statue wore it. Furthermore, the inscription records that something – perhaps the statue of Leto – was 'in' a linen cloth (*othonion*) that featured a purple-edged circle with decoration in gold thread and was bound with a belt interwoven with gold. The inscription also records the presence of dress-fasteners at the shoulders – a further indication that the garment was placed on a statue.[38]

In the Heraion, moreover, there was a garment of fine linen (*othonion*) for Hera (296 and 250 BC),[39] and the inventory of 155/4 BC records that two statues (*agalmata*) were clothed in linen.⁴⁰ Finally, it is possible that the statue of Dionysos also wore a garment. According to an inventory of 141/0 BC, it had a *chitōn* that belonged previously to the goddess Artemis.⁴¹

Official decrees

Temple inventories are not, however, the only source of information on the dressing of cult images. Several official decrees also describe this ritual act. For example, at Magnesia on the Meander, a decree from 197/196 BC concerning the establishment of a festival with thanksgiving sacrifices to Zeus Sosipolis, Artemis Leukophryne, and Apollo Pythios reads:

> 'Let the Stephanophoros leading the procession have the wooden statues (*xoana*) of all the Twelve Gods carried (in the procession) in the finest clothes (*esthētes*) possible and have a round tent (*tholos*) erected in the Agora next to the altar of the Twelve Gods.'⁴²

There is no doubt that these statues were clothed in real garments.

A honorary decree dating to 62/1 BC from Mantineia on the Peloponnese praises Nikippa Pasia for

> '... her cooperation in the adornment of the Goddess with the priests in office on each occasion, on whose request she has furnished each group with what was required for the worship and adornment of the Goddess'

and

> '... offered a sacrifice to the Goddess and obtained good omens ..., contributed a robe (*peplos*) for the Goddess, ...'⁴³

The law of the Delphic Amphiktyons (380/79 BC) indicates that the statue of Athena Pronaia at Delphi was washed annually and provided with new 'adornment' (*kosmos*), including a mantle (*ampechonon*) with gold fibulas (*porpāmata*) and a gold diadem (*stephanē*):⁴⁴

> 'The bath (*lōtis*) of [Athena] Amphi[ktyonis (?), x] Aiginetan [staters;] her mantle (*ampechonon*), 150 [Aiginetan] staters, [and] 100 Aiginetan staters for [its shoulder fastenings (*porpāmata*);] for her diadem (*stephane*), 100 [Aiginetan] staters; ...'.⁴⁵

Further down the text the gold dress-fasteners (*porpāmata*) are mentioned again.⁴⁶ Finally, the *Lexica Segueriana* refer to the *gerarades* as 'the women who clothe the statue of Athena at Argos',⁴⁷ which is explicit evidence that the statue of Athena was clothed in some sort of garment.

Literary evidence for the dressing of cult images

Several literary sources confirm that cult statues were dressed in garments. In particular, Pausanias' *Description of Greece* records several examples in which the images of deities are clothed.

In book 1 (1.18.5), Pausanias reports that

> 'The Athenians are the only people whose *xoana* of Eileithyia are draped to the tips of the feet.'[48]

Furthermore, at Megara (1.43.5), he reports that

> 'Polyidus also built the sanctuary of Dionysus, and dedicated a wooden image (*xoanon*) that in our day is covered up except the face, which alone is exposed.'[49]

Certainly, the image might have been concealed with something else, but textiles seem to be the most likely choice.

In book 2 (2.11.6), he relates that the image of the god Asklepios in the sanctuary at Titane near Sicyon was clad in a white woollen *chitōn* and a *himation*:

> 'As for the statue of the god (*Asklepios*), (…) only the face, hands and feet of the statue are visible, for it is clothed in a white woollen chiton and mantle (*himation epibebletai*). There is also a similar statue of Hygeia, which is also barely visible, so completely is it surrounded by the locks of hair, which women cut off to give to the goddess and with the strips of Babylonian garment (*esthētos Babylonias*).'[50]

In book 3 Pausanias describes the Sanctuary of Apollo at Amyklai, the Amyklaion, where the festival of *Hyakinthia* was celebrated. According to this description, the sanctuary was shaped like a throne, in the middle of which was an altar that served as a base for the column-shaped semi-iconic image of Apollo.[51] He states (3.16.2) that the (Spartan) women weave a *chitōn* for Apollo every year, and that the building where they weave it is called 'the *chitōn*'.[52] Pausanias does not mention the *Hyakinthia* in this passage, but it is generally accepted that the garment was presented during this festival, and that it was conveyed in a procession on the Sacred Road called *Hyakinthis*.[53] Also in book 3 (3.15.10–11), Pausanias reports that, at Sparta, there is a hill on which there is an ancient temple with a wooden image (*xoanon*) of an armed Aphrodite. The statue sits and wears a veil.

In books 5 and 6, Pausanias refers to a garment made for Hera at the sanctuary in Olympia. In book 5 (5.16.2), he writes:

> 'Every four years the Sixteen women weave a peplos for Hera at Olympia …'[54]

This description relates to a section in book 6 (6.24.10) stating:

> 'In the Agora at Elis a building has been built for the women who are called "the Sixteen", in which they weave the peplos for Hera.'[55]

Thus, a special house was assigned to the weavers, which was located in the Agora of Elis, a city about 20 km northwest of Olympia. Of the 16 poleis that originally sent the women to weave Hera's dress, there remained at the time of Pausanias only eight tribes (*phylai*), who each therefore sent two women.[56] It is generally assumed that the *peplos* was draped on the cult statue in the Heraion, but this is not mentioned by Pausanias. The statue was probably seated, and the *peplos* may therefore have been placed on the lap of the statue or draped around it.[57] It has also been suggested that the *peplos* may have been a veil rather than a robe.[58]

In book 6 (6.25.5-6), Pausanias describes a statue of the god Poseidon standing in the most densely inhabited section of Elis:

> 'In the most thickly-populated part of Elis is a statue of bronze no taller than a tall man; it represents a beardless youth with his legs crossed, leaning with both hands upon a spear. They cast about it a garment of wool (*esthēta de erean*), one of flax (*apo linou*) and one of fine linen (*byssos*).'[59]

In Mansfield's translation, the statue is dressed in only two garments: 'They clothe it in a woollen garment (*esthēs*) and one of linen (*byssos*)'.[60] On the basis of the Greek, however, it is impossible to determine whether the text is describing only one garment made from three different kinds of fibre or two or three different garments, one of wool and one of linen and *byssos*.

Book 7 (7.23.5-6) mentions the sanctuary of Eileithyia at Aigion, Achaia, where the wooden statue of the goddess 'is covered from head to foot with a fine woven cloth (*hyphasma*) ...'[61], and the temple of Demeter at Boura, Achaia, where the image of the goddess 'has a garment (*esthēs*)'.[62]

Book 8 (8.42.4) describes the cult of Demeter Melaina at Phigalia in Arcadia, where the old wooden cult image had the head and hair of a horse and was clad in a black *chitōn*,[63] and book 9 (9.3.7) describes a ritual at Plataia, Boeotia, where the wooden cult statue of Hera was dressed as a bride. Pausanias describes the garment as an *esthēta*.[64]

Finally, book 10 (10.24.6) mentions that the *baitylos* (*lithos*) at Delphi was decorated with shining wool (*eria epititheasi ta arga*) at each feast. The wool may simply have been unworked wool, or perhaps woollen ribbons,[65] and the shining effect could have been caused by using olive oil.

Further literary sources attest to the dressing of statues of gods and goddesses. For example, the writer Polybios (2nd century BC) records that cult statues at Antioch were clad in garments (*stolai*) interwoven with gold during a procession at the games held in 166 BC.[66] Athenaeus (2nd century BC) describes the famous Dionysiac procession in Alexandria during the reign of Ptolemy Philadelphos (283-246 BC), where the participants included men dressed as *silēnoi* wearing purple cloaks, 120 boys and 500 girls in purple tunics, and one or two statues of Dionysos, 4.5 m tall, garbed in a purple *chitōn* and *himation*.[67]

The orator Hyperides (4th century BC) records that Dodonian Zeus, through an oracular response, ordered the Athenians to send Dione a new dress, ornaments, and a mask (*prosōpon*):

> 'For Zeus of Dodona commanded you through the oracle to embellish[68] the statue of Dione. (...), you embellished the statue of Dione in a manner worthy of yourselves and of the goddess.'[69]

There are also several literary sources that testify to the dressing of cult statues in the Roman world. A famous example is the statue of Hercules in the Forum Boarium at Rome, which was clothed in full triumphal garb during processions,[70] and Cicero reports that the statue of Zeus Olympios at Syracuse wore a woollen mantle (*pallium*).[71] According to Ovid (43 BC–17/18 AD), the identity of a statue in the temple of Fortuna was unknown as a consequence of the fact that it was covered in robes.[72] Lucian (125–180 AD) mentions in his description of cult at Hierapolis in Anatolia that the temple treasuries include many robes (*esthēta*).[73] Furthermore, he states that the statue of Zeus resembles the god in all respects – facial features, garments, and throne.[74] However, whether this means that the statue actually *wore* actual garments or whether it implies that the statue's clothing *looked* real is impossible to say.

There are, as well, several literary sources attesting to the dressing of cult statues, but unfortunately not stating what deity or specifying the type of garment. For example, Tertullian (160–225 AD) describes how pagan idols are dressed in different garments: 'bordered and striped togas, and broad-barred ones, are put even on idols themselves',[75] and the Old Testament condemns the custom of adorning wooden idols of foreign cults with purple garments.[76] The early Christian author Lactantius Firmianus writes between 303–313 AD:

> '(...) to statues which have no use for clothing they consecrate robes (*pepli*) and costly garments,'[77]

and Valerius Flacchus writes:

> 'Hypsipyle placed her terrified father King Thoas in the quiet temple (of Dionysos) under the feet and right hand of the god; and he remained hidden, received by the sacred garments.'[78]

Official staff for dressing the cult statues

The act of clothing the cult images was an important responsibility assigned to particular individuals. Many female deities had official wardrobe mistresses, who changed the clothes of images when necessary and appropriate. The dressing and adornment of statues appears – at least at some sanctuaries and cults – to have been managed by officials and subject to regulations, in order to ensure that the practice was correctly observed, as well as to give the impression that images were in need of regular attention and maintenance.[79]

There were regulations/traditions that determined how the image ought to be dressed and who could perform this task. The latter point was perhaps considered to be especially important, because those who dressed the statue naturally had to remove the previous garment, and would therefore essentially see the image 'naked'.[80] Those responsible for these intimate interactions appear to have been officially categorised in the religious terminology, and in some sanctuaries were called *kosmophoroi*.[81] The term comes from *kosmos*, which can be translated as 'adornment' or 'adorning'. Another term connected with the official duty of dressing cult statues is the priestly title *stolizōn* or *stolistēs*, which can be translated as 'rober'.[82] The word is employed in several inscriptions, e.g. on statues and a herm, all in the Sanctuary of Isis. The majority date to the 2nd century AD.[83]

There are other related titles in the written record. One is the *hypostoloi*, who were probably assistants of the *stolistes* ('robers'). The term is recorded in two inscriptions from the 3rd century BC and 117 BC.[84] Another title for staff that dress a cult statue at Argos is *gerairades*, who are defined as 'those who dress the statue of Athena at Argos'.[85]

The *Praxiergidai*, who are mentioned in relation to the Sanctuary of Athena on the Athenian Acropolis, are also important in this connection. They were an Athenian clan, whose members were responsible for keeping the statue and its garments clean, that is, washing the statue and its garments, an activity that took place during the *Plynteria*.[86] Members of the *Praxiergidai* also placed the new *peplos* upon the statue at the annual Panathenaia.[87] There were some cult personnel titles that were probably filled by members of this clan. These include: the *Loutrides* ('bather-women'),[88] who bathed the statue of Athena; the *Plyntrides* ('washer/laundry-women'), who washed the statue's garments; and probably also the *Kataniptēs* ('washer'),[89] who likely cleaned and washed the whole statue and its garments throughout the year.[90]

Conclusions on the dress of cult statues

The written evidence thus confirms the custom of dressing cult statues in garments. According to these sources, the statues involved included both male and female deities, and a very wide range of garments was used, including not only common clothing items such as wraps, *chitōnes, himatia, chitōniskoi*, and veils, but also less common garments such as *tarantīna* and a *kandys* (Brauron), or even the more elusive garment terms *esthēs* and *othonion*. *Peploi* are also recorded, e.g. in the cases of Athena in Athens[91] and Hera in Olympia. In both of these cases, however, the *peplos* was made specifically for the image, which might imply a different meaning. Often, the sources merely inform us that the cult statue was dressed – not in what or how. It thus seems, according to these sources, that cultic images could be clad in almost anything. The written sources indicate that the practice of dressing cult statues existed at least from the 5th century BC, as evidenced by the inventories from Brauron and Samos, and continued into the Roman Period.

6. Written evidence for the dressing of cult statues

What, in summary, can we infer from the iconographic and written sources on the dressing of cult statues? Rituals involving the dressing of Greek cult images and the offering of clothing are well attested, and it is reasonable to conclude that the practice of dressing cult images – wooden *xoana*, as well as stone and acrolithic statues[92] – in real clothing was probably common. The majority of the iconographical evidence, however, consists of Classical, Hellenistic, and Roman examples, and the written evidence is comprised of Post-Archaic Greek and Roman literary sources and Hellenistic inscriptions.[93] This can produce a somewhat skewed picture, since we do not have any useful illustrations or testimonia from the early periods. Nevertheless, some sources indicate that this custom possibly occurred already in the early periods. As an example, we know from Homer that the presentation of garments to a deity (but not necessarily the dressing of cult statues) was practised at least as early as the 8th century BC in Greece.[94] Scholars thus tend to agree that the practice of dressing cult statues occurred at least from the 8th or 7th century BC, and that this ritual act continued for a very long period, at least into Roman times.[95]

The sources show that images of different gods and goddesses were dressed in garments, and that the practice was not reserved for one particular deity or one particular gender. Nor is the custom specific to a certain geographical area, but appears, rather, to have occurred across the Mediterranean region. According to the iconographic, epigraphic, and literary sources, goddesses are more often dressed in garments than gods – especially Athena, Artemis, and Hera, but also Aphrodite. There are nonetheless several examples of male gods being dressed in garments, as attested in both iconographic and written sources.

The cult statues were usually dressed in a single garment or a set of two or more clothing items, but in a few cases attested in the written sources the images were dressed in several garments at the same time. The use of multiple garments is rarely visible in iconography, except in the case of depictions of the cult image of Artemis of Perga, which was depicted either as a *baitylos*, or as an unshaped mass, in order to indicate that the statue was covered in numerous garments.

Colour is another interesting aspect of the garments used to dress cult images, but unfortunately, the iconographic material can rarely provide any information. Colour is mentioned only rarely in the written sources, and usually the only colours specified are purple and white, and of course also saffron in the case of the *peplos* of Athena at Athens (Appendix 1). Yet the garments used to dress the cult statues often feature heavy decoration, which perhaps imitates embroidery, applied decoration, or inwoven patterns. Another interesting feature is the tightly fitted garments for cult statues rendered in iconography.

It can be concluded, furthermore, that there was no differentiation in dress between gods and goddesses, since the type of garments used for male and female appear to have been the same, with the obvious exceptions of the *peplos*, veil, and girdle, which were reserved for women, and it seems that a wide variety of garments were deemed appropriate for the dressing of cult images. It is thus possible that the type of garment was not of the outmost importance, but instead its quality, decoration, and the way that the garment was arranged on the statue. In sum, the dressing of cult images in real clothing was a widespread practice, a fact testifies to the importance of this particular custom in Ancient Greek religion.

Notes

1. LSJ s.v. ekhō.
2. Culham 1986, 237.
3. IG II² 1514, 34–39.
4. IG II² 1514, 22–23.
5. IG II² 1514, 27–28.
6. IG II² 1514, 42–43.
7. IG II² 1524B, 204–206, 215–216, 224; IG II² 1523, 27–29.
8. IG II² 1524B, 207.
9. IG II² 1522, 28–29.
10. Paus. 1.23.7. Translation: Jones & Ormerod 1918.
11. Mansfield 1985, 459. Acropolis Museum, inv. no. Acr. 1352.
12. Linders 1972, 3, 15.
13. Linders 1972, 14.
14. Linders 1972, 14.
15. IG II² 1514, 37.
16. IG II² 1514, 34–36.
17. IG II² 1514, 38.
18. IG II² 1514, 22–23.
19. IG II² 1514, 27–28.
20. IG II² 1524B, 224.
21. IG II² 1524B, 227.
22. IG II² 1514, 42–43.
23. IG II² 1522, 28–29.
24. IG II² 1523, 27–28.
25. None of the statues mentioned in the catalogues was the Artemis of Praxiteles, since this statue, according to Pausanias (1.23.7), stood on the Acropolis. Linders 1972, 3, 15.
26. Ohly 1953, no.11.
27. The Euangelis was probably a third statue of Hera or a personification of 'good things'. Mansfield 1985, 484.
28. O'Brien 1993, 29.
29. Lactant. Div. inst. 1.17.8. Translation: O'Brien 1993, 34.
30. O'Brien 1993, 54.
31. LSJ s.v. amphiazō.
32. ID 1442B, 54–56. Mansfield 1985, 475–477.
33. IG XI,2 161B, 62.
34. IG XI,2 203A, 73 and IG XI,2 204, 75–76.
35. LSJ s.v. amphiennymi.
36. ID 1417A, I, 100–103.
37. ID 1450A, 200–201.
38. Mansfield 1985, 478–479. ID 1428, II, 53–58.

39. *IG* XI,2 154A, 21–22; 287A, 120–121.
40. Mansfield 1985, 482. *ID* 1417A, II, 22.
41. Mansfield 1985, 480. *ID* 1444Aa, 38; Furthermore, the *ID* 1442B, 146/5 BC, 54–56, report that the old garment of Artemis was put on the statue of Dionysos.
42. *LSAM* 32. Translation: Mansfield 1985, 485.
43. *IG* V², 265, 4–26. Translation: Mansfield 1985, 468.
44. Mansfield 1985, 463.
45. *CID* I, 10; *SEG* 28:100, 26–32. Translation: Mansfield 1985, 462.
46. Translation: Mansfield 1985, 468.
47. *Lex. Seg. Gloss. rhet. s.v. gerarades.* Translation: Mansfield 1985, 467.
48. Translation: Donohue 1988, 366.
49. Translation: Jones & Ormerod 1918.
50. Mansfield 1985, 464; Kleijwegt 2002, 107. Translation: Jones & Ormerod 1918.
51. Petterson 1992, 11; Paus. 3.19.1–5.
52. Mansfield 1985, 467.
53. Petterson 1992, 11; Dillon 2002, 137.
54. Translation: Mansfield 1985, 470.
55. Translation: Mansfield 1985, 470.
56. Scheid & Svenbro 1996, 11.
57. Mansfield 1985, 470.
58. Marinatos 1967, 43.
59. Translation: Jones & Ormerod 1918.
60. Mansfield 1985, 470–471.
61. Translation: Mansfield 1985, 471.
62. Translation: Mansfield 1985, 471.
63. Radke 1936, 18; see also Jost 1973, 249–250; Bruit 1986.
64. Mansfield 1985, 461.
65. Jones & Ormerod 1918.
66. Polyb. 31.3.
67. Reinhold 1970, 32; Ath. *Deipn.* 5.27–28; Blum 1998, 90, 106.
68. *LSJ s.v. epikosmeō*: add ornaments to, decorate.
69. Hyp. *Eux.* 4.24–25. Translation: Burtt 1954. Goldman 1942, 63.
70. Weddle 2010, 54; Plin. *Nat.* 34.16.
71. Cic. *Nat. D.* 3.34.83.
72. Ov. *Fast.* 6.569ff.
73. Luc. *Syr. D.* 10.
74. Luc. *Syr. D.* 31. Kleijwegt 2002, 107.
75. Tert. *Idol.* 18. Translation: Thelwall 1969. Reinhold 1970, 56–57.
76. Reinhold 1970, 21; *Jeremiah*, 10.9: 'Their idols (...) are dressed in violet and purple cloth woven by skilled weavers.'
77. Lactant. *Div. inst.* 2.4.15. Translation: Mansfield 1985, 450.
78. Val. Fl. *Argon.* 2.257–259. Translation: Mansfield 1985, 451.
79. Weddle 2010, 54.
80. Weddle 2010, 54.
81. Weddle 2010, 53.
82. Vidman 1970, 62. 'Bekleider'. *LSJ s.v. stolistēs*: = *hierostolos*: 'those who had charge over the sacred vestments'.
83. *IG* II/III², 4771, 4772, 3564, 3644, 4818.
84. Mansfield 1985, 455; *IG* XII, suppl. 571; *IG* IX, 2, 1107b.

85. Robertson 1996, 51. Hsch. *s.v. gerērades*. *LSJ s.v. geraira*: priestesses of Athena at Argos.
86. On the *Plynteria*, see Robertson 1996, 48–52; Dillon 2002, 133–134; Sourvinou-Inwood 2011, 135–224.
87. Mansfield 1985, 366.
88. Hsch. *s.v. loutrides*; Phot. *s.v. loutrides*.
89. *LSJ s.v. kataniptēs*.
90. Mansfield 1985, 367–368; Sourvinou-Inwood 2011, 265.
91. See Appendix 1.
92. Mansfield 1985, 445.
93. Romano 1988, 129.
94. Romano 1988, 129.
95. Mansfield 1985, 445.

Chapter 7

Discussion: Dressing of cult statues

Why did the Ancient Greeks dress their cult images? To the modern mind, statuary is composed of inanimate material, so what purpose does clothing serve? Ancient sources indicate that cult images do not need to be highly decorative or ornate,[1] so there is no obvious point to adornment from an aesthetic perspective. There is thus no modern logical explanation as to why cult statues would need elaborate wardrobes,[2] which perhaps also explains why the entire topic has been rather neglected in scholarly literature.

The question of why cult images were dressed has been left largely unaddressed by classical scholars, perhaps because it is a very complicated issue, but also, of course, because the subject of the use of textiles in cult and ritual has so far been overlooked. When addressed, the question is usually passed by as something secondary and less important in comparison with other rites. Mansfield has even suggested that often 'the adorning of cult statues with garments is not a ritual act at all, but part of the normal devotion paid to the statues as objects of worship.'[3] Romano suggests that the custom of dressing cult images might have begun as a way to conceal the poor craftsmanship of wooden statues,[4] which, as we shall see, most probably has nothing to do with acts of dressing these images. Other possible explanations include: the need to appease the god by honouring him with gifts; the usefulness of textiles as a means of protecting the image from weather, whether rain or bleaching sun, or perhaps from wear and tear: and, of course, the need to conceal nudity. All these explanations, though, are more of a practical character, and no one has thus far performed a more thorough study of the underlying motives and effects of this specific act in Ancient Greece.

Anthropologists of religions in general and of religious images in particular have long grappled with issues of anthropomorphic representation and the treatment of images as though they are themselves divine or in some way alive[5] – in other words, whether the cult images simply *represent* the deity or whether they embody the deity, which is to say that spirits or deities are believed to 'take up residence' in statues and other religious objects.[6] Anthropological and sociological theories are thus of great help in dealing with such questions and will therefore serve as a point of departure in the following discussion.

Cult images and agency

Actor-network theory (ANT) focuses on the relation between human beings and objects. According to ANT, modern scientific theory has established an artificial distinction between human beings and non-human beings, by which is meant both organic and inorganic things such as animals, plants, minerals, objects, and so on. ANT argues for the rejection of this dualistic thinking, which has been dominant in the social sciences, and which creates a separation between subject and object. Instead, ANT theorists designate humans as well as non-humans as 'actors' if they cause action.[7] Thus, one of the main points of Actor-network theory is that both things and human beings have the ability to act.[8]

This leads us to the useful but difficult concept of 'agency'. One of the most influential scholars in this field is Gell, according to whom an 'agent' should be understood as one who 'causes events to happen' in their vicinity, i.e. that which has an effect on its surroundings.[9] This means that a conscious intention to act is not necessarily embedded in the agent, but that agency instead is defined as that which causes an act or event to take place.[10] Gell further distinguishes between primary agents and secondary agents. Primary agents are agents with intentions to act and affect their surroundings, which include not only human beings, but also objects (which from an animistic perspective are considered animated). Secondary agents are persons and objects that cause effects in their surroundings without having the intention to do so.[11] Gell also distinguishes between agents and patients: an agent is something that causes effects, while a patient is something that is affected by something else. This means that human beings as well as things (e.g. a cult image) can serve as both agents and patients.[12] Thus, we cannot assume that human beings always have the active and determining role in the way that events unfold, since material things often are co-creators of the circumstances under which human beings act.[13]

Gell puts forward several ideas about idolatry that are very useful for a study on Ancient Greek cult images. First, it is important to specify what is meant by idols. Idols come in many varieties, but it is conventional to distinguish between two polar types: purely aniconic idols versus iconic idols. Gell does not pay a great deal of attention to this distinction, since he considers all idols to be iconic, whether they resemble a familiar object or not. In this way, aniconic images are also realistic representations of a god. His theories as outlined below thus include both types.[14] Furthermore, idolatry has nothing to do with primitive religion. As claimed by Gell:

> 'The idea that only the uneducated or 'primitive' worship idols of stone, wood and metal fashioned to resemble the human form, is a consequence of the convergence between anti-imagistic forms of religiosity (such as Judaism, Islam, and certain forms of Christian sectarianism and Protestantism) and the rise of a more generalised religious skepticism, which has ancient antecedents.'[15]

One of Gell's important points is that, in the context of idolatry, the idol is not simply a depiction of the god, but rather the body of the god in artefact form.[16] Similarly, persons may be distributed, i.e. all of their parts not physically attached, but instead distributed around the ambience. This means that something can be both an image and part of something, e.g. a deity.[17] He further argues that:

> 'if appearances of things are material parts of things, then the kind of leverage which one obtains over a person or thing by having access to their image is comparable, or really identical, to the leverage which can be obtained by having access to some physical part of them.'[18]

This means that access to the cult image could imply some sort over power it, which could be enacted by the worshipper.

Gell includes many examples from different religions in his discussion of idolatry. One especially useful example for the issues being examined here is the concept of *Darshan*. In Hinduism, *Darshan* is a gift or offering made by a superior to an inferior, and it consists of a 'gift of appearance' imagined as a material transfer of some blessing, and is thus a mode of divine agency and intimately connected to the concept of the evil eye. To place oneself before the idol of the god, therefore, is to lay oneself open to the divine gaze and to internalize the divine image.[19] Yet *Darshan* is a two-way affair: the gaze directed by the god towards the worshipper confers his blessing, and, conversely, the worshipper reaches out and touches the god. According to Gell, 'the result is union with the god, a merging of consciousness according to the devotionalist interpretation.'[20] This indicates a form of reciprocity and inter-subjectivity in the relationship between the image of the god and the recipient/worshipper. Gell argues, further, that the image of the god is a manifestation of a 'Social Other', and that the god/worshipper relationship is a social one, absolutely comparable to the relationship between the worshipper and another human person. The worshipper, however, does know that the image of the god is only an image, not flesh and blood, and if it should do something such as move or speak, that is considered a miracle.[21]

Another of Gell's points is that ritual animacy and the possession of life in a biological sense are far from being the same thing. Nevertheless, a worshipper who addresses prayers to a stone must believe, somehow, that the stone, though not a living thing, sees and hears like the worshipper, thinks and reacts as he does, and also has the power to plan and execute actions.[22] So although these acts might appear irrational, they are far from without meaning:

> 'It is surely irrational, or at least strange, to speak to, offer food to, dress and bathe a mere piece of sculpture, rather than a living breathing human being. And so it is: those who do these things are just aware of the "strangeness" of their behaviour as we are, but they also hold, which we do not, that the cult of the idol is religious efficacious, and will result in beneficial consequences for themselves and the masters they serve in their

capacity as priests. It is not that the priests cannot distinguish between stocks and stones and persons, rather, they hold that in certain contexts stock and stones possess unusual, occult, properties.'[23]

He argues, further, that social agents can be drawn from categories that are very different, because social agency is not defined in terms of basic biological attributes, but is relational: it does not matter, in ascribing social agent status, what a thing is in itself. What matters is where it stands in a network of social relations. All that is necessary to make an image a social agent is that there are actual human beings 'in the neighbourhood' of these inert objects, not that they are biologically human beings themselves.[24]

The last of Gell's points to be discussed here is concerned with the question of external and internal conceptions of agency. The external (or practical) aspects of agency attribution are formed by outer actions or conditions: language, practices, routines, rules, etc., and are more related to the body. The internal concept of agency attribution is related to inner aspects of the mind.[25] According to Gell, the simplest solution to the question of idolatry is an 'external' theory of agency, since idols are social others to the extent that, and because, they obey the rules laid down for idols as co-present others (gods) in idol-form. So although idols may not produce much visible behaviour, they may be very 'active' invisibly.[26] The externalist theory of attributing agency is easily applicable to the cult of idols in Ancient Greece. The washing and dressing of the images of the gods imposed agency on them by making them 'patients' in social exchanges, which necessarily imply and confer agency. As Gell contends, there is no make-believe in such performances: these rituals would be pointless if they were not literal transpositions of the means by which we induce agency in social others in human form. These acts are not symbolic: there was no alternative way to dress the gods. Thus, the acts of e.g. dressing a cult image were real, practical services performed for divine social others in image-form, not symbolic acts.[27] In sum, according to Gell, the essence of idolatry is that it permits real physical interactions to take place between persons and divinities.[28]

The concept of 'relationality'

Gell's theory of agency has been criticised by Whitehead on the grounds that agency, as it stands, is conceptually incomplete in terms of accounting for the roles of religious objects specifically, because the weight of its emphasis relies upon human intentionality.[29] Furthermore, she finds that agency in the way that Gell understood it is not capable of addressing the full relational potentialities of religious statues due to its emphasis on representation.[30] Instead, Whitehead proposes the concept of 'relationality', which builds on or is a form of Harvey's new animism,[31] which suggests that animals, objects, and other persons are animated when they are in relationship with another person.[32] Relationality thus asserts that the 'personhood' of religious statues is dependent on relational engagements where objects and subjects bring each other into forms of co-relational being through encounters.[33] Whitehead argues, further, that it is only in moments of active relating that statue persons come into being. The concept of relationality allows statues to contain 'spirits', be embodied with divine presences, be representative of divine presence, or be inert matter.[34] Thus, according to Whitehead, cult images are not just representations of the divine, but (can) embody participants in ceremony and ritual.[35]

God and image in Ancient Greece

The question of how the Ancient Greeks thought of their cult images is not easy to answer, since we obviously cannot ask them directly: was it a case of representation or embodiment?

It has been claimed that the equation between god and image appears to have been prevalent in the Archaic and earlier periods. Some scholars such as Hölscher, however, have maintained that there was a shift from an Archaic equation of god and image towards a separation of statue and god in later periods. In this view, the 'distant' god is separated from his statue in the 5th century BC, and is supposed to be an example of 'Götterferne' – a disturbed relation between gods and their worshippers.[36] This distinction between statue and god was often understood as a differentiation in a much wider sense, which would affect the history of religion: as a sign of the distance between statue and god that was to be equated with a separation of the human (in the cult statue) and divine spheres.[37] Yet I do not find Hölscher's view entirely satisfying. It seems that Greek literary sources tend to ignore the distinction between deity and image of deity,[38] and there are many examples in which gods appear to their worshippers in the form of their images.[39] Gods frequently appear in the form of their cult images especially in dream narratives, in part because the dream world facilitated the coalescence of deity and image that was always implicit in the form of cult statues themselves, allowing the living statue to be fully animate.[40] This perception of a correspondence between immaterial divine bodies and material images is exemplified by Artemidoros (2nd century AD), who states:

> 'Statues of the gods have the same meaning as the gods themselves.'[41]

Artemidoros also remarks, in relation to the goddess Artemis:

> 'It makes no difference whether we see the goddess herself as we have imagined her to be, or a statue of her. For whether gods appear in the flesh or as statues fashioned out of some material, they have the same meaning.'[42]

Similarly, in Aristides, the connectedness of transcendent deity and material image comes through in the *Sacred Tales* on occasions when the god takes the form of his statue.[43] In this work, statues of the gods, particularly Asclepios, repeatedly appear either as a focus of worship, as the identifying feature of a specific sanctuary, or as animated forms of the deities that they represent.[44] During incubation Aristides converses with Asclepios, who appears 'in the posture in which he is represented in his statues', and he receives surgical treatment from Serapis 'in the form of his seated statues'.[45] Another example can be found in Kallimachos, who in his seventh *iambos* plays with the idea of equating the god with his statue.[46] This text features an ancient cult statue of Hermes Perpheraios that speaks of its origins and the institution of its cult in the city Ainos in Thrace.[47] In ancient Greek, moreover, temple statues were frequently referred to as 'the deity' rather than 'an image'. This tendency not to distinguish between god and image on the level of language is found in many texts: for example, the Trojan statue of Athena in the *Iliad*, which is referred to as Athena herself, and references to Phidias' Athena as 'the goddess'.[48]

But, most importantly, in consideration of the theory of agency, especially Gell's theory of idolatry and Whitehead's arguments, and since it is clear from the archaeological, iconographical, and written evidence examined in this and the previous chapters that the Greeks did interact with the cult images (e.g. by washing and dressing them), it seems to me that there is no doubt that the Greeks thought of the cult images as embodying divinity.

According to Schnapp, for the Greeks, cult is inconceivable without images,[49] which underlines the importance of the cult statues. He argues, further, that, in Greece, the image is not only a category of figurative art, but also a means of approaching the divine and communicating with it, since the image materializes the divine.[50] And as Freedberg argues, acts of washing, dressing, and adorning an image might be habitual or conventional in nature, but, even if this is the case, they are still not sufficiently explained. Thus, people do not interact with cult images simply out of habit:

> 'they do so because all such acts are symptoms of a relationship between image and respondent that is clearly predicated on the attribution of powers which transcend the purely material aspect of the object.'[51]

As Platt argues, images thus have the potential to be viewed as 'epiphanic embodiments of the divinities they represent,' and they can simultaneously symbolise and constitute divine presence. Yet there appears to have been a 'double consciousness' whereby the viewer of a work of art moves between an apprehension of the entity in the image (the deity) and an awareness of the image's created nature and its status as an object (a statue). Thus, Greek viewers/worshippers 'were fully aware of the materiality of cult statues, but they were rarely impious enough to think of them as 'just' statues.'[52] Platt further argues that the most insistent, direct, and powerful means of experiencing divine encounter was the viewing of a physical cult image, usually a cult statue in a temple. In these cases, representation was not merely a symbolic actualisation of divine form, but a living embodiment of the divine.[53] The relation was never constant, however, and the bond/correspondence between gods and their cult statues was continually shifting and elusive, and the ritual model of image as epiphany (in which viewing a cult image was akin to encounter with the deity it represented) existed alongside other epiphanic models.[54]

Following upon these arguments, Ancient Greek cult images appear to have been treated and addressed in ways that indicate that they housed power. The ancient belief that ancient wooden cult statues, such as *xoana* and *bretades*, in particular, possessed special powers was widespread. For example, Lucian describes how the statues in a Syrian temple could sweat, move, and utter oracles.[55] They were sometimes believed to be dangerous and able to inflict blindness, madness, or sterility on their viewers.[56] Thus, Plutarch recalls that there was a *bretas* of Artemis at Pellene that was kept covered for most of the year, but on specific days it was carried in procession. During the procession, no one could look directly at the image, since its eyes caused terror and death and made trees sterile.[57] A further example is the statue of Artemis Orthia at Sparta, known as Lygodesma ('bound with willow'), which was considered extremely dangerous, and the two men who found it were afflicted with madness because of the encounter with her gaze.[58] This also illustrates that these cult statues were believed somehow to have the capacity to act.[59]

As already discussed, dressing could have the practical functions of protecting the cult statue from wear and tear and preserving its modesty. But dressing a statue could also have the opposite function of protecting the viewer.[60] As these examples demonstrate, one could be struck by several calamities simply by looking at a cult image. By covering the image, potential viewers would be protected.[61] In this way, the textiles serve the dual purpose of protecting cult image as well as viewers, and potentially function as a symbolic means of protecting against, and controlling, the supernatural world.[62] In this way, dressing can also be perceived as a way to achieve a form of control over the cult images and hence the deities themselves.[63]

In comparison, Fletcher has shown that, in Greek tragedy, garments can be associated with curses, an example being that the garment in which Agamemnon is captured and he dies in is what Clytemnestra calls 'an inescapable net … an evil wealth of clothing.'[64] Hittite texts demonstrate that spells can be cast by throwing a cloak over a person and that an individual can escape an oath or break a spell by removing his or her cloak.[65] Curses can therefore be understood as means of exercising control over somebody, in this case by the use of garments. If this pattern of thought is applied to cult images, the use of dress may represent an attempt to control the deities by 'binding' them with garments. This may be expressed in the depictions of cult images dressed in tightly wrapped garments, for example the Hellenistic and Roman reproductions from Asia Minor as well as in vase-paintings and on coins – all of which could be understood as a means of controlling or binding the deity. But the act of dressing also reflects an element of co-dependency. Although statues are relationally more powerful than devotees since they have supernatural power, devotees potentially have power over statues in terms of their care, restoration, and regeneration.[66] If the statues indeed needed garments, they would be dependent on human beings to provide them.

Dressing as concealment or revelation? The question of visibility and invisibility

A question arises as to whether the textiles used to dress the cult statues were intended to make the statues visible or invisible – were they meant to conceal or to reveal? This might seem like a strange question to ask, since the obvious interpretation of the act of dressing is that it has the effect of concealing, primarily the naked body. Yet dressing can also be used to enhance, highlight, or emphasize certain aspects of a body. For example, a textile wrapped tightly around a specific part of the body may, instead of concealing, have the effect of attracting attention to this particular area. In this way, dressing may make the contents (i.e. the statue) clearer or conceal it to the extent that it ceases to exist,[67] for example by dressing it in numerous garments or wrapping it in many layers of textile. Perhaps the reproduction of the cult image of Artemis from Perga discussed above is an instance of this.

In her study on the use of textiles for shrouding the dead, Gleba draws a clear distinction between items that were made invisible by wrapping and urns that were made more visible by being dressed like the dead person or decorated with ribbons, and thereby challenges the concepts of visibility and invisibility.[68] In other words, the act of wrapping or dressing may have been either a means of drawing boundaries between the living and the dead or a way of communicating between the living and the dead – or both.[69] Perhaps this was also the case with regard to the dressing of cult images, in that this particular act not only served to conceal the 'nakedness' of the statue, but also had other functions, which will be examined below. When dealing with the use of textiles to dress the cult images, it is therefore important not only to think of visibility/invisibility or naked/dressed, but to also consider the other potential functions of textiles.

The potential functions of textiles in cult and ritual

The preceding discussion on the dressing of cult images leads to six further points on the potential functions of textiles in ritual. First, when textiles were placed around the cult image, they literally came in touch with the deity represented, and thus became like a second skin.[70] Thus, textiles were potentially the closest thing to the god or goddess, thereby automatically becoming something special.

Second, textiles have certain properties that make them a particularly useful medium in cult and ritual. For example, they are light and flexible, and therefore can be wrapped around an image, such as a large cult statue. Rudy explains further:

> 'The formlessness of textiles is precisely what makes them so multivalent. Like liquids, they take on the forms of solids that contain or define them. Other properties of textiles, furthermore, contribute to their multivalence: textiles report and mark (sweat, blood, mutilation). They are usually opaque, a fact that makes them candidates for performing functions of ceremonial concealing and unveiling.'[71]

This leads to the third point, which is the ability of textiles to shape and transform what they surround. This ability can change the expression and thus the message of the cult image to the observing worshipper. The addition of a garment or a piece of textile can thus be used to alter the content it enwraps, to manipulate its perception by others, and to provide a surface with which to contain or convey emotions.[72] In this way, dressing may involve not only physical transformations, but also psychological and symbolic transformations of objects, places, and people – in this case cult images.[73] This is very important, since textiles were the only medium available to do this in antiquity, with the possible exception of the jewellery embellishments, which can, however, be considered part of dress.

A fourth and related point is the interchangeability of textiles. In contrast to permanent forms of adornment or decoration, such as, for example, the sculpted dress of a bronze or stone statue, textiles were impermanent. They could be used to dress the cult image to signal specific occasions, perhaps by employing specific garment types, colours, dress arrangements, or textiles with certain patterns or figurative decoration. At the end of a particular occasion, the textiles could be removed and/or replaced with other textiles. In this way, textiles may have served as a means of creating an impression that the cult image was transformable and changeable; this practice also speaks to the usefulness of textiles as a dynamic means of communication.

A fifth point is the ability of textiles to communicate identity. That dress was employed as a means of communicating and signalling aspects of identity has long been established in scholarship. In the case of human beings, the addition and removal of clothing allows different aspects of the person to be revealed and concealed according to the situation. Clothing can thus materialise aspects of a person, whether these are aesthetic, economic, or moral values, or other qualities such as charisma, power, or gender.[74] This was so for cult images too, for which the adding or removing of specific textiles or clothing items could signal specific properties or aspects of the divinity in question.

The sixth and final point is the functionality of textiles as a mediator between the sacred and the profane. Textiles are often said to mediate between the personal and the social, for which reason it has been argued that they can also be used to define the liminal passage toward sanctity.[75] And since textiles can mediate between categories such as the physical and the visionary, it can be argued that they also can denote the borders between the sacred and the secular.[76] Conversely, textiles can also construct boundaries to create interfaces not only between objects, subjects, and the world,[77] but also between the sacred and the secular.

This last point about the functionality of textiles as a mediator is related to a further possible potential of textiles, which is their ability to transmit power.[78] By this it is meant that the textile can act as a kind of medium through which certain characteristics or powers can be transferred. There are Judeo-Christian examples attesting to this phenomenon. In this respect, the Biblical story of the *haemorrhoissa* is quite interesting. A woman sickened by constant bleeding from the womb is healed by touching the garment of Jesus: 'She heard about Jesus, and came up behind him in the crowd and touched his cloak, for she said "if I but touch his clothes, I will be made well".'[79] What is most interesting is that she touches the textile, and there is no skin-to-skin contact between her and Jesus. The touch of the woman with the haemorrhage is also peculiar because the touch seems to have a strong effect on Jesus, and it seems that something 'flows away' from him.[80] The text even implies a power over which Jesus has no control. Baert proposes that this might be an expression of an archaic magical effect.[81]

Healing via garments is not an unprecedented phenomenon. For example, in the Acts of John, many brothers were healed in Ephesos by touching the garments of John,[82] and the Old Testament makes reference to leprous garments.[83] There are also Greco-Roman examples, such as Epidauros 7, which states that a disease can not only be transmitted to a cloth, but can also be transferred in full from the sufferer to the cloth.[84] In this way, the textile can also function as a sort of vessel. According to Lalleman, healing by touch is less typical of Hellenistic cults than of Judaeo-Christian healing practices.[85] Instead, Greek and Roman healing practices appealed to therapy, such as the bath therapies and dreams of the Asklepios cults.[86] This does not mean, however, that it did not take place, only that we do not have any solid evidence for this particular practice.

The transfer of power through textiles occurs not only between human individuals and the divine sphere. There are also examples of textiles transmitting power between people. An example is provided by Plutarch, who describes how a woman named Valeria Mesalla attempted to share in Sulla's power by taking a bit of nap from his mantle:

> 'As she (Valeria Mesalla) passed on along behind Sulla, she rested her hand upon him, plucked off a bit of nap from his mantle, and then proceeded to her own place. When Sulla looked at her in astonishment, she said: 'it's nothing of importance, Dictator, but I too wish to partake a little in thy felicity.'[87]

This indicates that Plutarch actually believed that the textile would transfer – would share with her – the power of the body that wore it.[88]

These testimonia have led Baert to conclude that 'Garments and by extension textiles are mysterious and multisensory threads connecting the glory of god to mankind.'[89] This obviously concerns Judeo-Christian religion, but it can also be said to apply to Ancient Greek religion in the sense that garments may provide a link between humans and the divine. In fact, it has been claimed that garments help to link the natural and supernatural world.[90] And, as the previous examples illustrate, textiles appear to have the capacity to transfer essences of the owner's body, such as the healing powers of Christ or the desirable power and fortune of the Dictator Sulla.

This means that the clothing of a body somehow constitutes that body.[91] This is a very important point in relation to the phenomenon of dressing Greek cult statues in garments, since this implies that the garment or textile surrounding the cult statue *IS* the cult statue and thus, as argued above, the deity it represents. In this way, the action of wrapping creates a relationship between the wrapping material and the contents, whether they are bodies or objects.[92] As Gell argues in relation to images of gods on the Marquesas Islands: 'whatever forms the wrapping, is itself made sacred as a result of this intimate contact'.[93]

Textiles also have the potential to serve as containers, since a wrapping keeps whatever has been wrapped apart from the external world, and keeps qualities or properties from being dissipated. The intent of wrapping a cult image in textiles, however, is not necessarily only to prevent certain qualities from being dispersed, but also to allow it to be dispersed into an enclosing form, a container – in this instance a textile.[94] Textiles used to dress cult images, then, became sacred in themselves, since they absorbed properties of the divinity that they enwrapped.

Textile as narrative

There is one more important property of textiles that merits discussion. As previously noted, ancient textiles could bear detailed figural scenes illustrating myths and stories of the glorious past, as did, for example, the *peplos* of Athena on the Acropolis, which, according to literary sources, bore scenes of the Gigantomachy,[95] as well as several of the Hellenistic copies of cult images from Asia Minor that are dressed in garments decorated with figurative scenes in friezes. This figurative decoration means that it was possible, quite literally, to read clothing based on the stories illustrated thereupon.[96] Such figurative messages could transform the bodies or cult images that they clothed or were understood to clothe,[97] thus providing the viewer with further clues as to how to understand, contextualize, and possibly approach the images. This, combined with their ability to be in direct contact with the cult image, makes textiles an obvious choice as e.g. a means of disseminating propaganda/a political ideology – like a giant picture book for all visitors to the sanctuary to see, readable to all, whether illiterate or not. In addition, the use of specific symbols could be a further way of expressing power through decoration. In this way, the garment used to dress a cult image could serve as an expression of politics and common identity. For example, the *peplos* of Athena on the Acropolis had great political relevance in the Panathenaic procession because of the images depicted on it, which were manifestations of the self-representation of the *Polis*.[98] The *peplos* was thus an expression of identity, which was also reflected in the city's building program, and served as a symbolic bond between the city and its main deity.[99] For this reason, the decision about what images should be depicted on the garment was not left to chance, but was probably made by political organs.[100]

The power of touch

The term 'sensual religion' refers to religious expressions that can be touched, felt, smelled, etc. Sensual religion is, however, problematic because of the fact that (modern) conceptions of religion are often defined in terms of transcendence or metaphysics. Therefore religious objects (such as cult images) are often defined as symbols or representations.[101] Yet, as demonstrated above, Greek cult appears to have been different, in that cult images had agency and were somehow understood as embodying the deity. This is very important as far as interactions between human beings and cult images are concerned, since this means that being close to the cult image also meant being close to the deity. By touching the image, one would actually, by proxy, touch the deity. But what does that mean for our understanding of the interaction between Greek cult images and the ancients? According to Whitehead, a statue that is touched frequently might play a more animistic, fetishist role to the devotees than one who is not, since touch aids in fostering relationships.[102] Furthermore, whether or not a statue can be touched and physically interacted with has an impact on the role that it plays contextually.[103] This means that the way in which the statue is displayed and whether or not it is physically accessible to worshippers must play a significant role in any understanding of how performances and interactions with the statue occurred. Unfortunately, we do not know whether or not the ancient Greek cult images were placed in ways that made them accessible to the visitor, or whether such close encounters were reserved for specific persons or cultic personnel.[104] We do know, however, from written sources, especially Pausanias, that at some Greek sanctuaries only the priest or priestess might enter the temple or see the statue of the deity,[105] while other temples were open to the public only once a year.[106] This indicates that access to temples was generally restricted. And even when we find bases for the original cult image *in situ*, we cannot tell whether this spot was accessible to worshippers, since the fact that worshippers were allowed into the temple does not necessarily mean that they were allowed close enough to the cult image to touch it.

This leads us back to the cultic personnel who dressed and/or washed the cult images, since they were those closest to the cult images, sometimes even seeing them 'naked', which made dressing the cult image a significant point of interaction. As Douny and Harris argue, undressing is not simply the reversal of dressing:

> 'the act of unwrapping is significant in itself and has its own outcomes. Unwrapping may refer either to a physical or conceptual revelation whereby knowledge is gained or secrecy exposed. The removal of wrappings and their application elsewhere may be a devise to accumulate and store the power of their contents.'[107]

Therefore, these individuals were, because of this privilege, possibly among the most important cultic personnel. This does not, however, appear to be the perception in the scholarship. This may be, again, a problematic result of modern perceptions of gender and ecclesiastical hierarchies, in which washing and dressing are women's tasks and thus, by extension, not as important as other tasks, such as performing the sacrifice of a bull, which are considered men's. On the contrary, this study has established that women were the closest to the cult images in their interactions, and thus that their role in cult and ritual was far more important than earlier believed. This changes the current, usually unspoken, belief that priestesses were less important and powerful than priests.

In sum, these chapters have laid out comprehensive evidence for the phenomenon of dressing cult images in textiles. This means that it is important to consider the possibility that statues which we are used to seeing undressed may once have been dressed. If this was indeed the case, then their appearance and their impact would have been much altered;[108] in fact, textiles may have been so integral to the cult images that their removal would have changed their nature entirely.[109]

This also means that – at least to some extent – we have to reconsider our conceptualization of Ancient Greek cult and religion, since the common perception seems to be that the relationship with the gods was 'distant' and that the interaction with and involvement of the cult image was limited. This relationship can be considered distant because rituals, especially sacrifices, took place at an altar placed outside the temple and hence at considerable distance from the cult statue and thus the deity. This is also the situation for the dedication of votive offerings, which could be placed in the sanctuary, often in considerable distance from the cult image. Conversely, the current study has shown that the Greeks DID in fact interact with cult images through the use of textiles, and that this interaction carried particular significance. Consequently, textiles appear to have been among the most important material agents in the performance of Ancient Greek ritual.

Notes

1. E.g. Paus. 5.17.1; Plin. *Nat.* 35.45.
2. Weddle 2010, 60.
3. Mansfield 1985, 442.
4. Romano 1988, 130.
5. Weddle 2010, 69.
6. Whitehead 2013, 28.
7. E.g. Latour 2005; Callon 1986.
8. Bille & Sørensen 2012, 57–61.
9. Gell 1998, 16, 19.
10. Bille & Sørensen 2012, 62.
11. Gell 1998, 20–21; Bille & Sørensen 2012, 63.
12. Gell 1998, 63; Bille & Sørensen 2012, 63.
13. Bille & Sørensen 2012, 63.
14. Gell 1998, 97–98.
15. Gell 1998, 115.
16. Gell 1998, 99.
17. Gell 1998, 106.

18. Gell 1998, 105
19. Gell 1998, 116.
20. Gell 1998, 117.
21. Gell 1998, 117.
22. Gell 1998, 122.
23. Gell 1998, 122.
24. Gell 1998, 123.
25. Gell 1998, 126–127.
26. Gell 1998, 128.
27. Gell 1998, 134–135.
28. Gell 1998, 135.
29. Whitehead 2013, 111.
30. Whitehead 2013, 139.
31. Harvey 2005.
32. Whitehead 2013, 181.
33. Whitehead 2013, 5.
34. Whitehead 2013, 181.
35. Whitehead 2013, 184.
36. Hölscher 2010, 112.
37. Hölscher 2010, 118.
38. Tanner 2001, 262.
39. Platt 2011, 12.
40. Platt 2011, 258.
41. Artem. 2.39. Translation: Schnapp 1988, 573. According to Platt, however, this relationship between god and image was not always so straightforward, and we should not take Artemidoros' statement at face value. She argues: 'despite that fact that image and prototype have the same general import for the dreamer in Artemidorus' interpretative model, any details that draw attention to the god's status qua image necessarily generate a further layer of signification that, as he fully recognises, must then be subjected to further analysis.' She further argues that Artemidoros introduces a Platonic notion of mimetic hierarchy (the statue is only an imitation of the deity, after all, and thus less efficacious in prophetic terms), which contradicts the model of phenomenological unity suggested by his statement that statues of the gods have the same meaning as the gods themselves. Platt 2011, 259, 283–284.
42. Artem. 2.35. Translation: Platt 2011, 284.
43. Platt 2011, 263.
44. Platt 2011, 261
45. Platt 2011, 262; Aristid. *Or.* 50.50, 49.47.
46. Petrovic 2010, 211.
47. Petrovic 2010, 208.
48. Platt 2011, 78. Hom. *Il.* 6.263–311; Ar. *Av.* 667–70; *Eq.* 1168–70; Thuc. 2.13.5.
49. Schnapp 1988, 569.
50. Schnapp 1988, 568, 570.
51. Freedberg 1989, 91.
52. Platt 2011, 47, 50.
53. Platt 2011, 77.
54. Platt 2011, 122.
55. Luc. *Syr. D.* 10.
56. Petrovic 2010, 210.
57. Plut. *Arat.* 32. Freedberg 1989, 33.

58. Paus. 3.16.7–11. Freedberg 1989, 75.
59. See e.g. Mitchell 2010, 266.
60. This was also so for divinities revealing themselves to human beings. As an example, the Homeric *Hymn to Aphrodite* (181) tells the story of Aphrodite and Anchises. When he awakes from sleep, Aphrodite is dressed once more after their amorous encounter. Though her beauty is conveyed only by her neck and beautiful eyes, Anchises must hide his face even from this partial revelation of her form. This implies that textiles could conceal the deity and thus protect the viewer. Platt 2011, 69.
61. An example of the use of garments for supernatural protection is in Homer, where Aphrodite uses her shining *peplos* to shield Aeneas in battle. Hom. *Il.* 5.311–317. Fletcher forthcoming, 25. Similarly, Apollo uses the aegis to cover Hektor's body when it is being dragged by Achilles. In this case, the aegis functions as a protective shelter, equivalent to a shield. Hom. *Il.* 24.18–21. Deacy & Villing 2009, 115.
62. Douny & Harris 2014, 29.
63. Douny & Harris 2014, 9.
64. Aesch. *Ag.* 1382–1383. Translation: Fletcher forthcoming.
65. Fletcher forthcoming, 2.
66. Whitehead 2013, 184.
67. Douny & Harris 2014, 4.
68. Gleba 2014, 143.
69. Douny & Harris 2014, 31.
70. Baert 2007, 242.
71. Rudy 2007, 20–21.
72. Douny & Harris 2014, 6.
73. Douny & Harris 2014, 6.
74. Douny & Harris 2014, 24.
75. Rudy 2007, 22.
76. Rudy 2007, 35.
77. Douny & Harris 2014, 4.
78. Baert 2011, 324.
79. Mark 5:24b-34.
80. Baert 2011, 311.
81. Baert 2011, 312.
82. Baert 2011, 312.
83. *Leviticus* 13.47–59. Fletcher forthcoming, 3.
84. Baert 2011, 312; Theissen 1983, 134.
85. Lalleman 1997, 360–361.
86. Baert 2011, 312.
87. Plut. *Sull.* 35. Translation: Perrin 1920.
88. Baert 2011, 324.
89. Baert 2011, 311.
90. Baert 2011, 310.
91. Baert 2011, 312.
92. Douny & Harris 2014, 3.
93. Gell 1993, 179.
94. Gell 1993, 179. Similarly, during the annual cleaning of the niche containing the statue of Christ in the Cathedral of Valetta, cotton wool wadding is dabbed onto Christ's bleeding wounds, so as to 'soak up' his power. The sick then keep the wadding to dab on their own wounds in the hope of being healed. Mitchell 2010, 272.

95. See Appendix 1.
96. A further example of the use of textiles as narrative is the myth of Prokne. Her sister Philomela was raped by Prokne's husband Tereus, who tore out her tongue to prevent her from revealing the crime. Philomela, however, wove a tapestry illustrating what she had suffered. Ov. Met. 6, 412, 504.
97. Warr 2010, 15.
98. Reuthner 2006, 300.
99. Reuthner 2006, 323.
100. Reuthner 2006, 301. Arist. [Ath. Pol.] 49.3, tells us that at one time the designs of the *peplos* had been judged by the Council (*boulē*), but that in his own day this was done by the jury court (*dikastērion*).
101. Whitehead 2013, 117.
102. Whitehead 2013, 146.
103. Whitehead 2013, 145.
104. According to Hewitt, sometimes the cult image and presumably the chamber in which it stood were only accessed by the priests, except at a certain time or from a distance, while sometimes the restriction was waived for a certain class or applied to one sex only. Hewitt 1909, 84. Corbett (1970, 154) argues that there was a wide range of variation in the extent to which the temples were accessible to ordinary citizens and that we cannot accept as a general rule that entry by ordinary people into the *cellas* of temples was exceptional.
105. Corbett 1970, 156. Paus.:
    2.10.2: Only priests were allowed in the inner part of the Temple of Apollo Karneios at Sikyon;
    2.10.4: The Temple of Aphrodite at Sikyon was only entered by a woman who was barred from sexual intercourse and a virgin with the title of bath-bearer (*loutrophoros*) who held the office as priestess for one year.
    2.35.11: In the temple of Eileithyia, only the priestess might see the statue;
    3.20.3: In the Temple of Dionysos at Bryseia in Laconia, only women might see what was inside, since they were the ones performing the secret rites;
    6.20.3: The inner part of the Temple of Sosipolis at Olympia was entered only by the woman who served the goddess;
    8.36.6: Only women were permitted to enter the Temple of Demeter near Megalopolis.
106. Paus. 9.25.3: the Precinct of the Didymaean Mother was only open once a year;
    8.41.4–6: the Shrine of Eurynome was only open once a year;
    10.35.7: the Shrine of Artemis at Hyampolis was open twice a year.
107. Douny & Harris 2014, 3–4. See also Gell 1993, 89.
108. Douny & Harris 2014, 17.
109. Douny & Harris 2014, 24.

Part IV

Sacred dress codes

Chapter 8

Introduction: Sources and methodological discussion

In addition to the temple inventories and the different testimonia on the dressing of cult statues, another important aspect of dress in sanctuaries is the garments worn by priests and priestesses, as well as those worn by regular people when visiting a sanctuary. This subject has not yet received significant treatment in the scholarly literature for two reasons: the limited sources available and the difficulty of drawing conclusions from these sources because of their geographical and chronological spans. Thus, previous studies have focused primarily on a single inscription that describes the prescribed attire for entering a sanctuary or briefly refers to singular aspects (e.g. the colour) of priestly garments. Attention has focused especially on a famous inscription from the Sanctuary of the Great Goddess at Andania on the Peloponnese.[1] Thus far, only few studies has attempted to conduct a thorough study – including relevant inscriptions, literary sources, and iconography – to determine what conclusions can be drawn about the garments worn by individuals in a sanctuary. This chapter will fill this gap in the scholarly literature by examining diverse sources in order to determine what forms of dress were considered appropriate for a person visiting a sanctuary or performing a rite.

First, I will examine the evidence for priestly garments in iconographic, then epigraphic and literary sources. Second, I will discuss iconographic evidence for the dress of sanctuary visitors, as well as clothing regulations in epigraphic and literary sources that are aimed at sanctuary visitors. Finally, these diverse sources will be analyzed with an eye to the evidence that they can provide for sacred dress-codes and their effect on the ritual experience of the individual when in a sanctuary.

Priests and priestesses in Greek religion

Before we come to the evidence, it is necessary to discuss nomenclature and definitions. Scholars of antiquity and Greek religion have tended to designate some or all Greek religious officials, functionaries, and agents of the divine indiscriminately as 'priests' or 'priestesses'.[2] As Henrichs explains, part of the problem with this designation lies in the fundamental discrepancy between the word 'priest' and its modern connotations of personal religious vocation or sanctity and the wide range of cult-related public and private offices known from antiquity.³ In English 'priest' is thus used as a translation for all of the different Greek terms for cult agents, which implies 'a potentially misleading unity of conception and an analogy with the roles of priesthood in later religions',⁴ but in fact, our term priest does not translate into a single Greek word.⁵ Since, furthermore, the term 'priest' is 'derived from a monotheistic religion with an exclusively male and hierarchical clergy, the very concept of priest suggests a uniformity of identity and function that contrasts sharply with the inherent diversity of Greek religion.'⁶ The modern priesthood differs greatly from Greek religious offices, which were not restricted to men and commonly included priestesses.⁷

The French philologist Dumezil and the linguist Benveniste introduced a theory that Indo-European cultures exhibited a tripartite social division into priests, warriors, and peasants⁸ – a theory that has had a great influence on our perception of ancient Greek religion and the idea of a priestly 'caste'. Other scholars subsequently argued against this tripartite division of ancient Greek society. Thus, in contrast to Dumezil and Benveniste, Bremmer contends that the Indo-Europeans had neither a separate priestly class nor a specific term for priests or priestesses, and that every city could develop its own organisation and titles. He argues, furthermore, that, despite the fact that an established order (priestly class) was lacking in Greek society, we (i.e. modern scholars) still tend to impose our own Judeo-Christian ideas of priesthood on the Greeks. Thus, we are inclined to think of a priest as one particular kind of person, while the ancient Greeks would have thought of several different kinds of religious officials that we may designate 'priests' or 'priestesses', but who cannot, in fact, be equated with present-day priests, who have different duties and training.⁹ So, while the Catholic and Protestant priesthoods have remained identifiable and fairly strictly defined offices, ancient Greek priesthood was characterised by a flexibility that could adapt continuously to new circumstances.¹⁰

Modern studies of Greek religion ordinarily refer only to *hiereis* as priests and *hieraiai* as priestesses.¹¹ The title *hiereus* is thought, first and foremost, to denote one who is in charge of the *hiera*, that is, the sacred objects stored in a sanctuary *hieron* and the sacred rites connected with a cult.¹² There is, however, a wealth of words in Greek that designate cult personnel,¹³ and to consider only this designation would therefore be wrong. There are dozens of designations of religious personnel in local cults: Chaniotis mentions a few, including the *kistiokosmos* ('adorner of the basket') of Athena in Messene,¹⁴ the *purphoros hieras nyktos* ('torchbearer of the sacred night') in the Sanctuary of Apollo Maleatas in Epidauros,¹⁵ the *diabetria* of Artemis at Mopsuestia, the *kynēgoi* ('hunters') of Artemis in Ephesos, the *loutrophoros* ('water bearer') in the service of Artemis Boulaia at Miletos and Artemis Kindyas at Bargylia, and the *neokoros* ('temple warden') in the cult of Aphrodite at Sicyon.¹⁶ These examples are listed here simply to give an idea of the diversity of the many different titles and functions of cult officials in ancient Greek religion. These offices were usually not life-long, and priests and priestesses were appointed for limited periods. Furthermore, it should be kept in mind that each sanctuary could have different cult officials with different titles and functions, and that the administration and control of the sanctuary was thus not necessarily reserved for only one person.

It has been claimed that one of the most fundamental functions of the *hiereus/hiereia* was to conduct sacrificial rituals, and some scholars therefore interpret them primarily as slaughterers or performers of animal sacrifice. Greek cult officials were, however, more than mere butchers, and they also play other important roles.[17] This is reflected in the many different designations for cultic personnel described above, which demonstrate that, for example, bearers of different objects for the cult, temple wardens, etc. also played a very important part in several cults.

The question of nomenclature remains, though: what is the best designation for these officials? One argument that speaks in favour of a continued use of the term 'priest' is that it, like similar key words such as religion, ritual, and cult, has become part of established scholarly parlance and conventional terminology.[18] I will therefore use the words 'priest' and 'priestess,' as well as 'priestly personnel' (as formulated by Chaniotis), in a general sense and not only as the translation of *hiereus* or *hiereia*, but as the designation for the personnel primarily responsible for a cult.[19] I will also employ the more general category of 'cult personnel,' which includes all persons responsible for cult activities who held a certain office that was not solely administrative in function.[20]

Notes

1. E.g. Gawlinski 2011; Themelis 2004, 2007; Deshours 2006.
2. Henrichs 2008, 1.
3. Henrichs 2008, 1.
4. North 1996, 1245.
5. Dignas 2003, 40.
6. Henrichs 2008, 4.
7. Henrichs 2008, 2.
8. Dumezil 1958; Benveniste 1973.
9. Bremmer 2008, 37.
10. Bremmer 2008, 52.
11. Henrichs 2008, 7.
12. Garland 1990a, 77.
13. Henrichs 2008, 5.
14. *SEG* 39:381.
15. *SEG* 39:358.
16. Chaniotis 2008, 27.
17. Henrichs 2008, 3.
18. Henrichs 2008, 1.
19. Chaniotis 2008, 19. See also Mylonopoulos 2013, 123.
20. Hoff 2008, 108.

Chapter 9

Priestly garments

One method of identifying a representation of a priest or priestess is to search for certain priestly accessories and attributes or inscriptions that securely identify the individual portrayed as a priest or priestess.[1]

Iconographic representations of priestly dress

Priestesses

The *korai* from the Athenian Acropolis are often mentioned in connection with cult identity, and it has been suggested that they may represent *arrēphoroi*, *kanēphoroi*, or priestesses.[2] They do not, however, have any attributes or epitaphs identifying them as cult personnel, and for this reason it has also been suggested that they simply represent female votaries.[3] Their contribution to our knowledge of priestly dress is therefore limited.

Representations of women who can be identified as priestesses with some certainty appeared sporadically in Greek art in the Archaic Period, but from the Classical Period onward they became more frequent.[4] According to Pilz, civic bodies only reluctantly honoured female priestly personnel with portrait statues, and publicly decreed honorary statues of such individuals first appeared in Athens and elsewhere during the 2nd century BC. Privately erected statues of priestesses, though, are attested as early as the first half of the 4th century BC.[5]

The symbol of female priestly office par excellence is the so-called temple key, a large metal bar bent at right-angles (Fig. 76).[6] Temple keys appeared as a priestly attribute in the late 6th century BC,[7] and they are attested in numerous sanctuaries and burials in southern Italy, with a particular concentration in Lucania.[8]

Fig. 76. Temple key of bronze. 5th century BC. Photograph: © Boston, Museum of Fine Arts, inv. no. 01.7515.

Fig. 77. Woman by the name Polystrate with a temple key. Attic funerary stele, ca. 380–370 BC. Kerameikos Museum, Athens, inv. no. I.430. © Hellenic Ministry of Culture, Education and Religious Affairs. Ephorate of Antiquities of Athens. Photo: S. Mavrommatis.

Priestesses with temple keys are depicted in several media: for example, on ten or 11 Attic grave reliefs and two marble funerary *lekythoi* from the 4th century BC (Fig. 77).[9] The priestesses are identifiable solely because of the temple key that they carry, but there are no attributes or inscriptions that indicate which divinity they served. Limestone statues of priestesses with temple keys have also been recovered from sanctuaries of Aphrodite on Cyprus.[10] The priestesses carrying temple keys are usually dressed in the same type of outfit: a *chitōn* belted high beneath the bosom and tied with a Herakles knot, over which is worn a *himation* pulled from under the right arm, across the front, and over the left shoulder. Cypriot priestesses are veiled, while Attic are not.[11] Identification of these women as priestesses is somewhat complicated by the fact that goddesses are also depicted with temple keys, as on a Greek votive relief of unknown provenance depicting a seated Demeter with Kore behind holding a temple key (Fig. 78).[12] In some instances, then, it can be uncertain whether the woman portrayed is a priestess or a divinity.

Some vases depict priestesses with temple keys, often in a setting inspired by theatrical and dramatic performances. Such scenes were popular in South Italian vase-painting. Here the priestesses with temple keys are often rendered as old, with white hair and wrinkles, and are typically clad in a *chitōn* and mantle, but sometimes in a *peplos* (see Fig. 55).[13] The combination of *chitōn* and mantle is also known from representations of

Fig. 78. *Goddess with temple key. Votive relief, ca. 425–400 BC. J. Paul Getty Museum, inv. no. 73.AA.124. Digital image courtesy of the Getty's Open Content Program.*

mythical priestesses such as Theano, Iphigenia, and Pythia in South Italian vase-painting.[14] The type of dress is always white and often decorated with patterns in a dark colour along the hem or in a band down the front (Fig. 79).[15] This type of decoration with a stripe along the hem of the garment is, however, not restricted to priestesses, but also appears in the dress of women in e.g. wedding scenes. There are also examples of key-bearing priestesses wearing elaborate decorative garments in Attic, Apulian, and Campanian vase-painting, but the women portrayed in these instances might be goddesses, not priestesses. This

Fig. 79. Boreas seizing Oreithyia at and altar. Priestess fleeing and temple key on the ground. Red-figure South Italian krater, ca. 350–340 BC. © Trustees of the British Museum, inv. no. 1931,0511.1.

identification is supported by the fact that these women seem to be placed in mythological settings: for example, in the company of other gods recognisable from their attributes.[16] For example, on a Campanian amphora a woman holding a temple key and wearing highly ornate garments – including a *polos* and a veil – is depicted in front of a temple, and is surrounded by other divinities (Fig. 80).[17]

To return to the Attic funerary *stelai*: only one depiction of a priestess is securely identifiable as belonging to a specific cult. The woman, who is named Chairestrate, is securely identified as a priestess of Kybele on the basis of both her attributes and her epitaph. She holds a temple key, and a young girl stands in front of her holding a *tympanon* (a 'kettle-drum'), which is considered the most important attribute of Kybele (Fig. 81).[18] On the basis of this stele, scholars have identified other female figures with *tympana* on contemporary *stelai* as priestesses of Kybele, and have come to consider the *tympanon* a priestly attribute. But *tympana* are also depicted with other figures, divine or not, and may not always point to the cult of that particular goddess.[19] There are three such examples of priestesses with *tympana* on Attic gravestones.[20] They are all dressed in *chitōnes* and *himatia*, like the priestesses with temple keys.

Xoanon-bearers are another possible visually recognisable representation of priestesses. They are described in literary sources, and depicted in vase-painting and sculpture, but they are absent from Attic funerary art. One example is a Boeotian grave stele, dating to the beginning of the 4th century BC (Fig. 82). It depicts the priestess Polyxena holding a small statuette, probably the *xoanon* of the divinity she served. She is clad in a *peplos* with the overfold pulled over her head so as to create a veil.[21] A final example is a marble *kore* wearing a belted *chitōn* from Miletos, which is equipped with a *lituus*-like staff to illustrate her cult function.[22]

If recognisable attributes such as temple keys and *tympana* are not present, it can be impossible to identify a statue as a representation of a priest or priestess. In some rare instances, however, we find marble portrait statues together with their inscribed bases, which enable us to identify the status of the person represented. These statues primarily belong to the Hellenistic Period. This is due to several factors: first, statues from the Classical Period were often made in bronze and have therefore not survived, while marble statues were common in

Fig. 80. Woman with temple key. Campanian amphora, 4th century BC (after Quercia forthcoming, fig.7).

the Hellenistic Period. In addition, inscriptions specifying the titles of priest or priestess did not come into wide use until the Late Classical and Hellenistic Periods, and the term *hiereia* is not used in dedicatory inscriptions for portrait statues until the first half of the 4th century BC.[23]

One of the most famous examples of an identified priestess is the marble statue of Nikeso, dating to ca. 300–250 BC, which was erected at the entrance to the Sanctuary of Demeter and Kore at Priene (Fig. 83).[24] The inscription on the base informs us that Nikeso was a priestess (*hiereia*) of Demeter and Kore.[25] She is dressed in a *chitōn* and a double-hemmed *himation*.

At the Sanctuary of Nemesis at Rhamnous, Attica, a marble statue of the priestess Aristonoe from the 3rd century BC was found, together with its inscribed base. The inscription designates the statue as a dedication to Themis and Nemesis (Fig. 84). The priestess wears a *chitōn*, a voluminous *himation*, and a rounded *strophion* in her hair.[26]

Fig. 81. Attic funerary stele depicting a woman by the name Chairestrate and young girl holding a tympanon. Archaeological Museum of Piraeus, inv. no. 3627. © Hellenic Ministry of Culture and Sports.

At the Temple of Artemis at Aulis, Boeotia, several statue bases with inscriptions recording the names of young women who served in the sanctuary were recovered in their original placement in the cella. In one instance, a statue dating to the Roman Period was found with its base.[27] This statue represents Zopyreina, a priestess of Artemis, who is clad in a *chitōn* and *himation* (Fig. 85).[28] Other contemporary female statues clad in similar *chitōnes* and *himatia* have been found at the sanctuary, and possibly also represent priestesses.[29] It is worth noting that these women wear their *himatia* in two different ways: Zopyreina and another statue wear their *himatia* like the 'Small Herculaneum Woman Type', while the others wear it under their right arm, across the abdomen, and over the left arm and shoulder.

Other statues identifiable as priestesses and female initiates have been recovered at the Sanctuary of Artemis Orthia at Messene.[30] Eleven statue bases (five of them inscribed) – to which eight statues can be matched – were found *in situ*, placed in a circle in the cella. Themelis has classified the eight statues in two groups: 1) at least five life-size statues, primarily dated to the 1st century BC and representing young girls; and 2) three smaller than life-size statues of adult priestesses.[31] Four of the statues depicting girl initiates date to the 1st century BC. They all wear a *peplos* with a long overfold fastened by a belt, which is tied beneath the bosom (Fig. 86).[32] One of the statues might be of a slightly later date, and is clad in a *chitōn* under a heavy *peplos*, which is fastened with a belt tied in a Herakles knot. Three of the statues of the young girls can be associated with inscribed bases, which inform us that the statues were dedicated by the girls' parents to Artemis Orthia.[33] The three statues of the mature priestesses date to the 2nd and 3rd centuries AD. They stood on cylindrical bases bearing inscriptions with their names.[34] The statue of the priestess

9. *Priestly garments* 279

Fig. 82. Xoanon-bearer. Boeotian funerary stele, ca. 400 BC. © Staatliche Museen, Antikensammlung, inv. no. SK 1504. Photo: Johannes Laurentius.

Fig. 83. Statue of the priestess Nikeso from Priene, ca. 300–250 BC. © Staatliche Museen, Antikensamlung, inv. no. SK 1928. Photo: Johannes Laurentius.

Fig. 84. Statue of the priestess Aristonoe from Rhamnous, 3rd century BC. National Archaeological Museum, Athens, inv. no. 232. © Hellenic Ministry of Culture, Education and Religious Affairs/ Archaeological Receipts Fund. Photo: Eliades.

Eirana Nymphodotos wears a *chitōn* with buttoned sleeves and, over that, a *himation* (Fig. 87).³⁵ The two others wear a sleeved *chitōn* under a *peplos* and *himation* (Figs. 88–89).³⁶

From Asia Minor, we know of (at least) six funerary *stelai* representing priestesses of Demeter, which are recognisable because of the large torch and the poppies and/or ears of corn that they carry in their hands. They are all probably from Smyrna and belong to the 2nd century BC.³⁷ The priestesses all wear a *chitōn* belted high beneath the bosom underneath a *himation*. In one case, however, the *chitōn* has long sleeves and a neck seam (Fig. 90).³⁸ In addition, one of the priestesses wears a small veil.³⁹

Priests

The representations of priests appear in the 6th century BC. The most well-known attribute of the male priesthood is the sacrificial knife, rather than the temple key, since men were usually responsible for the leading and butchering of animals in sacrificial rites.⁴⁰ Several specimens of such knives have been preserved in Greek sanctuaries,⁴¹ and they are also attested in, for example, Athenian temple inventories.⁴² Priests are thus occasionally identifiable on the basis of this attribute, especially on Attic grave *stelai* and funerary *lekythoi*, where the priests are dressed in long short-sleeved ungirded *chitōnes*. An example is the grave stele of the priest Simos, dated to ca. 400 BC, who is depicted holding a sacrificial knife and wearing precisely such a

Fig. 85. Statue of Zopyreina, priestess of Artemis from Aulis. Archaeological Museum, Thebes, inv. no. BE 66. © Ephorate of Antiquities of Boeotia/Archaeological Museum of Thebes, Greece.

Fig. 86. Four statues of young girls from the sanctuary of Artemis Orthia at Messene, 1st century BC. Messene Museum, inv. nos. 241, 244, 245, 246. © Society of Messenian Archaeological Studies

Fig. 87. Statue of Eirana Nymphodotos. Messene Museum, inv. no. 242. © Society of Messenian Archaeological Studies

Fig. 88. Statue of Klaudia Siteris. Messene Museum, inv. no. 243. © Society of Messenian Archaeological Studies

Fig. 89. Statue of Kallis Aristokles. Messene Museum, inv. no. 240. © Society of Messenian Archaeological Studies

garment (Fig. 91).[43] In vase-painting, however, there are depictions of men standing by altars and holding sacrificial knives, but these men are clad not in a long ungirded *chitōn*, but in a piece of cloth wrapped around the lower body, leaving their torsos bare. An example is a cup dated to the middle of the 5th century BC, which depicts a man holding a large sacrificial knife and wearing a decorated cloth with a border down the front wrapped around his lower body (Fig. 92).[44] This indicates that the sacrificial knife did not necessarily go with a specific garment, as in a specific priestly outfit, but could be used by males dressed either in long ungirded *chitōnes* or in wraps around their lower body. In addition, it should be noted that there are also rare instances in vase-painting of female figures holding sacrificial knives.[45]

Fig. 90. Priestess of Demeter. Funerary stele from Smyrna, 2nd century BC. Staatliche Museen, Antikensammlung, inv. no. SK 767. Photo: Johannes Laurentius.

In contrast with the numerous examples of priestesses, there is only one example of a priest depicted with a temple key: a *krater* depicting a mythological and epic scene, possibly the priest Chryses imploring Agamemnon (Fig. 93).[46] The priest is clad in a heavily decorated garment with long sleeves and a *himation* with a stripe along its edge.

The long ungirded *chitōn* described above has been identified as the typical dress of Greek priests.[47] Priests dressed in such garments are depicted on several media beyond grave markers. Mantis provides a catalogue of 18 iconographic depictions of priests wearing long ungirded *chitōnes* dating to the Classical Period.[48] The priests are depicted on reliefs, grave *stelai*, grave *lekythoi*, and in rare instances in vase-painting.[49] Some of the priests carry a sacrificial knife,[50] while others hold a *kantharos*[51] or a wreath,[52] and yet others are empty-handed.[53] Finally, there is the very famous example of the male figure folding the *peplos*, interpreted as the high priest of Athena, on the east frieze of the Parthenon (Fig. 94).[54] Another example is a marble votive relief from Eretria dating to the 4th century BC, which depicts Herakles seated in a cult building and attended by a man – who is identified as a priest because of his long *chitōn*.[55] The priests' garments are usually undecorated, but in some instances they bear rich decorations (Fig. 95).[56]

Fig. 91. Funerary stele of the priest Simos, ca. 400 BC. National Archaeological Museum, Athens, inv. no. 772. © Hellenic Ministry of Culture, Education and Religious Affairs/Archaeological Receipts Fund. Photo: K. Konstantopoulos.

Seers

A seer was a professional diviner, an expert in the art of divination who performed 'the craft of divination' – the art of interpreting signs sent by the gods e.g. *augury*,

Fig. 92. Man with sacrificial knife. Kylix, ca. 475–425 BC. Museo Archeologico Etrusco, Florence, inv. no. 4224. © Soprintendenza Archeologia della Toscana - Firenze.

extispicy, and *empyromancy*.⁵⁷ The term for 'seer' in Greek is *mantis*, which is used to designate persons who dealt with a broad range of religious activities.⁵⁸ Seers are sometimes recognisable on the basis of certain attributes – such as the entrails of sacrificial victims. The reclining male figure on the east pediment of the Temple of Zeus at Olympia, dated to ca. 460 BC, has been interpreted as a seer (Fig. 96). He is depicted with a beard, a naked torso, and a lower body wrapped in a mantle. A portrait statue of a man from the family of the Branchids, recovered at Didyma, may also be a seer, but has a very different appearance.⁵⁹ The statue – which dates to the mid-6th century BC – depicts a seated, bare-foot man holding a sceptre and clad in an unusual mantle, which is closed down the front (Fig. 98).⁶⁰ The garment, the sceptre, and the find context have led to the identification of this figure as a seer from the Sanctuary of Apollo.⁶¹ The sceptre, however, was not restricted to seers, but was also used by royalty and divinities. Men wearing this type of garment are also represented in vase-painting: for example, on a cup dated to the first half of the 5th century BC, which depicts a bearded male standing by a cauldron in a setting that may indicate a sanctuary (Fig. 97).⁶² The figure wears a long, white, short-sleeved garment, which is closed down the front and features a coloured border along the neckline, opening, and sleeves.

Women could also be seers, and are also represented in iconography. A famous example is the 420–410 BC funerary relief from Mantineia that depicts a female seer holding a liver in her hand and wearing a belted *peplos* (Fig. 99).⁶³

Headgear

Certain types of headgear may also be associated with priestly status,⁶⁴ the most significant being the so-called *strophion* – a headband of twisted cloth. The usual

Fig. 93. Priest with temple key. Apulian volute krater, ca. 350 BC. © Musée du Louvre, inv. no. CA 227. Photo: H. Lewandowski.

colour was white, but it is also described as white with purple stripes and gold.[65] Priests wore it as a sort of wreath, while priestesses used it to fix their coiffure, but although the *strophion* very often was a priestly attribute, it does not necessarily define a priest, since secular individuals wore it as well.[66] It has been suggested that it was the Eleusinian priests in particular who were distinguished by the *strophion*.[67] According to preserved media, however, the headgear of the Eleusinian priests was different: in vase-paintings they usually wear a wreath,[68] and in votive reliefs they wear *strophia*. The large votive relief that the priest Lakrateides dedicated ca. 100 BC depicts the priest himself and the Eleusinian deities.[69] Unfortunately, the relief

Fig. 94. High priest, the Parthenon frieze. © Trustees of the British Museum, inv. no. 1816,0610.19.

is fragmented, and it is not possible to tell what Lakrateides was wearing, besides a *strophion* (Fig. 100). The early Antonine votive relief of the *hierophant* from Hagnous is more well-preserved (Fig. 101).[70] Here, the priest is dressed in a short *chitōn*, *himation*, decorated boots, and a *strophion* and wreath. Clinton suggests that figures represented wearing similar headgear can therefore be identified as priests.[71]

One final example is a votive relief from Kertch, dating to the 5th century BC (Fig. 102), which depicts the goddesses Demeter and Kore.[72] To the right of the deities stands an Eleusinian priest wearing a short *chitōn* and *strophion*, and carrying two torches. For Eleusis itself, we have further information about the appearance of the priests. The priesthoods here were filled by individuals from the Eumolpidai and the Kerykes – the two ancient clans of Eleusis, which until late antiquity had a monopoly on the two main priesthoods of the Eleusinian mysteries: the Eumolpid *hierophant* and the Kerykid *dadouchos* (torch-bearer).[73] These iconic priests are depicted in vase-painting. Often the mythical founders of the two clans – Eumolpos and Keryx – are depicted. On vases from the first half of the 5th century BC, these priests are usually illustrated as bearded men clad in a long *chitōn* and *himation* (Fig. 103),[74] while in later examples they are represented as younger beardless men wearing *ependytēs* and high Thracian boots (Fig. 104).[75]

Fig. 95. Man performing libation at altar. Amphora, ca. 450–400 BC. © Hessisches Landesmuseum, Darmstadt, inv. no. A1969.4.

Priestly dress in epigraphic and literary sources

Inscriptions and literary sources also provide important evidence on the garments worn by religious dignitaries such as priests and priestesses, in that they can contribute substantial information not present in the extant iconography, especially concerning the colours of priestly garments.

Fig. 96. Reclining male figure on the east pediment of the Zeus temple at Olympia, ca. 470-457 BC. © Joanbanjo 2011. Wikimediacommons.

Fig. 97. Bearded male standing by cauldron. Kylix, ca. 500-450 BC. Museo Nationale Etrusco di Villa Giulia.

Purple priestly garments

Scholars have dealt primarily with epigraphic and literary sources that treat the use of the colour purple,[76] and they have catalogued considerable evidence for the use of purple garments in liturgical costumes – which symbolise sacerdotal dignity – throughout the Hellenistic Period.[77] Examples of purple priestly garments include those of the dignitaries of the cult of Demeter at Ephesos, administered by a local family called the Basilides, who had the privilege of wearing purple garments (*porphyrai*).[78] Other examples include a 2nd century BC inscription from the city of Skepsis in Asia Minor, which describes the official garment of the priests of Dionysos as a purple *chitōn*, shoes that match the clothes, and a golden wreath.[79] According to Strabo, Anaxenor, who was priest of Zeus Sosipolis at Magnesia ad Meandrum, wore a *porphyra*.[80] Moreover, Athenaeus states that the priest of Herakles in Tarsos wore a purple and white *chitōn* (*porphyroun mesoleukon*), an expensive horseman's *chlamys*, and white Spartan shoes.[81]

There are no fewer than three inscriptions describing priestly dress from Cos. The first dates to the 2nd century BC

Fig. 98. Statue of seer from Didyma, ca. 550 BC. Archaeological Museum, Istanbul, inv. no. 1945.

Fig. 99. Female seer. Funerary relief from Mantineia, ca. 420–410 BC. National Archaeological Museum, Athens, inv. no. 226. © Hellenic Ministry of Culture, Education and Religious Affairs/Archaeological Receipts Fund. Photo: H. Goette.

and describes the dress of the priest of Nike.[82] During a procession in the month Petageitnos as well as during sacrifices and when in the temple, he wore a purple *chitōn*,[83] a gold ring, and a wreath of twigs. At all other times, however, the priest was required to wear white and to be considered pure. The inscription states that the other priests should follow this injunction, i.e. also wear white.[84] The second Coan inscription also dates to the 2nd century BC and is concerned with the priest of Herakles Kallinikos.[85] This inscription states that the priest is to sit in the first row during the sacrifice and is to pour libations during the choral competitions with the other priests, and that he must wear a white *chitōn* (*chitōn dialeukos*). He should be crowned with a white crown (a crown of white poplar), and he is to wear

Fig. 100. Votive relief, Lakrateides, ca. 100 BC. Eleusis Archaeological Museum, inv. no. 5079. © Arachne, DAI.

an *aphamma* (perhaps a chain)[86] and a golden ring.[87] The third Coan inscription dates to the 1st century BC, and is connected with the cult of Zeus Alseios. It states that the priest(s) should wear a purple *chitōn* (*chitōna porphyrion*), an olive wreath, and a gold *aphamma*.[88]

In his speech *Against Andocides* from 400/399 BC Pseudo-Lysias provides evidence that the Eleusinian priests and priestesses wore purple garments:

> 'And for such a deed priestesses and priests (...) shook out their purple vestments (*phoinikidas*) according to the ancient and time-honoured custom.'[89]

It thus seems that, in some cults, priests were specifically required to wear purple garments.[90] The majority of these examples occur in the Hellenistic Period, and are often concerned with either new regulations or privileges (e.g. Skepsis, Cos) or newly established cults,[91] especially in Asia Minor. This has led some scholars to conclude that the Greeks rarely used purple in sacred contexts before the Hellenistic Period.[92] This, however, fails to take into account Pseudo-Lysias' account of purple vestments.

9. *Priestly garments* 293

Fig. 101. Votive relief of the hierophant from Hagnous. Museum of the Athenian Agora, inv. no. 13114. © Hellenic Ministry of Culture, Education and Religious Affairs. Ephorate of Antiquities of Athens.

The scarcity of evidence for the use of purple in priestly garments in the Archaic Period has been interpreted as an indication that sacred purple was adopted from Anatolia and the Near East,[93] where the colour already had cultic significance, and that sacred purple from this period is only attested in cults strongly influenced by Anatolian cultures.[94] This argument, however, disregards the nature of the evidence. Inscriptions of this kind are much more common in the Hellenistic Period and especially in Asia Minor, which explains why the bulk of the material is connected with this particular period. And, to state the obvious, one should be careful about making conclusions on the basis of non-existent evidence.

As an interesting parallel, a high valuation was placed upon the colour purple among the Jews in antiquity, both as a ritual and sacerdotal colour:[95] for example,

Fig. 102. Votive relief from Kertch, 5th century BC. © State Hermitage Museum, St. Petersburg, PAN 160. Photo: Vladimir Terebenin, Leonard Kheifets, Yuri Molodkovets, Aleksey Pakhomov.

purple is specified in the Old Testament and later sources for various cultic and sacerdotal purposes, such as the curtain of the Tabernacle of the Temple in Jerusalem, the cloth of the showbread, and the garments of the high priest.[96]

As a final note on the sacerdotal use of purple, the use of this colour in these examples was not necessarily related solely to the priestly status of the wearers, and the colour could just as well be explained by their high social status.[97] Thus, as argued by Nosch, purple is the quintessential royal symbol over the centuries, and throughout Greek history kings have been associated with purple clothing. But whether this association comes from the religious duties and privileges of kingships is still unresolved, and it remains to be seen whether purple was originally the colour of a priest who had become king or a king who was head of a cult.[98] In sum, we do not have any conclusive evidence for the use of purple garments as the typical dress for all priests and priestesses in the Greek sanctuaries at any period of time. Yet it is clear that, inasmuch as the colour of the priestly garments was emphasised in the written sources, it was of importance to the cult.

White priestly garments

Purple is not the only colour mentioned in connection with priestly garments; indeed, there is ample evidence for the use of white in priestly garments. An inscription

Fig. 103. Eumolpos and Keryx. Red-figure skyphos, ca. 500–480 BC. © Trustees of the British Museum, inv. no. E140.

from the reign of Attalos I (3rd century BC) at Pergamon records that the priests of Zeus wore a white *chlamys* and a wreath (*stephanon*) with a purple ribbon.[99] And an inscription dating to the 3rd century BC on a doorpost at the main entrance to the Alexandreion at Priene records that 'Anaxidemos, son of Apollonios, received the priesthood to enter the holy temple in white clothing'.[100] Priestesses could also wear white, as is shown in an honorary decree from Thasos dating to the 1st century BC, which states that the priestess Epie must wear white clothes as custom dictates when performing religious rites at the altar of Demeter.[101] Finally, an inscription, dated to ca. 100 BC, from the city Demetrias at Magnesia records a resolution in which new regulations are established for the consultation of the oracle of Apollo Koropaios.[102]

Fig. 104. Eumolpos and Keryx. Red-figure pelike from Kertch, ca. 360 BC. © State Hermitage Museum, St. Petersburg, inv. no. 1792. Photo: Vladimir Terebenin, Leonard Kheifets, Yuri Molodkovets, Aleksey Pakhomov.

According to the inscription

> 'Those named should sit properly in the temple, in white robes, ornamented with laurel wreaths, in (cult) purity and sober'.[103]

Those named include the priest of Apollo Koropaios, a representative of each of the colleges of *strategoi* and *nomophylakes*, a *prytanis*, a treasurer, the scribe of the divinity, and the prophet.[104]

Furthermore, a number of ancient authors provide evidence that white was used for the garments of certain religious personnel. For example, according to Pausanias (6.20.3), the inner sanctuary of Zeus Sosipolis at Olympia could only be entered by the priestess, who was required to cover her head and face in a white veil. According to Lucian of Samosata (2nd century AD), the priests of Dea Syria wore white garments (*esthēs leukē*), whereas the high priest wore purple and a golden tiara.[105]

There is an assumption in some scholarship that priests wore white garments, and that purple was reserved for chthonic cults.[106] This is based on a passage in Plutarch, which states that the chief magistrate of Plataea usually could not wear any other colour than white, but that, during the annual procession that accompanied the funeral offerings for the Hellenes who fell in battle at Plataea, he was clad in a purple garment,[107] which would suggest an association between funerary rites and purple. Yet this passage is a slight basis on which to make claims about the use of certain colours in priests' garments or cults, and in general it is difficult to ascertain whether all priestly personnel were distinguished by special vestments based on the literary and epigraphic evidence alone. Nonetheless, it can be concluded that priests and priestesses generally were allowed costly and colourful garments, since they were often allowed to wear purple garments and gold accessories. This greater freedom of dress is supported by a law about the sale of the priesthood for Dionysos Phleos from the 2nd century BC, which states that the priest can wear whatever dress or costume he likes, as well as a golden wreath,[108] and by a 1st century BC inscription from Miletos, which states that the priest of Asklepios could dress in any type of garments he wished.[109] This suggests that dressing as one wished was something out of the ordinary and reserved for a select few. Even if we cannot map precisely what priests should wear, we can conclude that society often had a keen interest in priestly dress and sometimes felt the need to regulate it or to free it from social convention.

An inscription from Cos regarding the priesthood of Zeus Polieus, although it does not mention colour, states that the priest should sit at the table and preside over the selection of the sacrificial ox wearing his 'sacred garment' (*hiera stola*).[110] This is important because it indicates that some garments were considered 'sacred', although we do not know whether this implies a special type of garment, a special fibre, colour, cut, or some other quality.

According to Paul, clothing regulations are concerned with male priests, never priestesses, and, furthermore, these regulations follow sections in inscriptions about *proedria* and *spondarchaeia* in contests, privileges that never concern female priesthoods.[111] With the exception of the decree from Thasos honouring the priestess Epie, this appears to be true. It is not easy to say what the significance of this gender discrepancy is: perhaps it results from the chance survival of the written sources, or it may indicate that it was of greater importance to regulate male priestly garments, or that it was important to describe them since they were something out of the ordinary.

Summary and discussion of the evidence for priestly garments

This iconographic and textual survey demonstrates that priestesses wore a limited range of garments, and that, no matter the priestly attribute, they tended to wear the same attire in the form of *chitōn* and *himation*.[112] This exact combination of *chitōn* and *himation* is considered the standard dress of women in the Classical and Hellenistic Periods, which means that female priestly garb is very hard, if not impossible, to distinguish, since it follows secular female attire traditions. In this respect, there is an interesting contrast to the Roman Period, where certain priesthoods were recognisable on the basis of specific clothing items, for example the Flamines, who can be identified in iconography by their special headgear,[113] and the priestesses of Isis, who were clad in *palla* with fringes tied by a special knot – the Isis knot.[114]

In some instances, priestesses wear a second garment over the *chitōn*, for example a *peplos*. This costume is often considered to be the typical garb of priestesses, even though the combination of *chitōn* and *himation* is more common.[115] The *himation* can, however, be arranged somewhat differently. Some portraits render the *himation* in a way that resembles that worn by the Small Herculaneum Woman (e.g. the priestess Zopyreina from Aulis), which is to say that it covers a large part of the body, including the arms, but in many other examples the priestesses wear the garment hanging free under their right arm, and across the abdomen and over the left arm and shoulder in a way that leaves the lower calves uncovered by the mantle. According to Connelly, this arrangement of the *himation* was popular from the 5th century BC to the 4th century AD, and it occurs on around 200 female statues. Connelly also argues that this arrangement of the *himation* was at first used in depictions of goddesses, but that in the late Hellenistic and Roman Periods it also came to be used in depictions of upper-class women, and was thus not related to priestly status.[116] This indicates that priestesses were women given a special duty that was symbolised by e.g. the temple key, not by specific garment types.

Other priestesses are portrayed wearing *peploi*: e.g. the Mantineia seer. It has been suggested that the *peplos* was not part of everyday female dress in the Classical and Hellenistic Periods. After the Archaic Period, it was primarily a ritual garment, but could also be used to signify tradition, feminine virtues of chastity, and domestic labour.[117] This can be considered an example of Arthur's 'fossilised fashion', which is a sudden 'freezing' of fashion wherein a group continues to wear a style long after it has become outmoded for the general population. This phenomenon has been explained as an expression of dignity and high social status or of the group's religious, old-fashioned, sectarian identity.[118] The fact that a woman is clad in a *peplos*, however, does not alone allow us to identify her as a priestess, but only to establish that she is most likely represented in a traditional ritual sphere. From this it can be concluded that priestesses were not immediately identifiable from their garment type, but only from attributes that signified their priestly office.

The situation is somewhat different with priests. As previously discussed, priests in the Classical Period were depicted in different garments. The most common type is the long, ungirded *chitōn*, but in other instances they were clad in a *chitōn* and *himation*, or only in a *himation*. They are also depicted wearing more extraordinary garments such as the *ependytēs*, the long-sleeved decorated garment of the key-bearing Chryses, and long garment depicted on the seated statue from Didyma and on the red-figure Attic cup in the Villa Giulia. Furthermore, *strophiones* appear to have been used by ritual personnel. There is thus much more evidence of specific garments for priests than for priestesses.

Consequently, priestesses are defined by their administrative authority alone, whereas priests are defined by their garments, which are occasionally supplemented by specific attributes such as the sacrificial knife.[119] It is therefore difficult to determine whether priestesses, in particular, wore certain garments. This is not necessarily surprising, since they were not separated from the ordinary population, but were usually private men and women who took on or were granted religious duties for a period of time.[120] Thus, the depiction of their clothes actually highlights their double status as citizen and priest/priestess. As Garland observes, the ability to act in matters of religion in Classical Athens was not the preserve of any one class or caste, whether social or political, and priesthood did not debar a man from holding political or military office.[121] This situation is further complicated by the fact that in some cults the priests and priestesses acted out the roles of the deities themselves,[122] which makes it even harder to recognise them in iconography. Yet, the fact of this role-playing reveals that priests and priestesses were dressed as gods during certain rites, probably in colourful, decorated, fine garments, since this is how divinities are usually dressed in iconography.

Because of these difficulties, some scholars have concluded that priestly garments did not exist: Miller, for example, states that the very existence of priestly garments in Classical Athens is extremely doubtful.[123] This might be true in the case of priestesses, but not of priests, who, in several instances, can be shown to wear special garments or dress items that identify their priestly function. This does not imply unity in dress, however, but rather outstanding clothing. Furthermore, it should be kept in mind that there may have been considerable difference between the individual cults, and one cannot speak of a general priestly costume either in Greece or during a certain period.

It seems possible that in some cults there were no specific dress-codes for priests and priestesses, but in others special garments were required. As the inscriptions from Cos demonstrate, these required garments differed in accordance with what ritual act the priest or priestess was performing, and where the priest was located (e.g. inside or outside the sanctuary).

The written sources record that priestesses in some sanctuaries wore *chitōnes* and *himatia*, while priests are described as wearing *chitōnes*. They also wore mantles, either *himatia* or *chlamydes*. Yet it generally appears that the main objective of the inscriptions was to describe colour rather than actual garment type, shape, or fibre. This is in contrast to the temple inventories, which primarily identify garments by type.

The written sources thus provide an interesting juxtaposition with the iconographic material, inasmuch as sculptures were painted in a range of colours, mostly lost today.[124] It is therefore possible that priestly garments differed from secular garments especially in the use of certain colours. The priestly garments on *stelai*, sculptures, and reliefs could have been painted in specific colours or patterns that would have made them recognisable to an ancient audience, so that further attributes would have been unnecessary. This particular colour could have been a shade of purple, since purple is often specified in relation to priestly garments. In addition, as we shall see, several clothing regulations instruct that visitors to sanctuaries were not permitted to wear purple, colourful, or decorated garments. According to temple inventories, moreover, many garments with purple stripes or edges were dedicated in sanctuaries, which may indicate that garments of that colour had a ritual meaning.[125]

Notes

1. Bergemann 1997, 45.
2. Brulé 1987, 248; Turner 1983, 392–395; Keesling 2003, 100–120.
3. Von den Hoff 2008, 109; Richter 1968, 3.
4. Kosmopoulou 2001, 293.
5. Pilz 2013, 169.
6. Kosmopoulou 2001, 294; Connelly 2007, 92. An alternative interpretation has been put forward by Quercia, who suggests, based on the depictions of these objects on red-figure vases, that they were not temple keys, but distaffs, Quercia forthcoming.
7. Kosmopoulou 2001, 294.
8. Quercia forthcoming.
9. Connelly 2007, 229; Kosmopoulou 2001, 293; Bergemann 1997, 121. The earliest example in Attic funerary art is dated to ca. 380–370 BC, Kerameikos Museum, inv. no. I.430.
10. Connelly 2007, 93. Larnaca Museum, inv. no. 663; Louvre, inv. no. N3278.
11. Connelly 2007, 94. A key-bearer (*kleidoukhos*) was also depicted on a small terracotta figurine from the Artemis-sanctuary in Cercyra. Due to its fragmentary state, it is impossible to determine its exact type of dress, but it is definitely wearing a long garment and possibly a mantle. Von den Hoff 2008, 110; Mantis 1990, 30–31 pl. 5.
12. J. Paul Getty Museum, inv. no. 73.AA.124.

13. Connelly 2007, 98. *Chitōn* and mantle: Kunsthistorisches Museum, Vienna, inv. no. 724 (261). *Peplos*: State Hermitage Museum, St. Petersburg, inv. no. 298 (St. 1734); *peplos* and mantle: British Museum, inv. no. F209.
14. Kosmopoulou 2001, 297.
15. Connelly 2007, 98. British Museum, inv. no. 1931,0511.1; Staatliche Museen zu Berlin, Antikensammlung, inv. no. F3025.
16. E.g. Antikenmuseum Basel, inv. no. LU51; Lullies 1968, no. 46.
17. Mantis 1990, 54 no. 17; pl. 20a.
18. Archaeological museum of Piraeus, inv. no. 3627. Kosmopoulou 2001, 295.
19. Kosmopoulou 2001, 296.
20. Kosmopoulou 2001, 296. Ashmolean Museum, inv. no. 1959.203; Archaeological museum of Piraeus, inv. no. 217; National Archaeological Museum, Athens, inv. no. 3287. Women with *tympana* are also depicted on terracotta figurines from Tanagra, e.g. Louvre, inv. no. CA 3462; Museo Archeologico Nazionale, Taranto, inv. no. 51676. These women wear *chitōnes* and mantles with coloured borders. Becq et al. 2010.
21. Connelly 2007, 238; Kosmopoulou 2001, 296. Staatliche Museen zu Berlin, Antikensammlung, inv. no. SK1504.
22. The kore is now lost. Von den Hoff 2008, 110; Mantis 1990, 31–32, pl. 6a; Karakasi 2001, 35, cat. M10, pl. 45.
23. Connelly 2007, 132, 135.
24. Mylonopoulos 2013, 126.
25. Pilz 2013, 166. Staatliche Museum zu Berlin, Antikensamlung, inv. no. SK1928.
26. Von den Hoff 2008, 129; Connelly 2007, 146; 2008, 192; Pilz 2013, 161. National Archaeological Museum, Athens, inv. no. 232; Base: *IG* II² 3462.
27. Demakopoulou & Konsola 1981, 80.
28. Connelly 2007, 158. Archaeological Museum, Thebes, inv. no. BE66.
29. Archaeological Museum, Thebes, inv. nos. BE63, BE64, BE65, BE66.
30. Specifically in the so-called Cult House K located in the west wing of the Sanctuary of Asklepios. Themelis 1994.
31. Themelis 1994, 116–117.
32. Connelly 2007, 150–152. Messene Museum, inv. nos. 241, 244, 245, 246.
33. Themelis 1994, 115. *SEG* 23, 1968, 220–222.
34. Themelis 1994, 111. *SEG* 23, 1968, 215–217.
35. Archaeological Museum, Messene, inv. no. 242.
36. Archaeological Museum, Messene, inv. nos. 243, 240.
37. Pfuhl & Möbius 1977, 136–138, nos. 405–410.
38. Pfuhl & Möbius 1977, no. 405. Staatliche Museen zu Berlin, Antikensammlung, inv. no. SK 767.
39. Pfuhl & Möbius 1977, no. 407. Basmahane Museum, Izmir.
40. Connelly 2007, 93; Palagia 1980, 41; Miller 1989, 322.
41. For knives in sanctuaries see *ThesCra* V, 308–312.
42. *IG* II/III² 1481, 5; *IG* I³ 342, 16–17; *IG* I³ 343, 13; *IG* II/III² 1424, 20.
43. National Archaeological Museum, Athens, inv. no. 772.
44. Museo Archaeologico Etrusco, Florence, inv. no. 4224.
45. E.g. Museo Archeologico Nazionale, Naples, inv. no. 82922.
46. Palagia 1980, 41. Louvre, inv. no. CA227.
47. Von den Hoff 2008, 113.
48. Mantis 1990. Bergemann records 17 depictions of priests on Attic grave *stelai*. Bergemann 1997, 121.
49. Von den Hoff 2008, 113.

50. E.g. Staatliche Museen zu Berlin, Antikensammlung, inv. no. SK944; National Archaeological Museum, Athens, inv. nos. 4502, 772 and 3492; Mantis 1990, fig. 36a, 37, 38, 39b, 41, 42g.
51. E.g. on a red-figure *kylix*, Ashmolean museum, inv. no. 1911.617, and on an Attic funerary *lekythos*, National Archaeological Museum, Athens, inv. no. 4495. Mantis 1990, pl. 40a.
52. E.g. on a black-figure Attic *hydria*, Louvre, inv. no. F10. Mantis 1990, pl. 35b.
53. E.g. Staatliche Museen Berlin, Antikensammlung, inv. no. SK1708; J. Paul Getty Museum, inv. no. 72.AA.120.
54. Von den Hoff 2008, 113.
55. Archaeological Museum, Eretria, inv. no. 631. Edelmann 1999, 54, no. B32; Tagalidou 1993, 252, no. 46, pl. 19.
56. Hessiches Landesmuseum, Darmstadt, inv. no. A1969.4.
57. Attyah Flower 2008, 72.
58. Attyah Flower 2008, 22–27.
59. The Branchids were the hereditary controllers of the oracle at Didyma. Pasinli 2010, 10.
60. Archaeological Museum of Istanbul, inv. no. 1945.
61. Von den Hoff 2008, 110–111.
62. Museo Nazionale Etrusco di Villa Giulia, inv. no. 2272. See also Ashmolean museum, inv. no. 1911.617.
63. National Archaeological Museum, Athens, inv. no. 226.
64. It has also been suggested that the headgear represented on the heads of three male Archaic statues from the Sanctuary of Apollo at Idalion, Cyprus relates to priestly or kingly status. Senff 1993, 70–71.
65. Connelly 2007, 92; Miller 1989, 320. E.g. *LSAM* 38A and 38B.
66. Von den Hoff 2008, 139. Many portraits with crowns and wreaths do not signify a priest or priestly office. The identification of priestly crowns, ribbons, and wreaths is a complicated matter, because local traditions, iconographic particularities, and changes over time impact the choice of portrait iconography. As Horster and Schröder have argued, it is thus not possible to develop a comprehensive system of iconography for depictions of crowns or wreaths. Horster & Schröder 2013, 235.
67. Clinton 1974, 32–33.
68. Which is also the case for the Ninnion Tablet from Eleusis, National Archaeological Museum, Athens, inv. no. 11036.
69. Archaeological Museum, Elusis, inv. no. 5079; *IG* II² 4701. Clinton 2007, 347.
70. Agora Museum, Athens. Clinton 1974, 32.
71. Clinton 1974, 33–35.
72. State Hermitage Museum, St. Petersburg, inv. no. PAN 160.
73. Simon 1983, 27; Miller 1989, 317, 323.
74. E.g. a red-figure *skyphos*, ca. 500–480 BC, British Museum, inv. no. E140.
75. E.g. a State Hermitage Museum, St. Petersburg, inv. no. 1792 and a relief *hydria*, the so-called *Regina Vasorum*, 4th century BC, State Hermitage Museum, St. Petersburg, inv. no. 51659.
76. E.g. Blum 1998; Reinhold 1970.
77. Reinhold 1970, 36.
78. Blum 1998, 96, Strabo 14.1.3; Nosch forthcoming, 6.
79. *REG* 89, 1976; *BE* 572; *SEG* 26.1334.
80. Strabo 14.1.41.
81. Ath. *Deipn.* 5.54. Patera 2012, 41, note 17.
82. *LSCG* 163.

83. The word for purple is restored p[orph]yreon; another suggestion is a restoration as p[anarg]yreon.
84. I thank Giovanni Fanfani for help with the translation of this inscription. See also Patera 2012, 38; Parker & Obbink 2000, 425.
85. ED 180, lines 20–25; SEG 45:1129.
86. I thank Peder Flemestad for this information.
87. I thank Giovanni Fanfani for help with the translation of this and the following inscription.
88. Segre 1993, ED 215, lines 15–18.
89. Lys. [6] 51. Clinton 1974, 33.
90. Connelly 2007, 92.
91. Blum 1998, 100.
92. Blum 1998.
93. 'Anatolia and the Near East' is, however, a very broad generalisation, which does not take the different cultures of these areas into account.
94. Blum 1998, 118.
95. Reinhold 1970, 20.
96. Reinhold 1970, 20; *Exodus* 25.34–35; 26.1, 31, 36; 28.4–5; 35.23; 36.8; 35, 37; 39.1–9, 29; Joseph. *AJ* 7.7; *BJ* 5.4, 6.9, 7.6.
97. Blum 1998, 96.
98. Nosch forthcoming, 15.
99. LSAM 11; SIG³ 1018; SEG 34:1250; SEG 15:756.
100. LSAM 35; Hiller von Gaertringen 1906, no. 205.
101. SEG 18:343; Salviat 1959, 362–379.
102. LSCG 83; IG IX, 2, 1109.
103. Translation: Stavrianopoulou 2006, 14.
104. Stavrianopoulou 2006, 14.
105. Luc. *Syr. D.* 42. The use of white for priestly garments is also attested in Mesopotamia. The Enmerkar epic describes a priestess in the role of the goddess Inanna dressed in a white garment. Waetzoldt 2013, 203.
106. Miller 1989, 320.
107. Plut. *Arist.* 21.4.
108. LSAM 37.
109. LSAM 52 A, 14.
110. IG XII 4, 278, 9–10. Paul 2013, 262; see also Kaestner 1976.
111. Paul 2013, 262.
112. Kosmopoulou 2001, 296.
113. See e.g. Goette 2012, 23–25.
114. This particular type of dress originated in the Greek East, where it was used for statues of the Goddess Isis in the Late Hellenistic Period. Goette 2012, 27.
115. Kosmopoulou 2001, 297; Von den Hoff 2008, 127.
116. Connelly 2007, 157. Archaeological Museum, Messene, inv. no. 242.
117. Cleland et al. 2007, 143.
118. Arthur 1999, 5.
119. Von den Hoff 2008, 115–117.
120. Connelly 2007, 86.
121. Garland 1990a, 75, 78.
122. Connelly 2007, 104.
123. Miller 1989, 319.

124. See e.g. Stubbe Østergaard & Nielsen 2014; Skovmøller 2014; Liverani et al. 2004.
125. In the Roman Period, some priestesses become much more recognisable partly due to the wearing of specific colours. The Vestal Virgins did not wear any special garment type. Their garments, though, were pure white, and thus different from the dress of 'ordinary' Roman women. Their head gear was also different: portraits of Vestal Virgins often depict them with a type of veil known from ancient literature as a *suffibulum*, which was white with a purple seam. Goette 2012, 27. Another example is Isis *melaneimōn* or *melanostolos*. Priestesses of Isis are depicted in a mantle/*palla* with the *contabulatio* ('bar') decorated with stars and fertility symbols, but the most conspicuous characteristic is the black colour of the garment. Goette 2012, 30–31. Yet this specific Roman situation did not come from a Greek cult clothing tradition, but rather from Egyptian priestly dress.

Chapter 10

Iconographic evidence for the dress of sanctuary visitors

Dress in sanctuaries: iconographic evidence and methodological considerations

Vase-painting is an important source for representations of Greek ritual, and consequently also for the visual appearance of participants in these rituals. But Greek religious rituals could occur in various locations, not necessarily in a sanctuary, which of course must be taken into consideration. Greek ritual is, moreover, elusive and can be hard to identify. My method in dealing with the material is to focus on three iconic types of ritual settings:

1. Representations of persons standing by an altar and performing a ritual act, such as a libation or a sacrifice.[1]
2. Representations of the preparation or leading of sacrificial animals decorated with ribbons to sacrifice are a further example of clearly identifiable ritual scenes. Both types of scenes are highly likely to represent scenes from a sanctuary or other place connected with ritual.[2]
3. Ritual processions – most commonly in connection with animal sacrifice – are depicted on vases and votive reliefs. In vase-painting processions occur from ca. 570 BC into the 4th century BC, while on votive reliefs depictions of processions begin in the late 5th century BC and cease to exist in large numbers after ca. 300 BC.[3]

Still, even when a ritual scene can be clearly identified, we are faced with the problem of distinguishing priests or priestesses from initiates or 'ordinary' participants, and human beings from divinities. In order to minimize the latter problem, I leave out representations where divinities are clearly identifiable by their attributes. Of course, it is possible that this will result in the inclusion of examples where gods or goddesses are represented performing a ritual act, although they are not identifiable to a modern audience. The difficulty of separating priestly personnel from regular sanctuary visitors cannot be resolved. Since there was no requirement that only a priest or priestess was allowed to or capable of performing a ritual act, a person depicted standing by an altar pouring a libation, for instance, could be either. Nevertheless, these depictions still provide important information about what individuals could wear when performing a rite. In the cases of black-figure and red-figure vase-painting, we obviously cannot determine the original colours of the garments. Even so, these types of vase-painting still depict whether a garment is decorated or not – for example, with patterns, figural decoration, or coloured borders.

There are myriad representations of ritual that fall into the categories of scenes described above. The following discussion will therefore not be a complete analysis, but rather a survey of the different tendencies evident in the material. The material falls more or less into two distinct groups: 1) 'regular' clothing also worn in other scenes not related to ritual and, 2) more 'exotic' garments. In the case of regular clothing, the representations fall into five sub-groups or costume assemblages, determined by decoration: a) white garments, b) white garments with purple borders, c) white *chitōnes* and purple mantles, d) dark/purple *chitōnes*, and e) *chitōnes* with dots.[4] But these sub-groups (if they truly can be considered such) are fluid and are very similar, in that they include the same garment types: *chitōn*, *himation*, and more rarely the *peplos*. Exotic garments will be treated here according to type, and the discussion will include specific garment terms, such as the *ependytēs* and the dress of the *kanēphoros*.

Clothing for performing ritual acts

White garments (chitōn, himation)

Among the most commonly depicted outfits in sacrificial scenes in red-figure vase-painting are white, undecorated garments. This observation applies to both men and women, yet there are differences in the choice of garment types and the way that they are worn. Women performing ritual acts are usually depicted in a *chitōn* and *himation*, sometimes with a headdress or ribbons used to tie up the hair, and bare feet (Fig. 105).[5] In other instances the woman is clad simply in a *chitōn* without the *himation*.[6] Men are also often depicted in white garments when performing a ritual act at an altar. A very common form of male dress is a white *himation* and nothing else, except possibly a wreath. Such scenes are quite common in Attic red-figure vase-painting.[7] Other vases depict men at an altar in white *himatia*, but with a white *chitōn* underneath, and thus dressed identically to women performing such acts.[8]

White garments with a coloured border/stripe

A variation on the pattern of dress described above is the same type of white garments, but with the inclusion of a coloured stripe, primarily along the hem of the *himation*. There does not appear to be any difference between individuals wearing completely white garments or garments with a coloured stripe, and figures in either type of dress can be present on the same vase. The stripe is usually on the *himation*, and it is attested for both men[9] and women. A white-ground *kylix*, dated to ca. 460 BC, depicts a woman pouring a libation over an altar (one of two) (Fig. 106).[10] She is dressed in a pleated *chitōn* beneath a white *himation* with a reddish purple border, which on the lower hemline is further decorated with a line of dots. She has bare feet and her hair is held up by a broad band of cloth; she is also wearing a coiled bracelet on each arm and a long staff rests beside her. She has been identified as Kore because of the two altars, one her own and one dedicated to her mother Demeter.[11] Furthermore, according to von Bothmer, the name [K]ore can be read in faint letters.[12] But it is possible that she may be simply a mortal woman, perhaps a priestess, pouring a libation to a deity, perhaps Kore or Demeter. In other instances, the purple stripe is on the undergarment, either a *chitōn* or *peplos*, sometimes as a border along the opening of the garment (Fig. 107).[13]

White chitōn and purple mantle

For the third group there are several examples in white-ground vase-painting. One example depicts a woman clad in a white, tightly pleated *chitōn* with a long overfold; over this she wears a purple mantle (Fig. 108).[14] She has bare feet and is adorned with gold bracelets and a necklace, and her hair is tied up with a ribbon. It is impossible to determine the identity of the woman with any certainty, but it has been proposed that she is Hera because of the sceptre.[15] She might, however, just as well be a priestess or another sanctuary official performing a libation.

Another example is a *phiale*, dated to ca. 450 BC (Fig. 109).[16] The tondo bears a depiction of a ritual in which women dance around or towards an altar. The girls are identically dressed

Fig. 105. Woman at altar. Alabastron by the Syriskos Painter, ca. 500–450 BC. © Staatliche Museen Kassel, Antikensammlung, inv. no. T551.

Fig. 106. Woman pouring a libation at an altar. Kylix by the Villa Giulia Painter, ca. 460 BC. © Ashmolean Museum, University of Oxford, inv. no. 1973.1.

in white *chitōnes* and purple *himatia*. They all have bare feet and all except one wear their hair up, with a ribbon around their heads. In this case, it seems clear that the women are either priestesses or initiates participating in the cult of an unknown deity.

Dark/purple chitōn *and white* himation
The opposite colour pattern – although rare – also occurs in sacrificial scenes. An example depicts a woman dressed in a purple *chitōn* underneath a white *himation* (Fig. 110).[17]

Chitōn *with dots*
A fifth and final variation of this dress scheme is the dotted *chitōn*, usually worn under a *himation*. This type of garment is depicted on a *kylix* dated to ca. 475–425 BC, which features a procession of women, all with bare feet, walking towards an altar (Fig. 111).[18] The majority of the women are dressed in white *chitōnes* and *himatia* with

Fig. 107. Women preparing bulls for sacrifice. Amphora by the Nausicaa Painter, ca. 450 BC. © Trustees of the British Museum, inv. no. E284.

purple borders, but one of them is dressed in a dotted *chitōn* under her bordered *himation*. Except for the dotted garments, she does not differ from the others. Another interesting detail on this vase is the woman standing closest to the altar, holding two torches. Her long hair hangs down her back and she – alone of the women on the vase - is dressed in an unbelted *peplos* with a coloured border on the hem at the feet as well as on the overfold. Perhaps her different dress and her position at the altar with the torches indicate that she has a different status or function than the women in the procession. In other depictions, the dotted *chitōn* is worn with a purple or white mantle.[19]

The *ependytēs* and the *chitōn cheiridotos*

A garment type that deserves special attention is the so-called *ependytēs*. The Greek term can be translated as 'put over'.[20] It is commonly interpreted as a tunic-like garment of Persian or Oriental origin, either of wool or linen, and usually worn over a

Fig. 108. Woman performing a libation. Kylix by the Villa Giulia Painter, ca. 470 BC. © Metropolitan Museum of Art, inv. no. 1979.11.15.

chitōn.[21] It came in different lengths: to the waist, thigh, or knee. According to Lee, in the Classical Period, the garment appears in vase-painting in particular as a women's garment.[22] A possible depiction of an *ependytēs* appears on a red-figure *oinochoe* dated to ca. 420–410 BC: on the left, the woman emptying an *oinochoe* onto a fire wears a highly decorated, sleeveless garment over her *chitōn* (Fig. 112).[23]

Miller distinguishes between the *ependytēs*, which she – although she does not explicitly state this – considers a sleeveless garment, and the so-called *chitōn/chitōniskos cheiridotos*, which is a sleeved *chitōn/chitōniskos*. It can be either monochromatic or multi-coloured, and is often heavily decorated with different patterns or figural decoration. In Greek art the *chitōn/chitōniskos cheiridotos* is generally worn by women, and by men in festival contexts. According to Lee, several variations of this garment have been identified, but it is unclear whether they should be considered different types.[24] The grave stele of Hegeso from ca. 410–420 BC depicts a standing young woman, clad in a long-sleeved garment, which might be identified as a *chitōn/chitōniskos cheiridotos*, worn underneath another garment that reaches to her feet (Fig. 113).[25]

10. Iconographic evidence for the dress of sanctuary visitors 311

Fig. 109. Phiale by the Painter of London, ca. 450 BC. © Museum of Fine Arts, Boston, inv. no. 65.908.

Miller suggests equating the *ependytēs* with the *chitōniskos*, which occurs frequently in the Brauron catalogues, where it is often described as coloured and/or decorated.[26] Although both are short and often decorated tunic-like garments, there is nothing to indicate that they should be equated. Despite Miller's attempt to distinguish between *ependytēs* and sleeved *chitōnes*, there appears to be a tendency to identify almost any type of decorated or coloured tunic-like garment as an *ependytēs* – whether long, short, sleeved, sleeveless, featuring different decoration, etc.

Most of the representations of *ependytai* appear on women, but, in Miller's view, this is simply because of the apparent preference for depicting women in Attic vase-painting, and not because the garment was strictly female.[27] In addition to the fact that it was not associated with a particular gender, it also appears to have been used in different contexts and for different occasions, by aristocrats, actors, musicians, dancers, and gods and goddesses. After the Persian wars, heroes and divinities also start to wear this garment, and it becomes more common after the mid-5th century. In short, wearing the *ependytēs* was not restricted to a narrowly delineated range of

Fig. 110. Woman at altar. Alabastron *by the Syriskos Painter*, ca. 500–450 BC. National Archeological Museum, Athens, M. Vlastos Collection 20.2740. © Hellenic Ministry of Culture, Education and Religious Affairs/ Archaeological Receipts Fund.

activities, and it has been proposed that the garment served primarily to add a sumptuous element to an individual's dress.[28]

Despite these different contexts for the garment, the *ependytēs* has been associated specifically with cult. Already in 1936 this association was put forward by Thiersch, who argued that the *ependytēs* was a garment worn by the great goddesses of Anatolia that was subsequently imported to Archaic Greek cults, before finally disappearing from the divine sphere and becoming a cultic garment associated with the Jewish *ephod*.[29] Miller instead interprets the garment simply as 'Sunday best', as a form of display indicating that the wearer has attempted to look his or her best, and perhaps also as an indication of wealth.[30] I do not find Miller's explanation entirely adequate. As discussed in the previous chapter, there are several examples of cult images that are dressed in an *ependytēs*. Closer examination of the depictions of the *ependytēs* in identifiable ritual scenes in vase-painting allows for the identification of two tendencies: first of all, the *ependytēs* is worn by only one person – often the individual closest to the altar or cult statue (i.e., in the centre of the narrative). Accordingly, this individual can be considered if not the most, at least an important actor in the scene. Second, with regard to the dress of the remaining individuals in the scene, they are either more subtly dressed, e.g. in white garments, or everybody is wearing highly decorated garments. It appears that in the latter case the individual wearing the *ependytēs* is surrounded by divinities, identifiable by their attributes, but in the former the remaining attendants are human beings. These tendencies establish that the dress worn by the attendants in these ritual scenes is not random or coincidental. On the contrary, this garment must have had some special meaning when worn in ritual scenes. This does not necessarily mean, however, that it *always* had a special meaning – only that it did when used in cult. The two different scenarios for the wearing of the *ependytēs* both illustrate the same point: visual recognition. In cases where the individual in the *ependytēs* is the only one wearing the garment, and perhaps also the only one in decorated garments, he or she is clearly meant to stand out and to be recognised as somebody special by the

10. Iconographic evidence for the dress of sanctuary visitors 313

Fig. 111. Red-figure Kylix, ca. 475–425 BC. Museo Archaeologico Etrusco, Florence, inv. no. 3950.
© Photo courtesy of the Soprintendenza Archeologia della Toscana - Firenze.

Fig. 112. Oinochoe, ca. 420–410 BC. © Metropolitan Museum of Art, New York, inv. no. 75.2.11.

Fig. 113. Grave stele of Hegeso, ca. 410–420 BC. National Archaeological Museum, Athens, inv. no. 3624. © Hellenic Ministry of Culture, Education and Religious Affairs/Archaeological Receipts Fund. Photo: G. Patrikianos

audience. In the opposite case, the individual in the *ependytēs* is meant to blend in and to be recognised perhaps as a particular type of person or as a person with a specific function – he or she is clearly not just anybody wearing 'their Sunday best'. Thus, the *ependytēs* is a clear and functional identity marker, not of the occupation of the single person, but of the singularity of the person within a group of people. The same may be so for Miller's *chitōn cheiridotos* ('sleeved *chitōn*').[31] This garment also appears in different scenes and contexts, and is thus not an exclusively cultic garment, but when depicted in ritual scenes, it is usually worn by only one person, as for example the *kanēphoros* on the red-figure *krater* in Ferrara, who wears this heavily decorated garment alone among men dressed in white *himatia* (see below). Furthermore, it should be noted that these garments often look a lot like the priestly tunics described above and represented e.g. on the Parthenon frieze - a possible indication of a ritual connotation. But the garment on the frieze for example is undecorated, which may be connected with the state of the frieze's preservation. A further thing to note is that *ependytai* are never recorded in the temple inventories investigated in the previous chapters. Perhaps this indicates that ritual garments were not considered to be appropriate as dedications in these sanctuaries, or perhaps they were denoted under a different term.

Kanēphoroi and their dress

Only very few ritual roles or duties can be recognised in iconography. One such role is the *kanēphoros*. Attic religious processions were usually headed by one or more so-called *kanēphoroi* – 'basket-bearers' – girls of marriageable age chosen from among the high-born citizens.[32] The baskets (*kana*) contained items necessary for the sacrifice, such as the sacrificial knife, and fillets for adorning the sacrificial animal.[33] There appear to have been different regional types, and the baskets could be shallow trays or deep basins with handles of different sizes. They came in a variety of shapes and could be made of either wicker work or metal, such as gold, silver or bronze.[34] The *kanēphoros* was not connected with any particular cult and she participated in processions for many different deities at different sanctuaries in Attica.[35] *Kanēphoroi* are depicted in Greek art from the 6th century BC onwards, and are easily identifiable from the *kanoun* that they are carrying, usually on the head. They are therefore an obvious choice for a closer study of ritual dress.

Roccos has previously conducted a study on this topic, in which she concludes that the *kanēphoros* was always richly dressed with what she terms a 'festival' mantle, a garment whose arrangement changed over the years. In the Archaic Period, the *kanēphoros* wore a large, voluminous mantle that was pulled up over her shoulders and hung down in front and back, almost reaching the ground. In vases from this period, the *kanēphoros* can be recognized only from the basket that she carries on her head, since she wears the same garments as other female participants in cult scenes (Fig. 114).[36]

In the mid-5th century BC the 'festival' mantle changed to a shoulder mantle, which hung lower on the back, while only just covering the shoulders and upper-arms on the front, and worn over a *chitōn* (Fig. 115).[37] Roccos argues that, from the mid-5th century onwards, the *kanēphoros* could be recognized even without her basket because of this distinctive 'festival' mantle, which would thus represent the first and only identifiable female priestly garment.

In the 4th century, the 'festival' mantle changed to a pinned back-mantle.[38] This type of mantle appears on the *kanēphoros* on a red-figure *krater*, dated to ca. 440–430 BC, which depicts a sacrificial procession in honour of the god Apollo at Delphi (Fig. 116).[39] The procession is led by a *kanēphoros*, carrying a *kanoun* on her head, and dressed in a long, patterned garment, reaching mid-thigh, with long sleeves – identified as a *chitōn cheiridotos* by Miller – and a mantle pinned at the shoulders, falling down her back. Under the patterned garment she wears a long *chitōn*. The remaining participants in the procession are male, dressed in white *himatia*, under which they are naked, and a wreath. With the exception of the *kanēphoros*, all participants are thus dressed like the god.

Another example is a terracotta figurine of a *kanēphoros* carrying a *kanoun* on her head, dated to the 4th century BC (Fig. 117).[40] The figurine is dressed in a long garment and the distinctive back-mantle, which hangs down her back in wavy folds and is pinned at the shoulders. This latter example demonstrates that in different eras the *kanēphoros* appears in different media, and that the milieu in which she appears also changes: the *kanēphoros* is usually depicted in vase-painting in the Archaic Period, where she participates in cult scenes, whereas in the Late Classical Period, the *kanēphoros* is also depicted in the form of votive figurines. In red-figure vases from the 4th century BC, the *kanēphoros* appears in wedding scenes and on funerary monuments. Yet Roccos argues that all of the young women dressed in this kind of mantle, whether carrying a *kanoun* or not, should be identified as *kanēphoroi*.

According to Roccos, this 'festival mantle' may be what is often referred to as an *epiblēma*, and the pinned back mantle could be the *epiporpēma*, which literally means 'pinned on'. Alternatively, the mantle may be equated with the *chlanidion*.[41] In this instance, however, no firm link between the iconography and terminology can be established, and the uncertainty about the name of this garment must remain unresolved.

Besides the mantle, the *kanēphoros* usually wears a *peplos* or a *chitōn*, and she has bare feet. Her garments are usually undecorated, but there are examples of *kanēphoroi* who wear a *chitōn* with a purple stripe, or a dotted *chitōn* underneath the mantle,[42] or a mantle with a narrow coloured border. The only exception to this is the *krater* from Ferrara, where the *kanēphoros* wears a heavily decorated, long-sleeved garment over her *chitōn*.

Roccos argues, though, that not every woman holding a *kanoun* is a *kanēphoros*, and that those standing at an altar are more likely to be priestesses.[43] Whether these women are in fact *kanephoroi* or priestesses does not detract from the point that they occupy offices with a special ritual status and function, and that they are not regular women visiting a sanctuary.

Fig. 114. Woman pouring libation. Red-figure Kylix by Makron, ca. 480 BC. © Toledo Museum of Art, Purchased with funds from the Libbey Endowment, Gift of Edward Drummond Libbey, inv. no. 1972.55.

Heavily decorated garments

Other scenes feature an altar surrounded by attendants wearing heavily decorated garments. An example is an Attic *pelike* depicting a sacrificial scene carried out by two youths surrounded by divinities (see Fig. 51).[44] All individuals in the scene are dressed in heavily decorated garments. Both youths wear footwear, the youth to the left very elaborate knee-high boots, the other ankle-length sandals; some of the gods go bare-foot, while others, e.g. Ares, wear footwear. This type of scheme where all individuals in a ritual scene wear heavily decorated garments appears to be reserved for mythological scenes of gods and goddesses and their attendants, and thus does not necessarily provide direct information on what cultic personnel wore when performing a rite at a sanctuary. Here the garments illustrate the richness and power of the gods.

Fig. 115. Kanēphoros. Red-figure Oinochoe, ca. 480 BC. Staatliche Museen, Antikensammlung, inv. no. F 2189. Photo: Johannes Laurentius

Offering deities and divine imitation

As already mentioned above, it is often difficult to determine whether images depict gods and goddesses, priests and priestesses, or regular men and women. This problem is complicated further by the practice of divine imitation. According to Connelly, divine imitation entails what she terms 'cult agents' dressing up in

Fig. 116. Kanēphoros. Red-figure Krater, ca. 440–430 BC. Museo Archaeologico Nazionale, Ferrara, inv. no. 44894.

the costume of the divinity in order to perform a part in a ritual drama, not the priests or priestesses becoming the god or goddess. Thus, worshippers sometimes participated in rituals that were based on the re-enactment of the experience of the divinity, in this way bringing deity and devotee closer together.[45] Such divine imitations are attested for several divinities – e.g. Artemis, Apollo, and Athena – and involve female as well as male participants.[46] It is therefore possible that what appears to be a depiction of a god or goddess is instead a priest or priestess dressed as a divinity.

This practice has implications for the concept of 'offering gods'. Vase-paintings include scenes of what appears to be gods or goddesses performing ritual acts at altars, such as pouring libations. Several scholars have interpreted these scenes as epiphanies,[47] as examples of the humanization of divinities,[48] or as manifestations of divine reflexivity (since divine power depends on the reversal of canonical categories of sacrifice and devotee).[49] Other scholars, including Nilsson and Connelly, question the validity of the designation 'offering gods', and instead argue that the depictions might instead represent offering priests or priestesses.[50] Connelly convincingly argues, moreover, that vase-paintings that represent women at altars with the name of a goddess stated do not necessarily mean that the image should be identified as that particular goddess, but might rather be depictions of a priestess performing a sacrifice or libation to the goddess whose name is stated (Fig. 118).[51] This argument acquires further support from the fact that some of the stated divine names are given in the genitive case.[52]

Staffs and sceptres

This raises further questions about whether it is possible to identify figures as divinities on the basis of a single accessory. Perhaps we place too much emphasis on single objects, which should perhaps not be identified as divine attributes. In this connection, the sceptre is of particular interest, because it has been perceived as a divine attribute. This is, however, not a straightforward matter, as priests and priestesses as well as kings have been associated with this object. For example, in Homer, the priests Chryses and Teiresias are distinguished by the sceptres that they carry, and, according to Connelly, the golden sceptre recovered in a burial at Taras may have been an insignia of a priestess, who was buried with the attributes of her service.[53] The same can be argued for the crown, which might not have been used exclusively as a divine attribute for such goddesses as Hera, but also for priestesses.

Fig. 117. Kanēphoros, 4th century BC. Terracotta figurine. © National Museum of Denmark, inv. no. 10780.

Fig. 118. Woman at altar. Inscription: Artemis. Cup, ca. 475-425 BC. © National Museum of Denmark, inv. no. 6.

The expression 'to set up a sceptre' appears in epigraphy from Lydia and designates the erection of a symbol of divine power – probably in a sanctuary. This aimed at preventing future crimes as well as punishing offenders. According to Chaniotis, this ceremony was performed by priests, who are, in fact, occasionally depicted with such sceptres.[54]

Blending in or standing out

As the examination of the iconographic depictions of individuals performing rituals has established, there were many ways of dressing when performing a rite.

The majority of the depictions feature 'regular' clothing, primarily the *chitōn* and the *himation*, sometimes coloured or with decoration in the form of a coloured stripe or border, or in the form of dots. Other examples show more 'exotic' garments such as the *ependytēs* or festival mantle worn by the *kanēphoroi*, while yet others feature heavily decorated garments. These depictions reflect two scenarios: 'blending in or standing out'.

As previously discussed, in depictions of processions, attendants all wear more or less the same type of garments, perhaps with minor differences such as the addition of a coloured border, dots on the *chitōn*, or the like. In other scenes, which appear to take place in a sanctuary, everybody is dressed in heavily decorated garments. In both cases, the garments function to create uniformity among and indeed assimilate the individuals in the scene, in this way indicating that they are 'the same' or of the same status. Another distinction may also be made: in the case of individuals wearing regular clothing, the scene should most probably be interpreted as mortals attending a ritual, whereas scenes in which everybody wears colourful, decorated garments in a sanctuary should most probably be interpreted as depicting deities. In other depictions, though, single individuals stand out by reason of their exceptional dress, as for example the *kanēphoros* leading the procession on the *krater* in Ferrara. Here, the garments are used to make one individual stand out so as to be recognised by the ancient viewer as a special personage, probably some kind of cult personnel.

Notes

1. For the importance of the altar in Greek cult, see Ekroth 2009a.
2. According to Ekroth, an altar may show that the action is set in public space: a sanctuary, the agora, the palaestra, or along a road. But at the same time, the altar could also mark the private sphere, since altars were located in the courtyard of the house. In both cases, however, the context is religious activity. Ekroth 2009a, 99–100.
3. *ThesCRA* I, 19–20.
4. There are also examples of naked individuals performing rites at altars. These will however be excluded due to the focus of this study on textiles.
5. Staatliche Museen Kassel, Antikensammlung, inv. no. T551.
6. E.g. British Museum, inv. no. E324.
7. E.g. Museum of Fine Arts, Boston, inv. no. 95.25.
8. E.g. Musee Dobree, Nantes, inv. no. D974.2.8.
9. E.g. British Museum, inv. no. E53.
10. Ashmolean Museum, inv. no. 1973.1.
11. Connelly 2007, 111.
12. Von Bothmer (*non vidi*); Connelly 2007, 111.
13. British Museum, inv. no. E284.
14. Metropolitan Museum of Art, New York, inv. no. 1979.11.15.
15. Connelly 2007, 111; *LIMC s.v.* Hera pl. 412.
16. Museum of Fine Arts, Boston, inv. no. 65.908.
17. National Archaeological Museum, Athens, inv. no. 2740.
18. Museo Archaeologico Etrusco, Florence, inv. no. 3950.
19. Purple mantle: e.g. British Museum, inv. no. 1894.7-19.1; white mantle: e.g. Museum of Fine Arts, Boston, inv. no. 13.195; Man: *CVA* Fiesole, Collezione Costantini 1, 18–19.

20. Lee 2015, 123.
21. Cleland et al. 2007, 58.
22. Lee 2015, 123–124.
23. Metropolitan museum of Art, New York, inv. no. 75.2.11.
24. Lee 2015, 121.
25. National Archaeological Museum, Athens, inv. no. 3624. Another possible example of a *chitōn/chitōniskos cheiridotos* appears on a red-figure *hydria* from ca. 430 BC. Arthur M Sackler Museum, inv. no. 1960.342.
26. Miller 1989, 323.
27. Miller 1997, 176.
28. Miller 1997, 180–181.
29. Thiersch 1936.
30. Miller 1989, 322.
31. Cleland 2005, 109.
32. In the Archaic Period, there are a few depictions in vase-painting of men or boys carrying *kana*. *ThesCra* 5, 269. E.g. a neck amphora from ca. 520 BC, Antikensammlung, Munich, inv. no. 1441; a *skyphos* from the end of the 6th century BC, National Archaeological Museum, Athens, inv. no. 12531.
33. Connelly 2008, 218.
34. Palagia 2008, 32; van Straten 1995; *ThesCra* 5, 269–274.
35. Roccos 1995, 642, 645.
36. Roccos 1995, 665. Museum of art, Toledo, inv. no. 1972.55.
37. Newcastle University, Shefton collection, Newcastle upon Tyne, inv. no. 203. Staatliche Museen zu Berlin, Antikensammlung, inv. no. F2189.
38. Roccos 1995, 651.
39. Museo Archaeologico Nazionale, Ferrara, inv. no. 44894.
40. National Museum of Denmark, inv. no. 10780.
41. Roccos 1995, 650. *LSJ s.v. epiporpēma*.
42. *Chitōn* with stripe: National Museum of Denmark, inv. no. 6. Dotted *chitōn*: Museum of Fine Arts, Boston, inv. no. 13.195.
43. Roccos 1995, 647.
44. State Hermitage Museum, St. Petersburg, inv. no. BAK7.
45. Connelly 2007, 105.
46. Connelly 2007, 106.
47. Himmelmann-Wildschütz 1959, 27–31.
48. Fürtwängler 1881, 106–115.
49. Patton 1990, 326.
50. Nilsson 1957, 99–106, esp. 101; Connelly 2007, 109–110.
51. National Museum of Denmark, inv. no. 6.
52. Connelly 2007, 110–111. E.g. National Museum of Denmark, inv. no. 6.
53. Connelly 2007, 88.
54. Chaniotis 2008, 32.

Chapter 11

Clothing regulations in sanctuaries: The written sources

Sources and methodology

At a number of sanctuaries, inscriptions detailing clothing regulations for visitors have been found. It appears from the inscriptions that 'visitors' meant anyone who wished to enter the temple or the entire *temenos* of a deity, or to participate in a cultic procession. Often nothing certain can be said of the issuing body, proposer, date, exact context (e.g. location), or political associations of these texts.[1] Inscriptions detailing clothing regulations have been recovered in several sanctuaries and areas, e.g. Boeotia, the Peloponnese, Arcadia, Delos, Rhodes, and Asia Minor (Map 2). These regulations, which were inscribed on stone *stelai* placed at the entrance to the *temenos* or perhaps on the walls of the temple itself,[2] range in date from about the 6th century BC to the 3rd century AD,[3] but the majority belong to the latter part of the period, and only one can be dated to the 6th century BC.

Before we turn to the analysis and discussion, a word of caution is in order: Regulations concerned with clothing and other forms of adornment constitute only a small portion of the preserved inscriptions relating to ancient Greek cults. Regulation in the form of such inscriptions may have been the exception rather than the rule, and their scarcity seems to indicate that the majority of sanctuaries did not require regulation of dress.[4] Furthermore, the inscriptions should not be treated as a cohesive group because they derive from a wide range of locations and time periods.

These religious regulations include provisions for the cleanliness, fabric, colour, type, and design of the clothing, as well as the wearing of certain accessories.[5] The regulations are thus an important source of information on ritual dress and the use of garments and dress items in sanctuaries. The majority of the regulations contain prohibitions rather than injunctions, meaning that they often specify what visitors were *not* allowed to wear, rather than what they *should* wear.

As for the inscriptions, several are classified as 'sacred laws', *leges sacrae*. This category was invented and coined in the 19th century as a means of organising the inscriptions for publication. The sacred laws are comprised of different types of inscriptions: decrees and laws.[6] These inscriptions include three broad areas of subject matter: the sanctuary, the cult personnel, and the rituals, particularly those involving purification and sacrifice.[7] The modern term 'sacred laws' is problematic, however, since it is very broad and includes inscriptions that are not strictly speaking 'laws' or even 'regulations'.[8] Carbon and Pirenne-Delforge therefore suggest that we move away from the term 'sacred laws' and instead collect those inscriptions which contain sacrificial and purificatory ritual norms, and term these Greek Ritual Norms (GRN) instead of sacred laws.[9] As it turns out, though, Carbon and Pirenne-Delforge set up stricter rules for acceptance into this category. For example, they suggest leaving out the inscription from Arcadia (*LSCGS* 32) that prohibits access to the sanctuary to

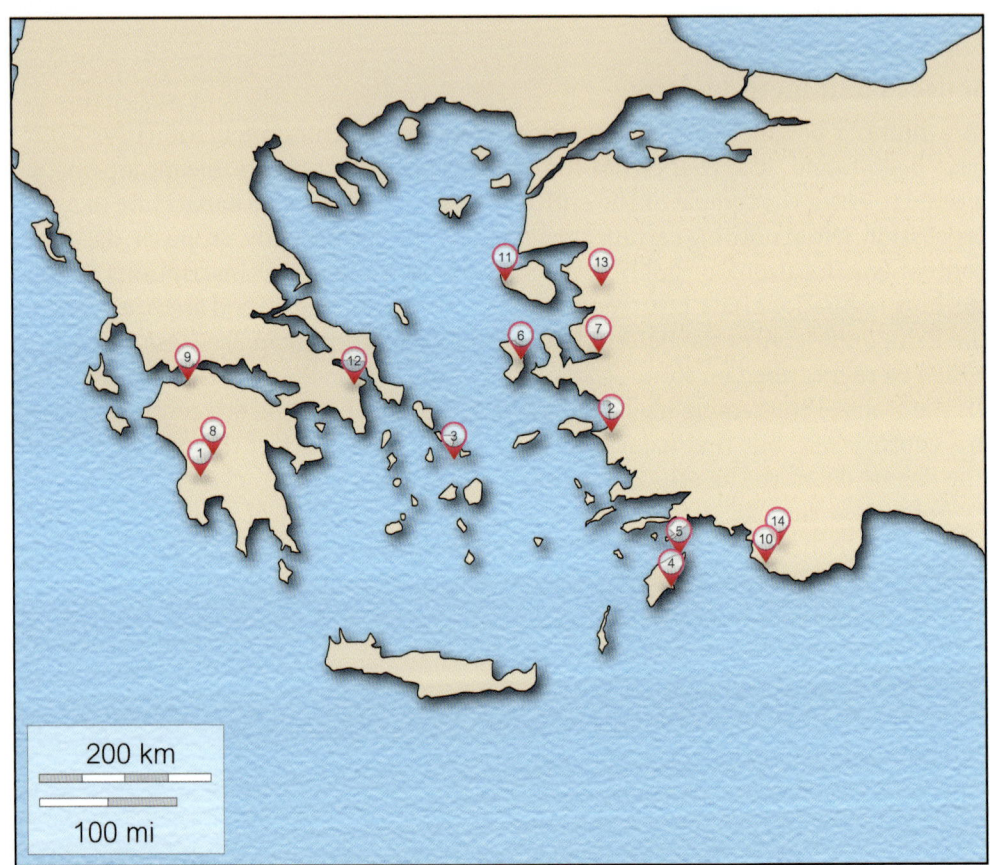

Clothing regulations

1. Andania (LSCG 65)
2. Priene (LSAM 35)
3. Delos (LSCGS 56, LSCGS 59, LSCG 94)
4. Lindos, Rhodes (LSCGS 91)
5. Ialyssos, Rhodes (LSCG 136)
6. Chios (LSAM 6)
7. Pergamon (LSAM 14)
8. Lykousoura, Arcadia (LSCG 68)
9. Patras (LSCGS 33)
10. Xanthos, Lycia (SEG 36, 1221)
11. Eresos, Lesbos (LSCG 124)
12. Marathon (SEG 36, 267)
13. Smyrna (LSAM 84)
14. Tlos, Lycia (LSAM 77)

those wearing brightly-coloured clothes.[10] Since this inscription is of great relevance, and therefore included in this study, I do not employ this new term GRN. Instead, I will simply use the term 'regulation' as a convenient way of indicating the nature of the inscriptions in question, but of course first and foremost so as not to exclude any inscriptions important to the subject of clothing and adornment in sanctuaries.

The clothing regulations under investigation here concern sanctuaries for different deities (Map 2):

'The Great Goddess' (Andania)
Demeter and Despoina (Lycosoura),
Demeter (Patras, Andania, Chios),
Athena (Lindos),
Alektrona (Ialyssos),
Pan and the nymphs (Attica),
Trophonios (Lebadaia),
Asklepios (Epidauros, Pergamon),
Dionysos (Smyrna),
Zeus and Athena (Delos),
the Egyptian divinities (Delos), and
Leto (Lycia).

In most cases the regulations are associated with the cults of female deities, especially Demeter (Patras, Andania, Lycosoura, Chios), in whose sanctuaries colourful garments are specifically prohibited.

According to Cleland, Greek cultic clothing regulations usually fall into two distinct types: one group simply stipulates that white clothing should be worn. None of these specify gender, and the prescription is brief and straightforward. They all date later than the early 3rd century BC. The other group of regulations range in date from the 6th to the 1st centuries BC. They originate in the Peloponnese, and are especially associated with cults of Demeter. These regulations are concerned with the use of particular colours in clothing, and often also with restrictions on the use of gold.[11]

As we shall see, however, even though the overall picture follows Cleland's simplified distinction between the two types, there are regulations that do not fit either of the two types exactly, and some inscriptions include prescriptions as well as proscriptions. The group of regulations requiring white clothing is a viable category, but the other group is somewhat more diverse than Cleland allows. This second group is not necessarily limited to the Peloponnese or the chronological period defined by Cleland. On the contrary, there are examples of a prohibition on colourful garments from sites in Asia Minor, e.g. at Chios and Tlos, where precious garments and flowery clothing respectively are prohibited. Other regulations do not fit in either group. For example, certain regulations are focused on prohibiting jewellery, metals, footwear, or specific textile fibres, which demonstrates that not just the clothing, but also the 'details' related to dress and adornment were of utmost importance.

In the following discussion, I will examine the evidence for clothing regulations recorded in epigraphic and literary sources. The material is divided into two main groups: prescriptive regulations, primarily regarding white or clean clothing, and proscriptive regulations, focusing on different aspects of clothing. The very complex regulations for the mysteries of Andania are prescriptive as well as proscriptive, however, and will therefore receive a separate initial treatment. The material will be presented in these three main groups, in chronological order as far as this is possible.

The regulations of the cult at Andania

A cult inscription connected with the mystery celebrations of the Great Goddess at Andania on the Peloponnese provides detailed clothing regulations for all entering the sanctuary (Appendix 3.1).[12] According to Pausanias (4.33.5), this festival was second in sanctity only to the Eleusinian Mysteries, but is otherwise barely known from literature. As the location of the mysteries has yet to be identified archaeologically, it would have remained practically unknown if it had not been for the discovery of this particular inscription.[13]

The beginning of the inscription is missing, but it is still the longest and most detailed regulation in existence, comprising 194 lines.[14] It comprises numerous paragraphs arranged by subject matter, and identified by appropriate sub-headings, and it covers most issues that the administration of the festival might entail.[15] The main part of the clothing regulations is concerned with the *hieroi* and *hierai*.[16] Fourteen lines of the inscription describe the type of clothing that the priests and priestesses, male initiates, female initiates and their daughters and slaves had to wear in the cult ceremonies (Table 33).[17] The inscription dates to 92/1 BC, the year of the reform of the cult, but the cult itself is much older and the clothing regulations might therefore have been in existence much earlier.[18]

According to the inscription, the board of ten in charge of the mysteries had to wear a purple headband during the mysteries. The adult priestesses were required to wear a *kalasīris* or a *hypodyma* and a *himation*, in total worth not more than two minas, while the girls were required to wear a *kalasīris* and a *himation*, in total worth not more than 100 drachmas, i.e. half the price of the priestesses'.

In the procession the priestesses had to wear a *hypodyma* and a woollen *himation*, with stripes not more than half a finger wide, and their daughters a *kalasīris* and a *himation* that was not diaphanous. In contrast with the details given for the priestesses, there does not seem to be any further reference to the clothing for the priests, except that they should wear a wreath.[19] The only clothing regulations for male initiates at Andania were that they should be dressed in white and go barefoot. The regulations are therefore often characterised as being concerned only with women, but it is notable that male participants are also subject to regulation, although their restrictions are less detailed.[20]

The female initiates were also required to wear a garment that was not transparent and that had stripes not more than half a finger wide. The prohibition on wide stripes is interesting, since it indicates that wide stripes were 'showy' and unsuitable for

Table 33. The clothing regulations of the cult at Andania (see also Appendix 3.1).

Wearer	Gender	Garment type	Decoration	Maximum price	Other
Hieroi	M				Wreath
Hierai	F				White felt cap
Recently initiated	M/F				Tiara – wreath of laurel
Initiated men	M	White clothing			Barefoot
(Initiated) women	F	No transparent clothes	Stripes max. half a finger wide (*himation*)		
Free adult women	F	Linen *chitōn* and a *himation*		Max. 100 drachmas (*himation*)	
Girls	F	*Kalasēris* or a *sindonites* and a *himation*		Max. 1 mina (*himation*)	
Female slaves	F	*Kalasēris* or a *sindonites* and a *himation*		Max 50 drachmas (*himation*)	
Hierai (adults)	F	*Kalasēris* or an undergarment and a *himation*	No coloured border (*himation*)	Max 2 minas (*himation*)	
Girls	F	*Kalasēris* and a *himation*		Max 100 drachmas (*himation*)	
Hierai (adults)	F	Undergarment and a woollen woman's *himation*	Stripes max. half a finger wide (*himation*)		
Girls	F	*Kalasēris* and a *himation* that is not transparent			
All women	F				No gold, rouge, make-up, hairband, plaited hair. No shoes unless of sacrificial leather or felt

both an initiate and a priestess. Even though the *hierai* have a wider choice of inner garments than other adult women, they were still not allowed much decoration.[21]

The women being initiated had to wear a linen *chitōn* and a *himation* worth not more than 100 drachmas, while their daughters were to wear a *kalasīris*, or a *sindonitēs*, and a *himation* worth not more than 1 mina.[22] The same provisions applied to female slaves, but in this case the value was reduced to 50 drachmas.[23] All of the women were to be unadorned. Gold ornaments, make-up, hair bands, and braided hair were forbidden, and their shoes could only be of felt or the leather of sacrificed animals. If these regulations were not respected, the garments were to be consecrated to the gods.[24] This could thus be evidence for a collection of garments in the sanctuary, like those recorded in the temple inventories.

Prescriptive clothing regulations for sanctuaries

The prescriptive clothing regulations are few and include only six examples, ranging in date from the 3rd century BC to the 3rd century AD, and originating in different geographical areas, including Epidauros, Delos, Chios, Rhodes, Priene, and Pergamon. They thus span a very large chronological period as well as geographical area.

The earliest example is the hymn of Isyllos to Asklepios, which is preserved in an inscription erected at Epidauros in 280 BC (Appendix 3.2).[25] The inscription states that participants in the procession of Apollo and Asklepios leading to the Temple of Apollo should wear their hair long or down and wear white garments and a wreath.

Another early clothing regulation, also dating to the 3rd century BC, concerns a phratry or family cult at Priene (Appendix 3.3).[26] The text is extremely fragmentary, and is thought to refer to a cult of a goddess.[27] According to the inscription, a person by the name Anaxideros obtained the priesthood or sacrifice and is required to wear white clothing.

The next example is an inscription probably dating to ca. 116/5 BC (Appendix 3.4). It records the regulations for entering the Sanctuary of Zeus Kynthios and Athena Kynthia at Delos. It is a reissue of a former ordinance by the priest of the two deities, who orders that visitors to the sanctuary should have pure hands and soul, wear white clothing and no shoes, belt/girdle, or iron finger ring.[28]

The last examples are all from the Roman Period. One is a *stele*, recovered at the Sanctuary of Athena Lindia on Rhodes, which bears an inscription with regulations regarding ritual purity (Appendix 3.5).[29] It stipulates that the sanctuary visitor must wear clean clothes, not bound by a belt/girdle, must not wear any headdress, and must be either barefoot or wearing white shoes that were not made of goat leather (items of clothing made of goat leather were generally not permitted at this site).[30] According to Blinkenberg, the inscription indicates that the clothes probably should be not only freshly washed, but also white.[31]

An inscription from Chios, dating to the 1st century AD (Appendix 3.6), requires that a participant in the rituals shall be barefoot and wear clean clothes and no gold.[32]

An inscription from Pergamon, dating to the 3rd century AD, declares that visitors who wished to undergo incubation at the Sanctuary of Asklepios were required to wear white clothing and wear a wreath of olive (Appendix 3.7). Furthermore, visitors are not allowed to wear a ring, a belt/girdle, anything of gold, or to bind up their hair, and they must go barefoot.[33]

The inscription from Andania also requires men to go barefoot and to wear white clothing.[34] Finally, even though it is not a clothing regulation, the decree of 108/7 BC from the Acropolis of Athens, treated in appendix 1,[35] should be mentioned, since it records the white garments of the *Arrēphoroi*. And as discussed above, the inscription from Demetrias at Magnesia stipulates that the priest of Apollo Koropaios, a representative of each of the colleges of *strategoi* and *nomophylakes*, a *prytanis*, a treasurer, the scribe of the divinity, and the prophet should all wear white.[36]

White garments

These examples of prescriptive clothing regulations demonstrate that the most important thing to emphasise was the colour of garments, not their type, fibre, or any other quality. In some instances, the regulations prescribe only white clothing, while in others this is only one requirement among several. This means that these regulations are often not simply prescriptive, but also proscriptive, prohibiting e.g. rings, girdles, gold, headdress, etc. Some of the regulations not only prescribe white clothing, but also require that sanctuary visitors wear their hair down, wear a wreath, or go barefoot.

In addition, this requirement to wear white can be viewed as an implicit ban against coloured garments. There are several literary sources that confirm the prescription against white clothing in specific sanctuaries: examples include the Athenian orator Aeschines (390–314 BC), whose speeches indicate that wearing white garments in sanctuaries was a common practice: 'And though it was but the seventh day after the death of his daughter, and though the ceremonies of mourning were not yet completed, he put a garland on his head and white raiment on his body, and there he stood making thank-offerings, violating all decency'.[37]

The requirement of white garments is also indicated by Plato, who writes:

> 'For woven stuff and other materials, white will be a color befitting the gods; but dyes they must not employ, save only for military decorations.'[38]

Apuleius (125–180 AD) writes that the initiates of Isis all wear linen clothes (*linteae vestis*),[39] which indicates light, possibly white clothing, since flax does not take dyes well, and Tertullian (*de Pallio* 4.10) informs us that the initiates of Ceres wear entirely white clothing.[40] Furthermore, the testimony of the apocryphal *Acts of John* in the Bible tells how John upsets the crowd by entering the Artemision wearing black clothes, while the worshippers are all wearing white.[41] Another written source is a fragment of the lost play *The Cretans* by Euripides, preserved for us by Porphyry in his treatise on

Abstinence from Animal Food. It describes an Orphic ceremony, and states that initiates of the Greek deity Zagreus wore white garments.[42] According to Athenaeus, at the Panegyris of Apollo Komaios during the festivities in the Prytaneion in Naukratis, the participants were to wear white garments.[43] Finally, Diogenes Laertius (3rd century AD) reports that Pythagoras' prescription for ritual purity included white clothing.[44] These examples have led scholars such as Gawlinski to infer that 'white clothes were generally used for marking festival time.'[45]

It has been proposed that white clothing emphasised the ritual purity of the worshipper.[46] This acquires support from the obvious fact that white clothing is delicate and easily shows dirt and stains, thus strengthening its association with cleanliness. That visitors should be clean when in a sanctuary is expressed in several inscriptions, e.g. the document from the mid-5th century BC inscribed on the wall of a public building in the agora of Thasos, which calls for a drastic purification including the washing (of clothes).[47] Yet the requirement was not necessarily solely based on physical cleanliness, but also on more symbolic meanings.

As Robertson observes, purity is often a requirement of Greek sacred regulations, which provide instructions on the purificatory procedures that an individual must complete before approaching a given deity to offer a sacrifice. Rules of purity appear in the 6th century BC and continue through the Late Hellenistic Period.[48] Certain terms are used to denote purity: *katharos* expresses purity as 'clean', while *agnos* means 'pure'. It is commonly assumed that Apollo and Artemis are especially associated with purity regulations. According to Robertson, however, the arguments for this are not very strong, and other gods are associated with purity as well. Nevertheless, it appears that certain forms of purity were particularly, but not exclusively, required by the Delian gods.[49] According to Robertson, the gods associated with *agnos* purity include only six of the Olympian deities: Zeus, Apollo, Dionysos, Artemis, Athena, Demeter, and also the younger god Asklepios.[50]

The requirement of purity could involve different practices: abstinence from certain foods, stipulations about cleanliness, a specific number of days of abstinence from sexual intercourse, birth, etc.[51] Some of these requirements can be explained by the entrance of Egyptian and Oriental deities into the Greek pantheon in the 4th century BC, since the traditional worship of these gods entailed unaccustomed food prohibitions and the exclusion of a woman during menstruation.[52] Nevertheless, these regulations clearly also existed in Greece before the 4th century BC, which means that this cannot be the entire explanation. As discussed above, rules of purity could also involve elements of a person's appearance, such as dress, adornment, and ornament, which were subject to regulation in some sanctuaries, as specific clothing regulations express, and as depictions in vase-painting may confirm.

Proscriptive clothing regulations in sanctuaries

There are more proscriptive than prescriptive regulations surrounding clothes and adornment in sanctuaries. The earliest example of a proscriptive inscription is a bronze plaque from the 6th century BC, probably from northern Arcadia, which is inscribed with regulations regarding the cult of Demeter Thesmophoros: 'If a woman wears a brightly coloured robe, it is to be consecrated to Demeter Thesmophoros' (Appendix 3.8).[53]

The cult of the chthonic goddess Despoina at Lycosoura also had specific clothing regulations.[54] At the sanctuary, which is close to Megalopolis in Arcadia, a stone inscribed with a regulation dating to the 3rd century BC was recovered stating that gold objects and flower-decorated or black clothing were not permitted, unless for dedication (Appendix 3.9).[55]

A regulation from the 3rd century BC connected with the cult of Demeter near Patras states that 'women may not (...) wear a brightly coloured garment or purple garment (porphyrean) (...)'. Furthermore, women are not permitted to wear gold ornaments worth more than one obol (Appendix 3.10).[56]

A marble stele inscribed with a regulation from ca. 300 BC describes the rules for entering the Shrine of Alektrona (the daughter of Helios and the nymph Rhodes) at Ialyssos. In addition, the inscription states that there is a prohibition against wearing shoes or anything made of pigskin (Appendix 3.11).[57]

An inscription found at the entrance of the Letoon at Xanthos, Lycia, Anatolia, dating to the end of the 3rd or the beginning of the 2nd century BC, states that persons entering the sanctuary are forbidden to wear a *petasos*, a *kausia* (essentially the Macedonian ethnic traditional headgear),[58] a dress-fastener (*porpē*), or any object made of brass or gold.[59] The inscription specifies that all clothes worn in the sanctuary must not have clasps, and that only a minimal outfit consisting of a simple garment and shoes is allowed (Appendix 3.12).

Several inscriptions from Delos include clothing regulations for entering temples, especially the Serapeion. Thus, an inscription dating to the 2nd century BC states that visitors are not allowed to wear woollen garments in the sanctuary for the Egyptian divinities (Appendix 3.13),[60] while another inscription from the 2nd century BC, found in the Serapeion, also relating to the Egyptian divinities, states that wearing bright-coloured garments is not permitted (Appendix 3.14).[61]

An inscription dating to the 2nd century BC from a sanctuary of an unknown goddess from Eresos (Lesbos) describes specific rules of purification before entering the *temenos*.[62] The inscription also records items prohibited from the *temenos*: e.g. no bronze, iron, and shoes, nor anything else made of skin (Appendix 3.15).

A dedication to Pan and the nymphs in the form of a pedimental marble stele has been recovered from the eastern entrance to the Cave of Pan at Marathon.[63] The stele dates to 61/60 BC and bears a prohibition against visitors wearing coloured or dyed garments (Appendix 3.16).

Another late inscription, this time from Smyrna and dating to the 2nd century AD, states that visitors are not allowed to approach the altar wearing black clothing in the Sanctuary of Dionysos Bromios (Appendix 3.17).[64] Finally, an inscription from the Lydian city of Tlos, of unknown date and associated with an unknown cult, records that flowery clothing and transvestism is forbidden (Appendix 3.18).[65] These proscriptive regulations appear, on first impression, quite diverse, but on closer scrutiny, they reflect five main types of proscribed clothing and accessories:

1. colourful or decorated clothing,
2. footwear,
3. headgear,
4. fibres or leather,
5. metals.

These prescribed items will be treated one by one in the following discussion, in which specific garment types will also be addressed.

Colourful garments

Especially prominent in these regulations is the prohibition against brightly coloured garments – either in general, sometimes emphasising the colour purple or, in one instance, black (*melas*) and flowery (*anthinos*). It is significant, however, that literary sources provide evidence to the contrary, since several sources describe how people in sanctuaries wore coloured garments: examples include the cult of Artemis Brauronia, Attica, where the young girls, according to Aristophanes, wore saffron-coloured garments.[66] According to the orator Polemon (2nd century BC), the women whom he saw in the Artemision of Perge had their heads covered and wore purple and white dresses as well as rich jewellery.[67] In both of these cases, it is impossible to be certain if this was a requirement for attending the cult. It is also possible that these ancient authors mentioned/described colourful clothing because it struck them as uncommon.[68]

A final example involves the old local chthonic god Trophonios who was venerated at an underground sanctuary at Lebadeia in Boeotia. This oracle cult was probably of pre-Greek origin and illustrates the importance of clothing as a ritual requirement.[69] The Greek philosopher Maximus of Tyre (2nd century BC) reports that those who wished to consult the divine power should be invested with 'a robe, reaching to his feet and a purple mantle (*phoinikis*)', indicating that purple was the appropriate colour to be worn when visiting the oracle,[70] while the Greek writer Philostratos (ca. 170–250 AD) states that the garments of visitors were required to be white.[71] It has been proposed that this indicates that the worshippers wore a white robe with a purple seam, or a purple mantle (*phoinikis*) over a white garment.[72] Alternatively, these testimonia could be interpreted as evidence that the ritual clothing for initiates changed over time from a purple to a white garment.

From the 2nd century BC to the 2nd century AD, then, regulations prohibit extravagant clothing in particular, such as garments embellished with ornaments and embroidered with woven patterns and figures. The specification of colour is not surprising, since colours signified value, because their application involved work and time and potentially very expensive dyes. These regulations could therefore be viewed as a reflection of a requirement of modesty and simplicity, as is expressed elsewhere in the price limits for the garments worn by visitors, for example in the regulations from Andania. In these regulations, the directives on the clothing and general appearance of females indicate that not only was there a concern with women appearing modest and without showy ornamentation, but also with the cost of their attire and how their costume signalled different levels of status within the cult.[73] The inscription from Andania also proscribes diaphanous garments, which has been interpreted as a requirement that women's bodies be covered, on the grounds that transparent clothing was seen as soliciting male attention, with the result that women thus apparelled could be associated with adultery and prostitution.[74] Alternatively (or additionally), though, this could be another example of a restriction placed on luxury, since see-through garments would imply a very fine – and thus expensive – quality.

The view of most scholars appears to be that the regulations were aimed at restricting the display of status expressed through dress and other types of adornment among the visitors to these sanctuaries, and that they thus served an instrument for imposing 'anti-luxuria' and modesty.[75] *Luxuria* was restricted because it had a dangerous potential to upset the natural order.[76] It could be argued, however, that the status indicators of the women would probably still be immediately apparent from the manner in which they dressed, even though none of them could wear makeup, gold jewellery, or have braided hair.[77]

Dress as hybris

In my view, the aim of these regulations was to prevent people from dressing too extravagantly in order not to 'outshine' the image of the god or goddess. It was thus probably not a coincidence that the sanctuary visitors could not wear the same colours or dress as extravagantly as the images of the deities or the priestly personnel. The regulations also created a way to visualise the difference between priest and community.[78] This was of course also the case for prohibitions against wearing gold. It is thus clear that there was a certain rhetoric of modesty towards the divinity, and what was permitted for gods was not allowed for humans.[79] In sum, there seems to have been a general conception of the attire that was proper for approaching a divinity. One exception is the cult of Asklepios at Epidauros, where the cult statue also wore white garments. This is also illustrated in iconography, for example on the Attic red-figure krater, which depicts a sacrificial procession in honour of the god Apollo at Delphi.[80] Here, the male participants are dressed just like the god sitting in his temple.

As the previous chapter demonstrated, the goddesses often received offerings of beautiful colourful garments, and purple was the most frequently described colour among the offered textiles. Furthermore, the preserved colours on e.g. Archaic sculpture and depictions in vase-painting illustrate that the images of the goddesses were richly ornamented and wore colourful garments that were embellished with figural decoration.[81]

Perhaps the ban against brightly coloured garments (or the requirement for white garments) cannot solely be explained by anti-luxury and the avoidance of outshining the deities in the sanctuary. In Mesopotamian myths brightly-coloured garments are given by gods to other gods (for example, in a myth, Enki gives Inanna a black and multi-coloured garment) and are said to possess numinous powers, indicating that even the gods were more powerful when wearing them.[82] It is perhaps possible that in Greece too such precious coloured garments were somehow imbued with special powers, and therefore prohibited to visitors and reserved for the gods.

Footwear
Regulations concerned with clothing or other types of adornment also often mention footwear. The regulations sometimes require specific footwear, as in the inscription from Lindos, which stipulates that visitors must wear white shoes not made of leather (Appendix 3.5), and in the inscription from the Sanctuary of the Great Goddess at Andania, where the shoes should be of felt or the skin of sacrificial animals (Appendix 3.1). Other regulations strictly prohibit footwear in the sanctuaries, and specifically require the visitors to go barefoot. This requirement occurs at the Sanctuary of Alektrona at Ialyssos (Appendix 3.11),[83] the Sanctuary of Zeus Kynthios and Athena Kynthia at Delos (Appendix 3.4), the Sanctuary of Asklepios at Pergamon (Appendix 3.7), the Sanctuary of Demeter at Chios (Appendix 3.6),[84] the Sanctuary at Eresos on Lesbos (Appendix 3.15),[85] and at the Sanctuary of Despoina at Arcadia no sandals are allowed (Appendix 3.9).[86]

The requirement that participants in cultic rites must go barefoot is also mentioned in literary sources, e.g. by Kallimachos (ca. 305–240 BC), in his *Hymn to Demeter*:

> 'And as unsandalled and with hair unbound we walk the city, so shall we have foot and head unharmed forever. And as the van-bearers bear vans full of gold, so may we get gold unstinted. Far as the City Chambers let the uninitiated follow, but the initiated even unto the very shrine of the goddess – as many as are under sixty years. But shoes that are heavy and she that stretches her hand to Eileithyia and she that is in pain – sufficient it is that they go so far as their knees are able. And to them Deo shall give all things to overflowing, even as if they came unto her temple.'[87]

Ancient Greeks of a certain status wore shoes when going out, and shoelessness was considered to be for the poor and slaves.[88] Perhaps the removal of one's shoes before entering a sanctuary (or a certain part of a sanctuary) was a way of signifying a transition from the secular to the sacred world.[89] This is also the view of Dillon, according to whom regulations of footwear are related to rites of liminality, and in this way stress the difference between the sacred and profane worlds.[90]

This act of removing shoes is depicted on the parapet of Athena Nike's Temple on the Acropolis of Athens, where Nike removes her sandal possibly as a sign that she is entering the sanctuary.[91] Furthermore, as previously discussed, in many depictions of ritual in vase-painting the participants have bare feet. A possible explanation for their inappropriateness in a sacred context is that shoes can carry dirt and thus pollute the sanctuary. Furthermore, if the shoes were made of leather, the tanning process could cause an even further pollution since it involved the use of filthy materials like urine.[92] Gawlinski suggests, as an alternative explanation, that being barefoot made one physically closer to nature, and was therefore the best way to approach a deity.[93] According to Wächter, the prohibition on shoes could also be explained by the binding of the shoes, which could have a hindering or obstructing force, as is also reflected in the prohibition in some sanctuaries against wearing a tied belt or girdle or braided hair.[94]

Yet it is also possible that the proscription against footwear was related to the avoidance of luxurious adornment, since an underlying purpose of this prohibition could be to keep out fancy foreign shoes – another control of display.[95] Ancient footwear is seldom preserved in the archaeological record, but we know of many different types from literary sources. Attested types include *baukides* – luxurious saffron-coloured Ionian shoes, which appear to have been especially popular with prostitutes - or the *blautē* – a man's sandal or slipper, sometimes with a turned-up toe, decorated with golden brooches and worn at banquets.[96] The *krēpis* was a leather boot reaching to the knee or shin and was used by soldiers and hunters, but could also be made of white leather and was, in such cases, considered effeminate.[97] These few examples serve to illustrate the possible richness of footwear and thus their association with luxury. This may also provide an explanation for Pausanias' remark (9.39.8) that the worshipper in the cult of Trophonios at Lebadeia who wished to consult the oracle had to wear boots that were of local manufacture (*epichōrias krēpidas*).[98] Perhaps this is an indication of a prohibition against luxurious footwear imported from e.g. the east. And in vase-painting ritual attendants often have bare feet, whereas deities often have footwear.[99] Perhaps, in some instances, footwear is a further indication of divinity or possibly a ritual status.[100]

That footwear and bare feet could have ritual connotations may receive some support from archaeological evidence. Sculpted feet with or without sandals or carved footprints, often accompanied by an inscription, have been found all over the Mediterranean, from the Greek islands to Asia Minor, and from Egypt (where they are especially common) to the Iberian peninsula. These sculptures occur from prehistoric periods onwards.[101] Many consist of sculpted imprints or outlines of bare or sandaled feet in a stone slab or in the pavement. Several such footprints have been recovered in sanctuaries, and the majority are associated with Egyptian deities and primarily belong to the Imperial Period, although some are earlier.[102] The mosaic from the upper terrace of the Sanctuary of Demeter and Kore on Acrocorinth is highly relevant in this connection, since it depicts an outline of a pair of feet, originally inserted in metal, pointing towards the sanctuary.[103] According to Dunbabin, because of their direction, the feet should be interpreted as belonging to the deity, which, on the basis of an inscription seems to have been the eponymous priestess Neotera.[104] The mosaic is placed near the entrance to the building.

Other examples include sculpted feet with or without sandals, of which many are directly associated with dedications to deities.[105] There are also examples of sculpted footprints, e.g. the pair of ca. 16 cm long stone imprints found in an Archaic level of the filling in the platform of the Temple of Nemesis at Rhamnous.[106] In yet other instances, they consist only of the footwear itself. As an example, three votive bronze sandals, dated to the end of the 6th or the beginning of the 5th century BC and dedicated to Apollo Corcyraeus, have been found in the god's sanctuary on Corfu.[107] They have been interpreted either as the feet/footwear of the worshippers that were dedicated to Apollo, or as representations of Apollo's feet and hence his presence in the sanctuary.[108]

Both the imprints as well as the sculpted types are usually interpreted as either the feet of pilgrims/worshippers or the feet of divinities.[109] Yet in some instances the imprints were placed in front of a cult statue, which has been interpreted as implying some specific ritual function, perhaps as an indication of the spot where the initiate had to stand for presentation to the god.[110]

In light of the evidence examined above, though, these depictions may instead have been references to the ban on footwear and to the presence of barefooted individuals. Of course, representations of footprints are also common in secular contexts such as baths, and their representations were always dependent upon the context.[111]

Headgear

A third sub-group of clothing regulations is concerned with different forms of headgear. This group includes both proscriptive and prescriptive regulations. The proscriptive regulations include prohibitions against particular or even any type of headdress, or against the binding/tying of hair. The regulations of the Sanctuary of Athena Lindia on Rhodes (Appendix 3.5) instruct visitors not to wear a headdress, and the regulations of the Sanctuary of Despoina at Lycosoura (Appendix 3.9) prohibit women from having braided hair and a covered head, which probably means that women should have their hair loose and completely uncovered.

The regulations of the Letoon at Xanthos prohibit visitors in specific terms from wearing a *kausia* or a *petasos* (Appendix 3.12). The *kausia* was a Macedonian broad-brimmed, flat felt hat, sometimes purple, which became part of Macedonian royal insignia; the prohibition of this item in the sanctuary could be an expression of male *anti-luxuria*.[112] The *petasos* was another type of broad-brimmed felt hat, part of the insignia of the ephebes and an attribute of Hermes, and its prohibition can likewise be considered an expression of *anti-luxuria*.[113]

The prescriptive regulations primarily consist of instructions that visitors should wear wreaths. This prescription is exemplified in the *Hymn of Isyllos*, which inform us that visitors should wear their hair long and a wreath (Appendix 3.2), as well as in the regulations from the Sanctuary of Asklepios at Pergamon, which instruct incubants to wear a wreath of olive (Appendix 3.7).

The regulations of the mysteries at Andania (Appendix 3.1) contain an entire section concerning wreaths. It states that the sacred men are to wear wreaths, the sacred women a white felt cap (*pilos*), and the first among the initiated a tiara (*stleggida*).[114] But when the sacred men give the order, they are to take off their tiara, and they all to be wreathed with laurel. Even though the heading of this part of the inscription refers to wreaths, the inclusion of a felt cap and tiara makes it clear that this heading is being used to refer generally to headwear, not just wreaths.[115] It is unclear exactly what the *pilos* would have looked like.[116] It does not appear, however, to have been strictly related to ritual, since it is also known from secular contexts.[117] The term *stleggis* usually refers to a strigil, but, according to Gawlinski, it indicates a tiara-like crown in this context. It is unclear what exactly it would have looked like, but it has been categorised with other kinds of wreaths, and it has been suggested that it might have been used in other religious rituals.[118]

It is clear from the regulations that wreaths especially were an important part of the adornment of sanctuary visitors, whether initiates or not. This conclusion receives support from iconography, especially vase-painting, where participants in rituals or processions often are depicted with wreaths. According to Gawlinski, the wreath was used in ritual as an outward symbol of the inward change created by initiation, marking the conclusion of a rite of passage. She compares the antique ritual use of wreaths with crowning ceremonies, which signify a change in status for those who begin political office or benefactors who are publicly honoured, in which cases the wreath is used as a symbol of initiation, as it was at Eleusis and other Mysteries.[119] Wreaths were thus a physical marker of religious ritual and festival time,[120] and were inexpensive and thus affordable for everyone.[121]

The prescription for long, loose hair and the prohibition against hairstyles such as braiding could be explained as another instance of *anti-luxuria*, since elaborate hairstyles required time and in some cases the use of jewellery such as hair pins, etc., and could thus denote high status. The prohibition against hairstyling thus speaks to the appropriate look for women at religious festivals; hair was just as important as clothing and accessories in this kind of context. This has contributed to the perception that long hair was the norm for Greek women, who in fact kept it bound up outside of festival contexts, when it was kept loose.[122]

Leather, skins, and fibres

A fourth sub-group of regulations prohibit leather, skins, or certain types of fibres to be brought into sanctuaries. Among the proscriptions against leather or animal skins are: the regulations from Andania, which prohibit leather sandals unless they are made from sacrificed animals (Appendix 3.1); the regulations from the Sanctuary of

Athena Lindia on Rhodes (Appendix 3.5.), which prohibits footwear or anything else made of goat skin; and the regulations from Eresos, Lesbos (Appendix 3.15),[123] which prohibit leather in the sanctuary. A regulation from the cult of Apollo, the nymphs, and the Charites at Thasos, dating to the 5th century BC, bars sheep and pig from the sanctuary,[124] and at Kos the priestesses of Demeter are forbidden to wear clothing made from dead animals.[125]

It has been suggested that leather was prohibited because it was ritually impure by reason of its association with death.[126] A hide used in manufactured goods may not have come from a sacrificial animal, which means that the product would be unclean and thus defile a sanctuary.[127] The ban against leather could – as is evident from a few of these examples – also be related to the prohibition against footwear, which was often made from leather.

Other regulations – although few in number – are more clearly directed at the specific fibres of the garments worn. An example is the regulation from the Sanctuary of the Egyptian divinities on Delos, which has a strict proscription against wool (Appendix 3.13). The regulations from Andania do not prohibit any fibres, but do require the female hierophants to wear a *hypodyta* and a women's *himation* of wool, while the free adult women are to wear a linen *chitōn*. Pausanias also reports that visitors to the cult of Trophonios must wear a linen *chitōn* (9.39.8). Furthermore, Apuleius, in his description of a procession honouring Isis at Cenchreae, Corinth, speaks of linen as worn only by priests and initiates.[128]

There is thus evidence that wool was deemed undesirable and that linen was required in some sanctuaries. The requirement for linen has been explained on the grounds that it was considered natural and 'organic', and that it had associations of purity and with the colour white.[129] Philostratos, for example, argues that although the gods love sheep, the purity of linen makes it preferable to other fabrics, including wool, especially among the Indians, Egyptians, and Pythagoreans, who use it in their rituals.[130]

As a parallel, in Judaism there is an intense preoccupation with the fibres from which religious garments are made. The Torah contains a general prohibition against cloth that combines sheep's wool and linen, which is referred to by the term *sha'atnez*:[131] 'A garment of two kinds, sha'atnez, shall not come upon thee'[132] and 'Do not wear Sha'atnez – wool and linen together.'[133] No explanation is given for this commandment.[134] *Sha'atnez* was in fact not entirely prohibited, but was accepted or even required in priestly garments:[135] This prohibition was designed to separate priestly from public practice and had the effect of reserving this fibre mix for holy purposes.[136]

In Judaism the garments of the priests were essential in order for them to fulfil their sacred functions. In fact, in the absence of these garments, the offerings made by the priests in the Temple had no validity. Thus, without his 'uniform', the priest serving in the Holy Temple was considered a 'stranger' and as a secular person.[137]

We should also include such considerations in the study of sacred dress in Greek sanctuaries, where it is possible that, in some cults, specific fibres were not considered appropriate or were restricted to certain people.

A number of Greek sources refer to the exclusion of sheep in Egyptian cults.[138] One of these sources is Herodotos, who provides information on animal sacrifices at Thebes: in the temple of Zeus at Thebes goats are sacrificed, but not sheep,[139] while in the temple of Mendes/Osiris sheep are sacrificed, but not goats.[140] According to Herodotos, the aversion from sacrificing or touching sheep can be explained by the fact that Zeus showed himself to Herakles displaying the head and wearing the fleece of a ram; because the Thebans therefore consider rams sacred, they do not sacrifice them.[141] Also, according to Strabo, the Thebans honoured sheep,[142] but once a year they flayed a single ram and put the fleece on the image of Zeus.[143] Herodotos, moreover, in his description of Egyptian ordinary clothing, emphasises the fact that the white woollen robe worn over the linen garment (*kalasīris*) was always removed before the wearer entered a sanctuary.[144] Plutarch also comments on the preference for linen and the aversion against sheep and goats in Egyptian sanctuaries: priests wore linen garments because flax was considered pure and afforded simple, clean, and convenient clothing, while wool (as well as the flesh of sheep) was avoided because it was considered unclean.[145] These sources indicate that the reason for prohibiting sheep and/or goats (and their wool) was either because they were considered sacred or because they were considered impure.

According to Seiterle, the Greeks believed wool to have a purifying, cathartic, healing, and generally apotropaic effect. He gives the examples of the mystics at Eleusis and other places, who wore white woollen threads around their wrists and ankles, as well as the custom of placing a tuft of wool on the door when a girl was born in the household.[146] However, this stands in contrast to the Greek clothing regulations, which appears to have preferred linen to wool. In addition, the temple inventories record only a few woollen garments, but they do attest to the dedication of unworked wool, which could provide support for Seiterle's statement.

Metals

A fifth sub-group of proscriptions remains to be discussed. Some regulations include prohibitions against bringing or wearing metals in sanctuaries. These prohibitions usually target objects of iron or bronze, but a few are concerned with objects of gold.[147] This could be interpreted as a prohibition against bringing weapons into the sanctuaries, but, in fact, such a prohibition would also prevent a visitor from wearing a garment fastened with a dress pin or fibula made of metal, and would thus affect the visitors' dress-codes. The only instance in which dress-fasteners are specifically prohibited is the inscription from the Letoon at Xanthos, which identifies dress-fasteners (*porpai*) or any object made of cobber (*chalkos*) or gold as forbidden objects in the sanctuary.¹⁴⁸ Further evidence for this practice is the inscription from Eresos (Lesbos), which prohibits visitors from bringing bronze or iron into the sanctuary, and a stele, dating to the 5th or 4th century BC, from a shrine of Apollo at Minoa on Amorgos, which declares that iron may not be brought into the sanctuary.¹⁴⁹ This prohibition is also attested in literary sources: for example, Kallimachos writes that iron is excluded from a shrine of the hero Menedemos on the island of Kythnos.¹⁵⁰ Plutarch also gives examples of the prohibition against bringing metals into sanctuaries.¹⁵¹

Several of the clothing regulations surveyed here mention the prohibition against wearing gold and restrictions on the maximum value of gold ornamentation: at Patras, women are not allowed to wear gold ornaments worth more than one obol; at Andania, women cannot wear gold ornaments at all; at Lycosoura, no gold is allowed in the sanctuary; and in the sanctuary of Demeter at Chios, women are to 'leave things of gold at home.'¹⁵² The prohibitions against gold could of course be considered as an expression of anti-luxury, targeted specifically at women since adornment was a means by which a woman could display her wealth publicly and express her femininity.¹⁵³

There are still further potential reasons for excluding gold rings: for example, they could be associated with magic and binding because of their form and the manner in which they are worn. Gold and gold-plated rings in particular are banned in the inscription from the Letoon at Xanthos (Appendix 3.12). Scholars have suggested several explanations, such as negative magical associations between gold and motherhood, or, if they are signet rings, a general intent to prevent worshipers from entering the sanctuary with their symbols of authority.¹⁵⁴

Garment types

Only a few of the regulations prohibit or prescribe specific garment types, which suggests that this was not of outmost importance in the sanctuaries in comparison with, for example, the proscription of certain colours, fibres, accessories, or materials. The only instance of a requirement that specific garment types should be worn is in the regulations from Andania, specifying attire for women, except for the women who are initiated – perhaps this was implicit and therefore did not need to be described. This fits rather well with the evidence from iconography, where attendants in rituals wear 'normal' garments. By way of comparison, in the temple inventories there is only one example of a *kalasīris*, in the inventory from Miletos.

Three of the prescriptive regulations specify – in addition to the requirement of white clothing – that the visitor is not allowed to wear a belt/girdle (*zōnē*).¹⁵⁵ This may appear in the iconography: for example the woman standing closest to the altar on a red-figure cup wears a loose garment (see Fig. 111),¹⁵⁶ and some priests wear an ungirded *chitōn*. Belt/girdles were often soft woven or braided belts tied in a knot rather than leather belts fastened by a buckle,¹⁵⁷ so the prohibition is not likely to be related to the ban on leather. Rather, the proscription could be connected with the fact that the girdle was tied in a knot, which is sometimes interpreted as having a hindering or obstructive force, as noted above.[158]

Gender: regulating women in the sanctuaries?

A further point should be made regarding gender in the regulations, which appear to focus on women's clothing to a slightly higher degree. The only instances of regulations that are specifically directed at women are at the sanctuaries of Demeter at Patras and Arcadia, but at the Sanctuary for the 'Great Goddess' at Andania there are clothing rules for both men and women, although the restrictions on female adornment are much more specific than those for males. In both of these Peloponnesian sanctuaries of Demeter, brightly coloured and purple garments are prohibited, and the Andanian inscription further limits the decoration of female garments. Only the regulations from the sanctuary of Zeus Kynthios and Athena Kynthia on Delos proscribe the clothing of men alone, and this does not involve decoration, colour, or the like, only the fibre type of the garments. This can be explained by the fact that women were banned from entering this sanctuary. There are other instances of cults that were exclusive to one gender, for which reason it was not necessary to specify the dress and adornment of the 'unwanted' gender at the entrance. Moreover, women were excluded from the Temple of Herakles at Delphi,[159] and a late 5th century BC inscription from Elatea associated with the Sanctuary of the Anakes disallows the presence of women.[160] Women are also excluded in one of the most substantial individual sets of sacrificial prohibitions, the inscription from Thasos, dating to the mid-4th century BC, which forbids the sacrifice of goats and pigs to Herakles.[161] Furthermore, Pausanias (8.31.8) reports that there were strict visiting rules in a sanctuary for the Great Goddess and the Maid (Demeter and Kore) at Megalopolis, where women were allowed to enter the sanctuary at all times, but men could only enter once a year.

Another possible explanation is the existence of different entrances for men and women, which would eliminate the need for regulations concerned with both male and female dress, since the inscription, being placed at one of the entrances, would only address one or the other gender. Furthermore, it might be significant that most ordinary Greek temples were opened only on special days,[162] in that the regulations would thus have been of importance only a few days of the year.

The majority of the regulations, however, appear to be 'gender-less' in their expression, since they do not specify whether they apply to men or women, which indicates that these rules and regulations were aimed at both genders. Still, even though the clothing requirements are usually not specifically stated to apply to women, several scholars assume that they are directed at them. For example, Dillon argues that the reference to e.g. 'brightly-coloured' garments is clearly more likely to apply to women than men.[163] According to Batten, the Andanian inscription expresses a concern about *luxuria*, which lies behind the directives for gold, clothing, and elaborate hair. For:

> 'if women were permitted to wear such things, they might continually pursue more and more adornment possibly to the financial and social detriment of the association. Although women in the association might have admired the hairbands and jewellery of their fellow initiates, the males in the association decreed that there must be limits.'[164]

It must be stressed, though, that these interpretations are rooted in typical gender stereotypes such as the assumption that 'real men' do not wear brightly coloured, decorated garments or jewellery, since these items are considered effeminate. It also has its roots in the perception that women are defined in a more visual, material, consuming, and outward way than men,[165] and thus run the risk of being anachronistic. Moreover, these interpretations are based on the assumption that regulation of female clothes is much more common than of male clothes, which is basically incorrect. In the Roman sources, there is often a tension between women's desire to wear jewellery and fine clothes and the Roman male stress on modesty and simplicity.[166] Yet it is unlikely that this situation is directly rooted in ancient Greece. So can we completely exclude the possibility that men could wear such garments? As an example, Alcibiades was a powerful man, but still wore luxurious and purple clothes,[167] although the descriptions of his clothes in the literary sources were probably intended as a form of ridicule and disassociation. My point here is simply that we must be careful before jumping to conclusions about the gendering of dress and adornment.[168]

The assumption that these restrictive regulations are directed exclusively at women (since men apparently would not wear brightly-coloured garments or jewellery) has caused scholars to consider them an attempt to exercise social control over women and a means to limit women's empowerment through the restriction of their display of wealth.[169]

Religious occasions could be unique opportunities for the public display of striking personal adornment and elaborate clothing. It has been claimed that these occasions would have been even more important for women than for men, since women were less engaged in the public life of the state and the market.[170] The clothing and jewellery restrictions would therefore have had their greatest real impact on daily conduct at the time of religious celebrations and at other times where women were travelling around to participate in religious worship. According to Culham, it is therefore possible to say that the Greek inscriptions do not 'reflect purely profane or exclusively sacred concerns and that neither was wholly sacred nor profane in effect.'[171] The restrictions 'reduced women's freedom to use and display wealth, and do this to maintain social control over women; and, indeed, the religious rites themselves appear to be another means to that end.'[172] Batten argues that limiting women's adornment with jewellery, fine clothes, and hairstyles would make them less visible in the public, in that they would blend in more and not attract attention.[173]

Yet the question is whether these regulations (in instances where they are specifically directed at women) are solely concerned with displays of social status. A further reason that the adornment of women, and not men, is the focus of attention in these cases may be that women play a more important role in the rituals of these sanctuaries: perhaps they were allowed into otherwise restricted areas such as the inner sanctum or were in closer contact with the cult statue, for which such colourful and precious items might have been reserved.

Penalties

Only a few of the regulations specify a penalty for a failure to observe properly rules of dress. Examples include the inscriptions of the sanctuaries of Demeter at Andania and Arcadia, which declare that the transgressor should dedicate the garment(s) that do not comply with the rules; likewise, at Lycosoura, forbidden items are only admitted for dedication. Because the penalty/provisions did not require that the offending objects be removed from the *temenos* or that the site be purified, but rather that the things stayed in the possession of the goddess, it is clear that no pollution of the temple can have resulted from such textiles or the negligence of their wearers.[174] This conclusion receives further support from the evidence of the temple inventories, where colourful and decorated garments are recorded. Dedication as a penalty argues against ritual impurity of the articles themselves. Instead, it has been suggested that the regulations perhaps functioned as a kind of tax on certain dress impulses.[175] Cleland suggests that it seems that those who could afford it simply wore the proscribed items and paid the penalty, in this way demonstrating that they had access to such items, and that they were indifferent to their loss or at least ready to pay a fine to the cult. She argues, further, that the clothing regulations were not an attempt to completely prevent the proscribed items from being brought into the sanctuaries, but instead represented 'an assertion of control over the behaviour of the participants.'[176] This is unlikely, however, at least if the purpose of the prohibitions was to ensure that visitors did not 'out-shine' the image of the present deity. This again can explain why the transgressing garments had to be dedicated, since they would please the deity, and were not at all related to impurity.

Clothing in funeral legislations

Regulations governing clothing were not limited to sacred contexts, but could also be found in private, sacred contexts such as funerals.[177] A number of Greek city states introduced legislation aimed at restricting the expense and extravagancy of the burial, in terms of the number of participants in the procession, the value and number of grave offerings made to the deceased, the duration of mourning, the size and costs of the grave monument – and some specifically address the value of textiles involved in the funeral rituals: for example, the garments worn by the participants as well as the textiles used to shroud the corpse.

One such example is the so-called law of Solon, which is also the earliest funerary legislation of which we have knowledge. It was supposedly passed by Solon at the beginning of the 6th century BC, and its existence is reported by three different much later literary sources: Demosthenes, Cicero, and Plutarch.[178] With regard to the specific regulation of clothing, Cicero and Plutarch are of particular relevance. Cicero tells us that the expense was to be limited to three veils and a purple tunic.[179] According to Plutarch, Solon introduced a number of laws, among which restrictions were also made on the use of textiles: female mourners were not allowed to wear more than three *himatia* and the deceased could not be buried in more than three *himatia*.[180] It is notable that Solon's legislation, or at least the three later sources, do not mention pollution as the cause for these regulations.[181]

Another example of funerary legislation that regulates the use of textiles is an inscription on a stone block at Delphi, which states the legal and religious obligations of members of the Labyad phratry.[182] The legislation dates to ca. 400 BC, and includes several rules, some of which concern textiles. The inscription stipulates that the shroud (*chlaina*) had to be thick and of light grey colour, and that only one bier cloth (*strōma*) and one pillow was allowed for the deceased.[183]

Funerary regulations from the 3rd century BC found at Gambreion near Pergamon also address the use of textiles.[184] In these regulations, women in mourning were to wear grey clothing, while men and children were also to wear grey, unless they preferred white. The inscription concludes by demanding that the next *stephanephoros* inscribe the law on two *stelai*, of which one copy should be erected before the gates of the Thesmophoreion and the other before the temple of Artemis Lochia. The regulations address post-burial rites exclusively. According to Garland, the obvious explanation for this is that already existing laws controlled conduct at the funeral, but post-funerary rites had only recently begun to constitute a public nuisance.[185]

From the city of Iulis on the island of Keos comes the most wide-ranging and extensive body of funerary legislation to survive from the Greek world.[186] It is inscribed on both sides of a stele dated to the second half of the 5th century BC.[187] According to the regulations, the corpse was to be wrapped in a maximum of three white *himatia* (specified as *strōma*, *endyma*, and *epiblēma*), worth a maximum of 300 drachmas. Furthermore, the corpse was to be borne on a bier with the head exposed, but the rest of the body covered with the *himatia* (or *othonia* according to another reading).[188] According to Garland, the primary intention behind this law appears 'to have been to prevent the death of a member of one's family from being exploited for political effect.'[189] Yet it is important to note that Greek law did not have the intent to protect the individual, but rather to regulate communities and ensure that people behaved in the right way, in this case limiting *luxuria*.

A final example of funerary legislation that addresses textiles is provided by Plutarch, who tells us that the lawgiver Lykourgos at Sparta forbade anything to be buried with the dead and ordered that the corpse was to be covered in a purple robe (*phoinikis*).[190] At Sparta, this type of garment was only worn by soldiers, and this regulation thus appears to have been aimed at a specific part of the Spartan population.[191]

These five examples show that textiles needed to be regulated not only in sanctuaries, but also at other ritual events such as funerals. The regulations are primarily restrictive, limiting the number or value of the textiles, but the regulations from Gambreion are more prescriptive in nature, specifying the colour to be worn by the attendants at the funeral, and the Iulis code appears unique in that it combines restrictive with prescriptive ordinances. This is not surprising, given the fact that on the whole Greek funerary legislation was restrictive in nature, and its primary objective was to set maximum limits, while they were rarely prescriptive or proscriptive.[192]

Generally, the regulations concerning funerals are notably similar to the regulations on sanctuaries. For example, they are primarily targeted at women's clothing, while there are no restrictions for men. Only the regulations from Gambreion stipulate the colour of the garments worn by men, but they are still allowed a greater freedom of choice than women.

Interestingly, there are also restrictions on the textiles used to shroud the corpse. That regulations were also aimed at the deceased has been explained on the grounds that the body was the focal point of the event, and thus the most obvious place for display through excess,[193] since an expensive, elaborate, and well-attended Greek funeral – as well as a religious festival – provided a perfect opportunity for the display of family wealth, status, and prestige in front of the entire community.[194] Thus, the main motive for Athenian funerary legislation appears to have been to impose sumptuary measures as an attempt to prevent invidious excesses in the display of luxury among the elite, which apparently was more likely to happen through the dress and adornment of women.[195] The reason that the dress of female participants in funerals was regulated was not that the women needed to be prevented from 'outshining' the deceased, who was after all the main actor of the event, as is demonstrated by the fact that the corpse was also subjected to limitations of display. This is in contrast to the situation in the sanctuaries, where the purpose of the regulations was both to restrict the display of luxury and perhaps to ensure that no visitor competed with the cult image of the deity.

Notes

1. Gawlinski 2011, 2.
2. Petrovic & Petrovic 2006, 175.
3. Mills 1984, 257.
4. Cleland 2002, 25.
5. Mills 1984, 257.
6. Gawlinski 2011, 3.
7. Carbon & Pirenne-Delforge 2012, 164.

8. Carbon & Pirenne-Delfore 2012, 170.
9. Carbon & Pirenne-Delforge 2012, 171–172.
10. Carbon & Pirenne-Delforge 2012, 175, note 58.
11. Cleland 2010, 8.
12. *LSCG* 65; *IG* V, 1, 1390.
13. Lupu 2006, 105.
14. For a detailed analysis of the entire inscription, see Gawlinski 2011.
15. Lupu 2006, 105.
16. Cleland 2010, 5.
17. Dillon 1997, 196. For translations see Mills 1984; Ogden 2002; Deshours 2006.
18. Sokolowski 1969, 130.
19. Dillon 1997, 197.
20. Cleland 2010, 5.
21. Cleland 2010, 7.
22. 100 *drachmae* = 1 *mina*.
23. Dillon 1997, 197.
24. Mills 1984, 259.
25. Stehle 1997, 132, 134. *IG* IV², 1, 128.
26. This inscription is also included among the evidence for priestly dress, since it refers to priesthood.
27. Cleland 2002, 31.
28. Sokolowski 1962, 114; *LSCGS* 59; *ID* 2529. Kleijwegt 2002, 122.
29. Blinkenberg 1941, 871–878, no. 487; *LSCGS* 91.
30. Mills 1984, 258.
31. Blinkenberg 1941, 871–878, no. 487.
32. *LSAM* 6.
33. *LSAM* 14. Mills 1984, 258; Radke 1936, 60.
34. *LSCG* 65.
35. *IG* II² 1060 and 1036.
36. *LSCG* 83/84.
37. Aeschin. *In Ctes*. 77.
38. Pl. *Leg*. 12, 956a.
39. Apul. *Met*. 11.10. Gawlinski 2011, 161; Patera 2012, 43, note 62.
40. However, he also writes that 'the (opportunity of) wrapping with a broader, purple tunic and of taking on a Galatic, red mantle commends Saturn (to others).' de Pallio 4.10.
41. Kleijwegt 2002, 122. John 38. The author of the *Acts of John* is unknown. It was probably composed in the 2nd century AD. Ehrman 2003, 94.
42. Eur. *Fr*. 475.
43. Ath. *Deipn*. 149D-E. Radke 1936, 60.
44. Diog. Laert. 8.1.19.
45. Gawlinski 2011, 117.
46. Dillon 1997, 199.
47. *LSCGS* 65. Robertson 2013, 234.
48. Robertson 2013, 195–196.
49. Robertson 1983, 393.
50. Robertson 2013, 237.
51. For example, for Artemis Kithōne the shrine announces ordinary intervals of keeping pure after death, birth, and intercourse. *LSAM* 51; Robertson 2013, 219.
52. Robertson 2013, 196.

53. Mills 1984, 258; Sokolowski 1962, 70; *LSCGS* 32.
54. Despoina ('mistress') is an old local goddess of Lycosoura who shares most of her characteristics with Artemis, but also with Hecate and Kybele. Her cult is associated with Demeter, who is considered her mother. Loucas 1994, 97–98; *LIMC* III, I, 383.
55. *LSCG* 68; *IG* V, 2, 514. Sokolowski 1969, 137; Loucas 1994, 97–98.
56. *LSCGS* 33; Sokolowski 1962, 71–2; Mills 1984, 258.
57. *LSCG* 136; *IG* XII, 1, 677; Sokolowski 1969, 233–234; Dillon 1997, 162.
58. *SEG* 36:1221; *LSAM* 6.
59. *SEG* 36:1221; Lupu 2006, 16.
60. *LSCGS* 56.
61. *IG* XI, 4 1300; *LSCG* 94.
62. *LSCG* 124 (shoes in line 17). Robertson 2013, 213, claims that the inscription beyond any doubt belongs to a sanctuary of Apollo.
63. *SEG* 36:267.
64. *LSAM* 84.
65. *LSAM* 77.
66. Ar. *Lys.* 645.
67. Polem. *Phgn.* 68.
68. Another example is Strabo 15.1.58 who writes that the worshippers of Bacchus wear robes and turbans, use perfumes and are dressed in dyed and flowered garments. Falconer 1903. This example, however, is problematic, since this is a description of 'the foreign/exotic' or 'the other', made by a Greek, and therefore is perhaps more representative of what and how the Greek author considered non-Greek clothing customs. See also Hartog 1988.
69. *Pauly-Wissowa s.v.* Trophonios.
70. Max. Tyr. XIV, 2.
71. Philostr. *V. A.* §19.
72. *Pauly-Wissowa s.v.* Trophonios.
73. Batten 2009, 485.
74. Batten 2009, 489.
75. E.g. Batten 2009; Cleland 2002; 2010; Mills 1984.
76. Batten 2009, 496.
77. Batten 2009, 485.
78. See also Shamir forthcoming.
79. Patera 2012, 39.
80. Museo Archaeologico Nazionale, Ferrara, inv. no. 44894.
81. See e.g. Brinkman 2004a; Brinkmann 2004b.
82. Waetzoldt 2013, 203.
83. *LSCG* 136.
84. *LSAM* 6.
85. *LSCG* 124.
86. *LSCG* 68.
87. Call. *Cer.* 118–133. Translation: Mair 1955. The requirement of bare feet in cultic contexts appears to have become common practice also in the Roman Period, and the custom is described by e.g. Philostr. (ca. 170–250 AD), *V. A.* 1, 8 and Iambl. *V. P.* (ca. 250–325 AD), 85.
88. Gawlinski 2011, 115; Blundell 2002, 146.
89. On a similar note, on his return from Troy, Agamemnon removes his shoes before he steps onto the textiles that Clytaimnestra had placed at the palace entrance. Agamemnon had religious reasons for this, since these textiles belonged to the gods and therefore entailed/signified a more sacred sphere. Blundell 2002, 147.

90. Dillon 2002, 263.
91. Blundell 2002, 149; Gawlinski 2011, 115.
92. Gawlinski 2011, 116. There are examples of tanning being banned near sanctuaries: IG I³ 257 lines 5-9 describes how tanning is prohibited upstream from the Sanctuary of Herakles near the Ilissos River because of the use of polluting materials - both literally and metaphorically.
93. Gawlinski 2011, 116.
94. An example of the prohibition against wearing a belt/girdle (zonē) is an Imperial inscription from the Asklepieion at Pergamon, Fränkel 1890-1915, no. 264, 10; Wächter 1910, 22.
95. Gawlinski 2011, 130.
96. Cleland et al. 2007, 169.
97. Cleland et al. 2007, 106. For Classical Greek footwear terms, see Morrow 1985, 175-184.
98. Paus. 9.39.8; Dillon 1997, 161.
99. For the representation of footwear in Greek sculpture, see Morrow 1985.
100. See e.g. Wortmann 1968.
101. Petridou 2009, 85.
102. Dunbabin 1990, 86.
103. The building dates to the 1st century AD, but possibly the mosaic is somewhat later. Dunbabin 1990, 85, note 1.
104. Dunbabin 1990, 95.
105. Dunbabin 1990, 86.
106. Touchais 1984, 751, fig. 30.
107. Archaeological Museum of Corfu, inv. no. MR 936,937.
108. Petridou 2009, 86; Dunbabin 1990, 86, 95.
109. Petridou 2009, 85. In his study of Graeco-Roman inscriptions with depictions of feet in honour of Isis, Takács interprets representations of feet, bare or shod, as either votive offerings, commemorations of a visit to a shrine, or as implying a wish for a successful return home, but he also suggests that such feet could belong to a god. Takács 2005, 354.
110. Dunbabin 1990, 93.
111. Dunbabin 1990, 97.
112. Cleland et al. 2007, 103.
113. Cleland et al. 2007, 147.
114. This type of headgear is related to the *mitra*, see Papadopoulou forthcoming.
115. Gawlinski 2011, 110.
116. According to Lee, the *pilos* is a close-fitting skull cap, which has been identified in different variations: craftsmen, especially metalworkers, wear a plain cap possibly of leather, while fishermen and shepherds often wear a softer version made of wool, fur, or animal skin. Elite men also wear a type of *pilos*, often with a narrow brim. Lee 2015, 160.
117. E.g. Hes. *Op*. 545-546.
118. Gawlinski 2011, 112.
119. Gawlinski 2011, 112.
120. Gawlinski 2011, 110.
121. Ar. *Th*. 443.
122. Gawlinski 2011, 127.
123. *LSCG* 124.
124. *LSCG* 114 A, line 2.
125. Parker 1983, 52; Dillon 2002, 263.
126. Cleland et al. 2007, 111.
127. Robertson 2013, 215.

128. Apul. *Met.* 11.4.
129. Gawlinski 2011, 122.
130. Philostr. *V.A.* 8.7.5. Gawlinski 2011, 122.
131. Milgrom 1983, 65.
132. *Leviticus* 19.19.
133. *Deuteronomy* 27.11. Translation: Morris & Brooks 2007.
134. Morris & Brooks 2007, 246.
135. Milgrom 1983, 65.
136. Shamir forthcoming, 9.
137. Shamir forthcoming, 11.
138. Lupu 2006, 211.
139. Hdt. 2.42.1.
140. Hdt. 2.42.2.
141. Hdt. 2.42.3–5.
142. Strabo 17.1.40.
143. Hdt. 2.42.6.
144. Hdt. 2.81.1.
145. Plut. *De Is. et Os.* 4.
146. Seiterle 1999, 251.
147. Wächter 1910, 115.
148. *SEG* 36:1221. Lupu 2006, 16.
149. *LSCGS* 60. Robertson 2013, 215.
150. Call. *Fr.* 663. Robertson 2013, 215.
151. Plut. *Prae. ger. reip.* 26. See also Plut. *Quaest. Rom.* 40 and *Aristid.* 21.4.
152. *LSAM* 6. Robertson 2013, 226.
153. Batten 2009, 490.
154. Gawlinski 2011, 126.
155. *LSCGS* 59 (Delos); *LSCGS* 91 (Rhodes); *LSAM* 14 (Pergamon).
156. Museo Archeologico Etrusco, Florence, inv. no. 3950.
157. Cleland et al. 2007, 80.
158. Wächter 1910, 22.
159. Loraux 1990, 25.
160. *LSCG* 82.
161. Lupu 2006, 58.
162. Lupu 2006, 74.
163. Dillon 2002, 263.
164. Batten 2009, 496.
165. Batten 2009, 493.
166. Batten 2009, 486.
167. Plut. *Alc.* 16.1.
168. See e.g. Brøns 2012.
169. E.g. Cleland 2002; 2010; Mills 1984, 261.
170. Culham 1986, 235; Mills 1984, 258.
171. Culham 1986, 238.
172. Culham 1986, 238.
173. Batten 2009, 498.
174. Günther 2013, 258.
175. Cleland 2010, 10.
176. Cleland 2002, 43.

177. Mills 1984, 261.
178. Garland 1989, 3. For the problems in ascribing these laws to Solon, see Garland 1989, note 3.
179. Cic. *Leg.* 2.59.
180. Plut. *Sol.* 21.4–5.
181. Garland 1989, 5.
182. *LSCG* 77. Garland 1989, 8.
183. Garland 1989, 9.
184. *LSAM* 16.
185. Garland 1989, 10.
186. Garland 1989, 11.
187. *LSCG* 97A; *IG* XII 5, 593, side A, 2–4.
188. *IG* XII 5, 593, side A, 6–8. Garland 1989, 11.
189. Garland 1989, 12.
190. Plut. *Lyc.* 27.1–2. For the Spartan *phoinikis*, see Campobianco forthcoming.
191. Lipka 2002, 191; Cleland et al. 2007, 148, 174. Xen. *Lac.* 11.3; Ar. *Lys.* 1140.
192. Garland 1989, 15.
193. Gawlinski 2011, 109, 121.
194. Gawlinski 2011, 109; Garland 1989, 2.
195. Cleland 2002, 44.

Chapter 12

Discussion: Sacred dress-codes in sanctuaries

Dress, collective visual appearance, and community in ritual

The clothing regulations, whether prescribing white garments or proscribing colourful garments, would have the same intent: to make people conform to similar patterns of dress and to prevent anybody from standing out because of extravagant adornment, with the aim of creating a collective visual appearance. As already argued above, these regulations could have derived from *anti-luxuria* ideals, and would have served to prevent people from displaying their wealth and status when in a sanctuary (or attending a funeral). Another result of such dress regulations, though, could have been to give visitors a sense of group identity that was emerged from the act of wearing similar dress and adornment. Adornment – or in this case especially the rejection of adornment – may therefore have served as a means for symbolically binding together a community and reinforcing a common sense of identity.[1] In this way, clothing could serve as a means of recognition and as a way of forming, advertising, and imposing group identity.[2] By way of comparison, studies of modern religious communities have shown that dress-codes can strengthen the sense of belonging that members of such communities feel by demarcating them within the larger population.[3] Therefore, the objective behind such regulations could be to create an outward display of conformity, which would serve to create a sense of community among the visitors in a sanctuary.[4] As Arthur argues:

> 'Symbols, such as dress, help delineate the social unit and visually define its boundaries because they give non-verbal information about the individual. Unique dress attached to specific cultural groups, then, can function to insulate group members from outsiders, while bonding the members to each other. Normative behaviour within the culture re-affirms loyalty to the group and can be evidenced by the wearing of a uniform type of attire.'[5]

Similarly, Joseph argues that the appearance of an entire group in similar dress affirms a common value or belief, and that dress may promote group cohesion, since members of a group can identify each other and draw comfort from knowing that they are not alone when all wear the same distinctive dress.[6]

Furthermore, religious dress – or perhaps simply obeying certain rules of dress – may protect the wearer from the disunifying effect of outside influences:

Roach and Eicher gives an example of the Hasidim, exemplified in the following quote of a Hasidic Jew:

> 'With my appearance I cannot attend a theater or movie or any other place where a religious Jew is not supposed to go. Thus my beard and my sidelocks and my Hasidic clothing serve as a guard and a shield from sin and obscenity.'[7]

This quote indicates how dress can influence the behaviour of an individual, preventing or encouraging certain acts. Similarly, dress used in ceremonial rites has also been shown to contribute to the creation of certain moods.[8]

This can be compared to social and military uniforms, which can be defined as 'standard appearances that are socially acknowledged to represent certain positions and the duties and rights that come with them.'[9] In this way, gaining the right to wear a specific uniform can become an expression of having become part of the group represented by this particular uniform. As Stig Sørensen explains, membership in the group is communicated visually and without involving direct contact to those who are knowledgeable or in possession of the code.[10] But what defines the uniform is not just the garment, but also the way it is worn, and the different constituent elements: in ancient Greek religion the headgear, the dress-fasteners, the footwear, and the belt, all these elements signal who the wearer is as well as his or her status within the religious community.

Sacred dress and rites of liminality

A crucial aspect of sacred dress is the capability to express, visualise, and materialise rites of liminality. The concept of liminality was first developed by anthropologist Arnold van Gennep, who introduced the 'liminal phase' of *rites de passage*.[11] According to van Gennep, all rites of passage or transition are marked by three phases: separation, margin (i.e. the liminal phase), and aggregation. During the first phase, the individual performs symbolic acts to signal detachment from an earlier point in the social structure. During the second (liminal) phase, the individual passes through a cultural realm that has few or none of the attributes of the past or the coming state – he/she is a *tabula rasa*, and has to follow prescribed forms of conduct. During the third phase, the passage is consummated and he/she returns to a stable state.[12] The concept of liminality has been taken up by Turner, who introduces the concept of *liminal personae* (or 'threshold people'). According to Turner, these individuals are ambiguous, since:

> 'this condition and these persons elude or slip through the network of classifications that normally locate states and positions in cultural space. Liminal entities are neither here nor there; they are betwixt and between the positions assigned and arrayed by law, custom, convention, and ceremonial.'[13]

This liminal phase is therefore likened to death, to being in the womb, to darkness, or to the wilderness.[14]

In ancient Greek cult, textiles were used to express such rites of liminality. For example, the inscription from Andania proscribes specific clothing for the initiates. Requiring that the initiates dress the same was a way of erasing their individual traits and anonymising them, thereby marking the liminal phase through which they were passing. This is further exemplified in the cult of Demeter and Kore at Eleusis, where initiates were expected to leave a specific offering behind after initiation: the clothes they had been initiated in,[15] thus indicating that they had left the liminal phase and successfully entered a new state. It is also possible that performing a sacrifice – or even simply entering a sanctuary – was a liminal phase, marked by specific attire. As discussed above, several regulations prescribe or proscribe certain types of clothing, footwear, or adornment, possibly as a way of indicating a difference between the sacred and profane worlds; it follows from this that the individual was in a liminal phase when in the sanctuary.

Religious hierarchies enacted through dress

This leads to the next important aspect of the function and meaning of dress worn in sanctuaries. These regulations did not necessarily serve to ensure that everybody present in the sanctuaries looked completely alike. On the contrary, they could also be used as a means of demarcating visitors or initiates as a group from e.g. priestly personnel. In this way, 'regular/ordinary' visitors or initiates could easily be distinguished from other actors in the sanctuary. According to Cleland, the effect of these regulations could be to erase the distinctions of profane social status usually expressed through dress, and to replace them with a different sacred/religious hierarchy articulated through dress provisions.[16] Consequently, ritual sub-hierarchies and status demarcations could still occur, and the position of each cult member in the external community could be reflected in what he or she was legally allowed to wear in internal cult activities.[17] Thus, specific adornment – or lack thereof – may have shown the individual's position within a religious group.[18] By way of comparison, the dress of the Roman Catholic clergy illustrates how clothing can be used to indicate position and rank within a religious structure: through the prescription of different dress for different clerical ranks such as monk, priest, bishop, cardinal, and pope.[19] This appears to have been the intention of the regulation from Andania, which distinguishes between the forms of dress and adornment permitted to priests, priestesses, initiates, slaves, etc. And, as demonstrated in the discussion on the dress of priestly personnel and visitors, priests and priestesses appear to have been allowed far greater freedom in their choice of dress and adornment in comparison to ordinary visitors. This is expressed through the fact that some religious personnel had the privilege of wearing decorated and/or coloured garments as well as gold jewellery, in contrast to ordinary sanctuary visitors. This would have made the priest or priestesses highly visible in the *temenos*, particularly because of the choice of colour. There are, however, several examples of priestly personnel wearing white garments. But perhaps at these sanctuaries white was not a requirement for visitors, which again would have rendered the priestly personnel distinct from the other people present in the *temenos*. In this connection, a word of caution is probably in order, since we cannot say with certainty that priestly personnel took advantage of this privilege or whether people actually observed these regulations, and we cannot be sure if all regulations had the same impact upon the recipients.[20] Obviously, such a regulation required either that people were able to read it or that somebody enforced the rules. Nevertheless, the evidence testifies to a religious hierarchy expressed through dress and adornment.

Colour as contrast

The written and iconographical evidence establish that the colour and decoration of dress especially were of great importance in sanctuaries. Also evident is the opposition between white and colourful or decorated fabrics – in particular, purple. Unfortunately, the sources are more or less silent regarding the function or meaning of the specified colours. Perhaps because it was obvious to the ancient Greeks or perhaps it was latent rather than conscious. Jones suggests three functions of colour in these texts: convention (what is agreed upon in society), association (the linkage a society makes between a certain colour and states or things), or contrast.[21] Yet these three functions cannot necessarily be discerned from each other if we do not have any further sources to rely on.

Nevertheless, contrast is an obvious explanation for the regulation of colour and decoration: when in a large area such as a *temenos*, an individual would be easily recognisable from a distance if he or she wore a specific colour such as purple or a heavily decorated garment. This means that it was easier to control, for example, who entered what areas in the sanctuary. The evidence should not therefore be interpreted as an indication that everybody in the sanctuary wore the same colour, but instead as an indication that some groups of people needed to be easily identifiable by their garments – perhaps so as to distinguish between men and women or priestly personnel and ordinary visitors. This receives support from Joseph's theories of visibility and dress, which argues that visibility in clothing is a social rather than a physical property. Thus, we respond not simply to the physical appearance of the clothing, but to the information it provides about the wearers' statuses or affiliations, the norms to which they are held accountable, their degree of conformity to these norms, and whether they are in the appropriate context.[22]

Thus, as shown by the evidence of priestly garments and the clothing regulations, in Greek processions white is almost always worn by groups, whereas purple appears to be the colour of individuals,[23] – a customary pattern that allows for a distinction between 'ordinary' participants (in white) and individuals placed higher in the religious hierarchy (in purple). This indicates that a kind of colour coding took place in Greek sanctuaries.

Dress as identifying with the divine

It appears, then, that priestly personnel were often allowed to wear purple garments and gold jewellery in contrast to the ordinary visitors, whose use of these items was restricted. This of course meant that the priestly personnel stood out. One reason for this could be that these highly marked garments illustrated the higher status and position of the priest or priestess, as well as his or her function within the cult. Yet in view of the conclusions made in the previous chapters concerning the dressing of cult images - namely, that these images were often dressed in very colourful garments bearing figural decorations - another potential purpose for priestly dress could be to associate the priests with the divinities whom they served. According to Paul, the essential characteristic of the priests and priestesses in the Greek world is his or her position as a mediator between the polis and the divine sphere. She argues that this intermediary position is most conspicuous in the sacrificial ritual, where the priest stands at the centre of the process of communicating with the divine.[24] The priest's outstanding position is defined by the ceremonial dress that he was required to wear and by the state of purity that he had to maintain. Paul further argues that this indicates 'if not an identification between the priest and the god whom she or he serves, at least a close connection with the divine.'[25] This implies that the dress worn by priests or priestesses was part of an identification process between them and the divinity whom they served. The cultic practitioner may even have been viewed as being transformed into a divine vessel of the spirits – a transformation made manifest by changes in dress.[26]

This conclusion gains strength from the fact that several cults required priestly personnel to wear items specifically identified with the deity in question. This is exemplified by a section in Plutarch in which he explains why the priest of Herakles at Antimachia wears a feminine garment and a veil when he sacrifices. Plutarch provides an explanation in the form of the myth of Herakles' arrival on the island of Cos and his escape from the inhabitants by dressing up as a woman.[27] According to Paul, this means that there can be a close link between a particular feature of the dress worn by the priest or priestess and a mythic event associated with the divine recipient.[28] This connection between items worn by priestly personnel and divinity is especially evident when it comes to wreaths, since the plant from which they were made could indicate a particular connection or association with the god or goddess. For instance, the crown of the priest of Herakles Kallinikos on Cos was made from white poplar, a tree which the demi-god had brought back from Acheron and was therefore associated with him.[29] Other examples include the priesthoods of Dionysos at Priene and Skepsis, where the priests were required to wear wreaths made of ivy, a plant closely associated with the god.[30]

This point can possibly be taken even further, by viewing clothing not only as an expression of religion, but as a form of religion itself. As Miller argues, there is a seamless connection between religion and clothing, and the designation of these two things as separate categories is artificial. Thus, 'when fibres, fabrics and ways of wearing are the medium for one's relationships to other people and to the gods, we cannot have "cloth" and "religion" we can only have the materiality of cosmology.'[31] This means that the dress worn by the priest or priestess or by the person performing a sacrifice in a way becomes religion, and not just an expression of religion. This again strengthens even further the potential and meaning of textiles in relation to cult and religion.

Symbolism versus recognition

Another important point is that men and women visiting sanctuaries were seen by others. This might seem obvious, but is actually of great importance for understanding the underlying function of these rules. By way of comparison, women and men who joined religious orders in the Medieval Period (and later) were increasingly enclosed within their convents, concealed from the gaze of outsiders – especially male. Thus, women who joined such an order, as for example the Poor Clares, did not encounter issues of recognition regarding their clothing.[32] Nevertheless, they were still subjected to strict rules of clothing regarding garment type, colour, and fabric, as well as of accessories such as belts. According to Warr, this indicates that these women's clothing carried an intimate symbolic meaning, that is to say, it represented their way of life and served as a symbol of their poverty not for others, but as a reminder to the wearer herself. This clothing symbolism was important in relation to their reward in heaven,[33] not in the relation to other human beings. This is in stark contrast to the situation in Greek sanctuaries, where people were seen, not hidden as nuns. And since Mediterranean communities are to be strongly ocular, i.e., allow for little privacy and exhibit a general anxiety about how individuals are observed in public,[34] this speaks to the importance of recognition and identification as a likely explanation for these rules of dress.

In sum, these chapters have established that clothing was an important part of attending and performing rituals in Greek sanctuaries. Not only priests, but also visitors appear to have been subjected to regulations and expectations as to what one should and could wear when in a sanctuary. This suggests that clothing played an important part in the non-verbal communication of the ritual environment, not least between suppliant or priest and the divinity. Clothing, moreover, may have played an important part in the communication with the divine, possibly even becoming part of cult itself. This illustrates how textiles are, in fact, central materialisations of Greek cult, by reason of their capacity to accentuate and epitomize aspects of identity, spirituality, and position in the religious system.

Notes

1. Roach & Eicher 1973, 18.
2. Warr 2010, 133.
3. Colburn & Heyn 2008, 6.
4. Gawlinski 2011, 109.
5. Arthur 1999, 3-4.
6. Joseph 1986, 50–51.
7. Roach & Eicher 1973, 19.
8. Roach & Eicher 1973, 9.
9. Stig Sørensen 2000, 124.
10. Stig Sørensen 2000, 124.
11. Van Gennep 1909. Liminality comes from the Latin word *limen*, meaning threshold.
12. Van Gennep 1909.
13. Turner 1969, 95.
14. Turner 1969, 95.
15. *ThesCra* I, 280.
16. Cleland 2002, 44.
17. Colburn & Heyn 2008, 6.
18. Roach & Eicher 1973, 17.
19. Roach & Eicher 1973, 17.
20. Petrovic & Petrovic 2006, 153.
21. Jones 1999, 252.
22. Joseph 1986, 50.
23. Jones 1999, 251.
24. Paul 2013, 273.
25. Paul 2013, 274.
26. Arthur 2000, 3.
27. Plut. *Quaest. Graec.* 304c–e.
28. Paul 2013, 263.
29. Paus, 5.14.1–2. Paul 2013, 264.
30. *LSAM* 37; *SEG* 26:1334. Paul 2013, 264.
31. Miller 2005, 7.
32. Warr 2010, 133–134.
33. Warr 2010, 141.
34. Batten 2009, 488.

Conclusion

This study combines the field of textile research with the study of Greek religion. The challenge is thus double, because both fields deal, for the most part, with aspects of life in ancient Greece without tangible remains: textiles have nearly all disappeared from the archaeological record and must be explored indirectly through images and tools; religion, beliefs, cult practices, too, can be accessed only through a few archaeological remains, images, and scarce textual materials that provide fragments of information about prescriptions and traditions. Such a study is therefore not without challenges and poses methodological problems. Despite these challenges it possesses enormous potential to increase our understanding of ancient Greek society by combining these two fields. Furthermore, I have not only introduced textiles into the study of religion, but I have also introduced religion, in turn, into the study of textiles. Traditionally, textile studies focus on functionality, production, and technique, while I have instead introduced the spiritual, symbolic, and religious value of textiles.

Textiles were an extremely important part of ancient life and economy, yet this vital part of material culture has usually been left out in studies of ancient Greek religion. The main purpose of the study has therefore been to introduce textiles into the study of this field, and especially into the research on votive practices, cult images, and dress codes. Through three separate, but closely linked studies, the present work has demonstrated that textiles were a far more important part of the ritual experience in Greek sanctuaries than hitherto anticipated. By combining the many relevant but various sources, such as iconography and other kinds of archaeological material (e.g., clothing accessories), epigraphy, and literary sources the present study demonstrates the importance and the specificity of textiles. I have investigated the roles, uses, users, and symbolism of textiles, and not least their function as material agents. My focus has been on three main aspects: dedication, the dressing of cult images, and dress-codes.

My examination of the ritual practice of dedicating textiles seeks to reopen and renew the discourse on votive offerings and the offering of textiles, primarily through a thorough examination of temple inventories. Despite the fact that the inscriptions from Brauron are still unpublished, the inscriptional material is a rich source that has more to offer than anticipated from the few previous and isolated studies. In contrast to what was previously believed, my survey and analysis of temple inventories establishes that textiles were an important category of votive offering in Greek sanctuaries. My analysis highlights the importance of the temple inventories as an immense source of information about all the different features of ancient textiles – not only their type, fibre, colour, and decoration, but also for clothing terminology, about which the inventories are probably more informative than ancient authors.

The study has laid out iconographic and written evidence for the phenomenon of dressing cult images in textiles, which appears to have been a more widespread phenomenon than suggested in previous studies. One further direction for my research on dressed cult statues is to consider the possibility that statues which we are accustomed to see without textiles may once have been dressed. If this was indeed the case, then their appearance and their impact would have been much altered. Yet the act of dressing the cult images was not mere decoration, but possessed immense meaning, in that the textiles acted as a material agent. I have found it useful to follow the theoretical lines of inquiry concerning the concept of agency developed by scholars, including Alfred Gell, in order to understand this phenomenon as more than merely textiles placed on cult statues. The study thus demonstrates the unique ability of textiles to serve as a means of communication with the divine through the connection with the divinity that the textiles acquire from their physical closeness to the cult image. This is further underscored by the fact that textiles can carry decoration or even letters, thus communicating specific messages. Furthermore, textiles could be expressions of extreme wealth, making them an obvious choice for display as well as expressions of power, and, at the same time, they had a close connection with identity.

The investigation of dress-codes establishes that dress was a significant element in the religious experience of the ancient Greeks. Iconography from the 6th to the 1st century and clothing regulations of the 3rd to 2nd centuries all reflect the fact that what one wore when in a sanctuary was far from unimportant. Indeed, there appears to have been a general concern with controlling what people wore, especially the colour of garments. A further clear tendency is the concern with restricting the sumptuousness of people's attire, by restricting people from wearing excessively lavish or valuable attire. Another conclusion is that purple clothing was much more widespread and used in cultic contexts by ordinary people than is recognized in traditional scholarship, which instead tends to focus on the relation of purple to monarchy in particular.

This concern with making people look a particular way, either blending in or standing out, also has a close relationship with rites of liminality, a central concept of ancient religion. Clothing and footwear were useful instruments by which to express liminality because of their closeness to the body, but also their visibility. This is evident from the focus on belts/girdles, and it seems that knots and leather were to be avoided in many sanctuaries. There also appears to have been a particular focus on headgear – again an easy way to signal an individual's state in specific rites or the religious hierarchy. Another item which was often restricted or prohibited in sanctuaries was footwear, which is also illustrated by the barefooted individuals performing rituals rendered in iconography. This is also reflected in the inventories, which only very rarely record footwear. These prohibitions can also express a possible taboo inherent in specific types of dress or accessories. Thus, much of the regulation on textiles in sanctuaries focuses on wealth and excess, but also on dangers and how one could adjust textiles to avoid these dangers. Clothing, moreover, was sometimes used as a means of identifying with the divine, by dressing as the gods. Since there is a seamless connection between religion and clothing, the designation of these two things as separate categories is therefore artificial. This means that the dress worn by the priest or priestess or by the person performing a sacrifice in a way becomes religion, and not just an expression of religion. This again strengthens even further the potential and meaning of textiles in relation to cult and religion.

This study has also demonstrated that the dedication of textiles was associated primarily with women, which underscores the role that women played in dedicatory practices, primarily in relation to female deities. Needless to say, but still important to remember, is that the female aspects of textiles in cult are yet another dimension of the invisibility of the subject since the female sphere is generally less frequently described in ancient texts and, until a generation ago, also quite overlooked in studies of ancient religion. The study highlights the importance and efficacy of textiles, which, as objects often made and worn by the donors themselves, strengthened the gift-exchange relationship with the recipient deity, since they were highly personal and can therefore be considered to be a part of the donor, who in this way gives part of herself. Furthermore, the act of dressing cult statues in textiles is demonstrated to have been a custom enacted in several ancient cults of primarily female divinities, which again may also be especially connected to women, underlining their importance in the performance of ritual. Yet, as I have shown, this does not mean that these ritual practices were exclusive to women, since there is evidence of also men dedicating textiles, as well as male divine recipients.

A further important point of this study is that the role of the *peplos* in Greek religion is restricted to Athens, where, however, it is of the greatest importance. Due to the rich sources on this particular phenomenon, it has dominated scholarship, which tends to assume that this was the 'norm' – a tendency I would term 'the *peplos* paradigm'. In fact, this study has shown that there is only written evidence for the making of a ritual *peplos* in a sanctuary in Athens and nowhere else.[1]

The phenomenon of dedicating textiles appears to have been widespread, and not restricted to any specific region. This is supported by the temple inventories and iconography, but also by the presence of dress-fasteners in several sanctuaries. The same appears to be the case for the ritual of dressing cult images. Here, most of the evidence pertains to Attica, which, however, is related to the fact that we possess an especially large number of testimonia from this region. The Hellenistic copies of cult statues from Asia Minor and the written sources clearly indicate that this was a far more widespread phenomenon and not exclusive to Attica. Ritual dress-codes also appears to have been widespread, as attested in iconography, but especially in the clothing regulations, and were not limited to any specific region.

With regard to chronology, there appears to be continuity in the use of textiles as votive offerings from the Bronze Age (attested e.g. in wall-painting) down through the Hellenistic Period. It is more difficult to reach a conclusion about the dressing of cult images. It is likely that this is a tradition that goes all the way back to the earliest cult images, but for now there is no solid evidence to back this up. The earliest proof of this ritual is the dedication of a garment to Athena in the Iliad. Further evidence includes the Brauron Catalogues (which date to the Classical Period), and there is evidence for the continuation of this practice down through the Roman Period.

By merging the study of Greek religion and the study of textiles, the current study has illustrated how textiles are central materialisations of Greek religion, by reason of their capacity to accentuate and epitomize aspects of identity, spirituality, position in the religious system, by their forms as links between the maker, user, wearer and tactility, but also as important material agents in the performance of rituals and communication with the divine.

This study has reached some new conclusions which challenge previous scholarship and which open up new lines of inquiry. Future research directions could include investigating how these results compare with textiles used in burial rituals of ancient Greece, which could provide further perspectives on the ritual use of textiles. As already discussed, there appears to be similarities between the dress of cult images and the dress of the dead, e.g. the dressing technique, such as the tightly fitted undergarments of some of the cult statues, which could resemble shrouding or mummy wrappings, as well as the use of gold appliques. Restrictions on the number and value of textiles used in burial parallel the concern in sanctuaries with regulating and prohibiting excessive colours or costly textiles. Additionally, investigation of textile tools that have been recovered in sanctuaries could provide interesting information on the possibility of ritual textile production, as testified in the Heraion at Foce del Sele and perhaps at Francavilla Marittima. Given that architectural textiles, such as curtains, were used in the temples, textiles must be factored into visualisations of temple architecture, which could potentially alter our perception of their appearance as well as their function.

Note
1. On this topic, see Brøns forthcoming b.

Appendix 1

The *peplos* of Athena at Athens

The most famous example of the custom of dedicating textiles and dressing cult statues in the ancient world is the offering of a *peplos* to Athena on the Athenian Acropolis. This is primarily due to the numerous epigraphic, literary, and iconographic sources for this ritual. Because of the abundance of sources, this religious occasion is especially important for our knowledge and understanding of the dedication of garments and their use in dressing cult statues. An overview of this case will be provided apart from the other examples.

Epigraphy and literary sources

The dressing of the statue of Athena at Athens is described in both epigraphic and literary sources.

One of the most well-known inscriptions referring to this practice is the so-called *Praxiergidai* Decree, usually dated to ca. 460–450 BC.[1] It was set up on the Acropolis behind or south of the Old Temple. Lines 10–11 state that 'Let them (the praxiergidai) clothe the [goddess] in the peplos'[2] and lines 24–25 records: 'The Praxiergidae are to dress with the peplos' or 'dress the goddess'.[3] Mansfield treats fragment c by itself, and interprets lines 10–11: '[The Ergas]tines (?) [is to clothe[the goddess in the robe]'.[4] But this translation has been criticised by Robertson, because *Ergastines* has no epigraphic parallels.[5] Furthermore, fragment b and c are badly worn near the bottom, and there is uncertainty about letters at both the beginning and end of line 25.[6]

Nicomachos' calendar of sacrifices was set up in the Royal Stoa in the Athenian Agora in 401 BC to record revisions of regulations that Solon had established in the early 6th century.[7] Lines 5–13 describe the expenses for the festival called *Plynteria*, and line 7 specifically mentions a *pharos*. Even though heavily restored, the inscription appears to record the name of the recipient as Athena Ergane (line 13).

Another important inscription is *IG* II² 1060 + *IG* II² 1036, which is one of a small group of Athenian inscriptions dated to ca. 100 BC that are concerned with the production of the *peplos* for Athena and its presentation on the Acropolis.[8] This inscription consists of a very fragmentary conclusion of one decree, which appears to contain regulations relating to the *peplos* and the Panathenaic procession (decree I), and a second decree honouring the *parthenoi* who had created the *peplos*.[9] 1036 is the largest of the two fragments and is, in fact, composed of two fragments, a and b. It speaks of the *Praxiergidai*, the *peplos*, and a *himation*, and the *peplos* is described, at its second occurrence, as the annual one.[10]

Aleshire and Lambert translate decree I as follows:

> (fr. a) '... the [agonothel]tes and the athlothetai[11] - - (for?) those (female) who have - - worked the peplos well - - - - - the People a crown of foliage - -
> (line 5) - - the peplos ... white raiment - -
> - - [agon]othetes for (or 'to') the procession ...'
>
> (fr. b) '... these ... proces[-] - the [Praxiergi]dai take over the year's peplos ... himation, they march out, they shall hand over to the (feminine plural, Praxiergidai?) ... (masculine subject) join in supervising the division ...
> (line 5) so that the Council and People may appear to di[vide justly - ?].'[12]

Mansfield translates the decree as follows:

> line 1-3: '[the arrephoroi] who finely made the [robe]',
> line 5: '[to dedicate to Athena along with the ro]be a white garment wh[ich they wore]',
> lines 8-10: '[in order that the praxiergi]dai may receive this year's robe and bring [it] up [to the temple? and] bring out [the old robe and the] himation, [the arrhephoroi (?)] are to turn it over..'[13]

The translations depend on different restorations/interpretations of the stone, but it is clear that it gives information on the garments of the statue. Decree I implies that the statue possibly wore other garments besides the *peplos*, since it also mentions a *himation*. According to Robertson, the *himation* is probably the same as the *pharos* of Nicomachos' calendar, which was presented every second year, since he interprets the Greek word *pharos* as mantle. He therefore concludes that the statue of Athena was dressed in both *peplos* and a newly woven mantle every second year.[14] There are, however, several potential problems with this interpretation. First, the texts refer to two different terms for mantles – *himation* and *pharos* – which might have been very different both in appearance and use. Second, the evidence for the dedication of a mantle on a biannual basis seems very slight. With regard to the white raiment, the absence of an article implies that it is the garment of persons who stand in some relation to the *peplos*, and not that it refers to the garment of Athena.[15]

Decree II is much longer and contains a prescript (lines 1–10), a body (11–22), provisions (23–27), and a name list (28–76). Aleshire and Lambert translates the first part of the body (which is the only part of relevance here) as follows:

> 'Since, having made an approach to the Council, the fathers of the maidens who worked for Athena the wool for (or into) the peplos make clear that they (sc. the maidens) have adhered to everything that was decreed by the People about these matters and have done what is right and have processed according to the stipulations in the most fine and seemly manner, (...)'.[16]

Also part of the group of inscriptions mentioned above are a number of fragmented inscriptions/decrees from 103/2 BC[17] and 100 BC respectively, all in honour of 120 *parthenoi* who worked the wool for the *peplos* of Athena.[18] These inscriptions were set up beside the Karyatid Temple, where the *Praxiergidai* decree (*IG* I³ 7) was also erected. The *peplos* in question is usually taken as the famous *peplos* given every fourth year.[19]

Among the literary sources for Athena's garments is Aristophanes (ca. 446–386 BC), who writes 'For whom shall we card the wool for the peplos?',[20] as well as two authors who specifically refer to the veiling of the statue of Athena: Xenophon (ca. 430–345 BC), who writes 'the city was celebrating the Plynteria, after the statue (*tou hedous*) of Athena had been veiled (*katakekalymmenou*)',[21] and Plutarch (ca. 46–120 AD), who describes how the image of Athena was disrobed and covered up:

> 'The Praxiergidai celebrate these rites on the twenty-fifth day of Thargelion in strict secrecy, removing the robes (*kosmos*) of the goddess and covering up (*katakalypsantes*) her image (*to hedos*).'[22]

Finally, the *Lexica Segueriana, Glossae rhetoricae* states that the *Plynteria*

> 'is so called because after the death of Agraulos (priestess of Athena) the sacred garments (*esthētes*) were not washed for one year.'[23]

Furthermore, the text refers to a man 'who washes off the dirtied spots (or garments?) under[24] the peplos of Athena,'[25] which confirms that the statue was dressed.

The festivals

The *Plynteria* were a festival celebrated on the 25th of Thargelion (May–June), 2 months before the *Panathenaia*. Athena Polias was the main deity honoured at the *Plynteria*. Other deities involved were Aglauros, Ge, Zeus Polieus, and Poseidon, the latter two in their capacity as a part of Poliadic pairs with Athena.[26] The name of the festival, *Plynteria*, derives from the word *plynō* 'I wash', which implies that a ritual washing took place.[27] During this festival, the *peplos* of the cult statue of Athena was removed and washed. Meanwhile, the cult statue was dressed/veiled in a temporary garment.[28]

The *Kallynteria* was another festival in honour of Athena, and was closely related to the *Plynteria*. There is still some discussion as to whether the *Kallynteria* took place before or after the *Plynteria*. According to Mansfield and Simon, it took place before, perhaps on the 22nd of Thargelion,[29] while Sourvinou-Inwood and Robertson suggest that it followed immediately after the *Plynteria*, perhaps on the 28th of Thargelion.[30] The name of the festival likely comes from the verb *kallynō* ('beautify', 'sweep clean').[31] Due to the word's multiple meanings, it is uncertain whether this festival was devoted to the cleaning of the sanctuary (and/or the cult statue) or the adornment of the cult statue.[32] Nevertheless, it is often assumed that the cult statue was somehow re-dressed during the *Kallynteria*, either with the old, newly washed *peplos* or its *kosmos* (adornment).[33]

These sources thus inform us not only that the ancient statue of Athena was clothed in a *peplos* woven annually by young girls, but also that it was provided with other clothing items. For example, at the *Kallynteria* the statue was decked out for three days in special finery, and at the *Plynteria*, before being re-dressed the following day, it was veiled or covered with a cloth or shroud.[34]

These diverse epigraphic sources need some further explanation: the presentation of the *peplos* to Athena occurred during the *Panathenaia*, which was the most important festival in the Athenian religious calendar, since it celebrated the traditional birth of Athens' patron goddess, Athena. This celebration took place towards the end of the month of Hekatombaion.[35] The high point of the festival consisted of the momentous events that took place on the Acropolis: the presentation of the *peplos* to Athena and the ritual slaughter of the hecatomb (100 heifers).[36] The festival was not identical each year, in that the Great *Panathenaia* occurred every 4 years, and the Lesser *Panathenaia* was held on the three off years.[37] Therefore, according to Mansfield, two different *peploi* are reflected in the sources, one offered annually, and a *peplos* woven by professional weavers every 4th year at the Great *Panathenaia*.[38] Yet Robertson rejects this scenario and argues for only one *peplos*, which was presented every four years at the Great *Panathenaia*.[39]

The cult statues of Athena on the Acropolis and their relation to the *peploi*

Which statue(s) received the *peploi*? There were many images of Athena on the Acropolis, but, unfortunately, the statues themselves have disappeared, and we often have only sporadic information on their appearance, size, material, etc. from literary sources and iconography – a fact which makes a study of these statues a very difficult task. Nevertheless, in trying to determine which statues were the recipients of the *peploi*, it is necessary to establish which statues were actually present on the Acropolis during the period in question, as well as what we know about them.

Athena Polias

The cult statue of Athena Polias ('of the city') was the most sacred on the Acropolis.[40] Pausanias refers to it as 'the most holy symbol', and claims that it had fallen from the sky (1.26.6). The image was already very old in antiquity, according to Philostratos, who writes that the image of Athena Polias was one of the most ancient cult statues in Greece,[41] and Plutarch, who states that Athena Polias was one of the first cult statues.[42]

Although, unfortunately, the statue of Athena Polias does not seem to have been depicted in Athenian art, a few ancient sources give hints as to its appearance. It was made of wood, and Tertullian compares it to a wooden log or a pole with no shape.[43] Plutarch mentions that the Gorgon's head was lost from the *agalma* of Athena when the Athenians fled with it to Salamis, which implies that it wore an *aigis*.[44] It was probably relatively small, able to be lifted and carried, since it was most likely removed from the Acropolis in 480, when the Persians invaded the city.[45]

We do not know whether the statue was standing or seated.[46] That it was seated was inferred from terracotta statuettes from the Archaic Period found on the Acropolis, but these do not necessarily represent a specific image. Ridgway states that these statuettes' arms are muffled under their clothing, as if a real garment had been added to a statue that did not allow proper draping.[47] Kroll, on the other hand, argues that the image of Athena Polias was standing.[48]

It has been suggested that the ancient *xoanon* of Athena Polias may have been carved into an anthropomorphic form by the sculptor Endoios in the 6th century BC. This is not certain, however, because this attribution is based on a corrupt text of Athenagoras,[49] and there is a great deal of debate among scholars as to the interpretation of this text.[50]

It is quite a puzzle where the statue of Athena Polias was kept, and there are several suggestions. We cannot determine, however, where it was housed during the Archaic Period. According to Ridgway, around 510 the statue was probably kept in the Old Athena Temple, which stood on the Dörpfeld Foundations, south of the Erechtheion.[51] For later periods (after ca. 420 BC), scholars have claimed the 'Erechtheion', understood as the Ionic structure with the Karyatid porch, as the Temple of Athena Polias.[52] A final suggestion is the *pteron* of the Parthenon. Restoration work revealed a rectangular opening in the paving stone of the north *pteron*, which was once surrounded by a *naiskos*. There are also signs of an altar in front of this small shrine. According to Ridgway, the shrine seems to have existed before the Parthenon and probably housed an early cult image – perhaps of Athena Parthenos.[53] Hurwit proposes that this pre-Periclean shrine housed a *palladion*.[54]

If the statue truly stood in the Old Athena Temple, we might be able to reach some conclusions about her appearance, since inventories from the 370s and 360s BC, recorded by the Treasurers of Athena provide evidence that the statue of Athena stood in this particular temple. In the sections of these inventories that catalogue the valuables in the Old Athena Temple (*archaios neos*), the Athena statue's ornaments are listed: 'a diadem that the goddess wears, the earrings that the goddess wears, a band that the goddess wears on the neck, five necklaces, a gold owl, a gold aegis, a gold gorgoneion, and a gold phiale that she holds in her hand.'[55]

Athena Nike

The *xoanon* of Athena Polias was not the only wooden cult statue on the Acropolis. Ancient authors describe a wooden image of Athena holding a pomegranate in her right hand, and a helmet in her left, but they do not record whether it was standing or seated, or when it was made.[56] These ancient sources include: an epigram describing a statue of Athena on the Acropolis holding an apple;[57] Pausanias, who states that the statue was wingless (3.15.7), and furthermore that a wingless statue of Nike at Argos was fashioned after the wooden image (*xoanon*) at Athens, which was called 'wingless Nike' (5.26.6); and, finally, in the 2nd century BC, Heliodoros, who describes the cult image of Athena Nike as a wingless *xoanon* holding a helmet in its left hand and a pomegranate in its right.[58]

The original statue has disappeared, but it has been suggested that it might have been rendered on a votive relief from the Acropolis, dated to ca. 420 BC (Fig. 119).[59] The relief shows a standing cult statue to the left behind an offering table in a small temple; the goddess Athena is seated nearby, and a line of worshipers approach from the right. The cult statue is clad in a *chitōn*, *himation*, and *polos*, and its elbows are close to the body, with its forearms held diagonally outward. Robertson and Beschi argue that the cult statue represents Athena Nike, primarily on the basis of Pausanias' meagre description.[60] Mark, on the other hand, identifies the statue as the cult statue of Athena Polias primarily due to its posture, contending that the discovered base for the cult statue of Athena Nike on the Acropolis shows that this particular image was seated, whereas the wooden image of Athena Polias was, in his view, standing.[61] Because our knowledge is so limited, however, it seems premature to conclude whether the cult statue depicts Athena Nike or Athena Polias. The only certain thing is that the relief depicts a female cult statue – most likely Athena.

Fig. 119. Votive relief, ca. 420 BC. © Acropolis Museum, Athens, inv. no. Acr. 2605+. Photo: Socratis Mavrommatis

Another marble votive relief from the Acropolis has been interpreted as a depiction of Athena Nike standing beside the three Graces.[62] Here, Athena is clad in a *peplos*, *aegis*, and *polos*, and she holds a *phiale* in one hand, and possibly a pomegranate in the other. We cannot tell, however, whether this particular votive depicts a cult statue, or even if it is actually Athena Nike.

The base for the cult statue of Athena Nike has been identified, as already noted, by Mark, who on the basis of its dimensions and shape suggests a seated or enthroned cult statue, ca. 1 m high and datable to 600–550 BC.[63] This is not, however, an undisputed fact, due to the iconographical representations, and a standing figure might seem more likely since literary sources describe it as a *xoanon*. At present, we can only state that we do not know its pose. Regardless of the statue's posture and exact appearance, we can conclude that there was a cult statue of Athena Nike on the Acropolis, probably already from an early age.

The cult image of Athena Nike stood in her sanctuary on the Acropolis. Pausanias (1.22.4) locates this to the right of the *Propylaia*, as one ascends the Acropolis. The site is presently occupied by an Ionic temple from ca. 427–424 BC, but beneath the Classical temple is an earlier shrine, which includes a *temenos* enclosure, a poros *naiskos*, a receptacle with figurines, a poros altar, an inscribed altar, and a poros base.[64] The earliest securely datable evidence for the cult of Athena Nike on the spot of the present temple is a fragment of a poros altar with an inscription reading 'tēs athe[naias] tēs Nikes Bōmos Patrokles epoiesen'. On the basis of the letter forms, the inscription can be dated to the mid-6th century, perhaps as early as 566 BC, thus possibly linking the cult of Athena Nike chronologically with the founding of the Panathenaic games.[65] The foundation of the Sanctuary of Athena Nike (i.e. the architectural remains) in this particular spot cannot be pushed back much beyond the mid-6th century BC.[66]

A last piece of evidence is an inscription, variously dated between 440 and 425 BC, which refers to a competition over the design of something made of ivory (perhaps a cult statue), probably for the Sanctuary of Athena Nike. Line 21 records *to archaion agalma* – 'the old statue' – at the end of a missing sentence.[67] This inscription indicates that an Archaic cult image stood in the precinct by the end of the 5th century BC.

Athena Parthenos

The most well-known of all the statues of Athena is the Athena Parthenos ('the virgin'), attributed to Pheidias and originally situated in the Parthenon.[68] This is the only statue of Athena from the Acropolis that can be identified with reasonable confidence, although the original statue has not survived.[69] Its appearance is known through several Late Classical, Hellenistic, and Roman reliefs, statues, intaglios, and coins, as well as from literary descriptions.[70] A famous example of such a copy is the so-called Varvakeion Athena, a marble statuette ca. 1 m tall, dated to ca. 200–250 AD, which is considered the most faithful and best preserved copy of the statue of Athena Parthenos (Fig. 120).[71] The most thorough ancient description is in Pausanias:

> 'The statue of Athena is upright, with a tunic reaching to the feet (*chitōni podērei*), and on her breast the head of Medusa is worked in ivory. She holds a statue of Victory about four cubits high, and in the other hand a spear; at her feet lies a shield and near the spear is a serpent. This serpent would be Erichthonius. On the pedestal is the birth of Pandora in relief.'[72]

He furthermore informs us that the statue was made of silver and gold, and wore a helmet with a sphinx and two griffins (1.24.5). Athena was thus rendered fully armed

Fig. 120. Varvakeion Athena, ca. 200–250 AD. National Archaeological Museum, Athens, inv. no. 129. © Hellenic Ministry of Culture, Education and Religious Affairs/Archaeological Receipts Fund. Photo: H. Goette.

Appendix 1. The peplos *of Athena at Athens*

with helmet, *aigis*, spear, and shield, and was of colossal size, perhaps some 10 meters tall.[73] Moreover, Pausanias is the earliest known writer to call the statue Parthenos.[74]

In Hurwit's view, it is unlikely that this particular statue was object of an official cult in Classical antiquity, since no public ritual or offerings were directed specifically to this statue. Furthermore, there is no record of any priestess to Athena Parthenos and there is no altar for this statue. He therefore concludes that the statue was not worshipped and thus not a cult statue.[75] Nevertheless, the statue depicts the most important goddess of the Athenians, and it was situated in a temple in a very sacred area – the Acropolis.[76] Nick argues that the Parthenon and the statue of Athena Parthenos both served to honour Athena Polias, and that the statue in the Parthenon was, in fact, a second cult statue to Athena Polias.[77]

Athena Promachos

The statue of the so-called Athena Promachos ('first-fighter' or 'who fights in the first line') was one of Pheidias' three famous Athenas on the Acropolis. The statue was very large and made of bronze, and was probably dedicated around 460–450 BC, which seems assured by virtue of the fact that it was a dedication made from the spoils of Marathon.[78] The only identified architectural remains are parts of the statue's marble base, which identify its original placement 40 m east of the later *Propylaia*, just in front of the ancient terrace-wall west of the Dörpfeld Foundations.[79]

Our knowledge of this statue is slight, and the reconstructions of its appearance are highly controversial, since there are no definite copies of this work, and no literary sources describe it sufficiently.[80] For example, Pausanias mentions only that its shield was engraved with a centauromachy, and that its spear tip and helm crest were visible to sailors coming from cape Sounion (1.28.2), thus implying that it was very large. It is, indeed, considered to have been colossal. Based on the testimony of Pausanias and the size of the base, scholars have reconstructed different heights for the statue, ranging from ca. 7–16 m.[81] Hurwit suggests a height of ca. 9 m,[82] whereas Robertson proposes a height of ca. 16 or 17 m.[83]

Images on the Panathenaic amphoras have been interpreted as depictions of an actual statue of Athena on the Acropolis, probably the Athena Promachos.[84] Here, the goddess is shown striding forward in a combative pose with raised spear; she wears a helmet and *aigis*, and is located occasionally in front of an altar. In some instances, the goddess is dressed in a *peplos* or *chitōn*,[85] while in others she is dressed in a long undergarment over which she wears a decorated *ependytēs* (Fig. 121).[86] Other vases depict the goddess in heavily ornamented garments featuring figurative decoration in friezes (Fig. 122).[87] It is difficult, however, to determine whether these are merely conventional representations of the goddess or depictions of a particular cult statue.[88] According to Hölscher, the image of Athena on the Panathenaic amphoras was depicted as an Archaic goddess, but the figure did not represent an existing statue of Athena. Instead, the type of Athena Promachos was closely connected with the function of the image on the Panathenaic vases.[89]

Fig. 121. Panathenaic amphora, ca. 525 BC. © National Museum Denmark, inv. no. ChrVIII 797.

Appendix 1. *The peplos of Athena at Athens* 375

Fig. 122. Panathenaic amphora, ca. 425–400 BC. © Trustees of the British Museum, inv. no. B605.

The epithet Promachos is a late designation, the earliest occurrence of which is in a dedicatory inscription recovered on the Acropolis and dating to the early 5th century AD.[90] The statue is named Promachos by only a single ancient source: the scholiast to Demosthenes.[91] Other writers concerned with the same monument describe it simply as the bronze Athena.[92] Hurwit therefore distinguishes between depictions of the Athena Promachos *type* (e.g. on the Panathenaic amphoras and bronze statuettes) and the bronze Athena – which he claims to have been its official title – by Pheidias.[93] Furthermore, according to Hurwit, it seems that the bronze Athena by Pheidias was not fashioned like a Promachos type, but stood at ease, perhaps holding a Nike and resting on her shield.[94] Another view is that of Lundgreen, according to whom the statue stood wearing a helmet and probably a *peplos*, extending one arm with a winged figure in her palm, while, on her opposite side, a spear rested against her shoulder.[95] It thus appears that there was a colossal statue of Athena just east of the *Propylaia*, but its appearance remains uncertain. Another potential point of uncertainty is whether this statue was a cult statue at all. The statue stood outside – not in a temple – and there are no signs of any altar in front of it. Furthermore, there is, to my knowledge, no evidence for a priestess of Athena Promachos. Perhaps this statue was intended solely as a dedication to Athena, not necessarily as a cult statue.

Athena Hygieia

Pausanias mentions a place of worship for the goddess Athena Hygieia ('Health') on the Acropolis. This goddess, whose cult can be traced back to the 6th century BC, had an altar near the east façade of the *Propylaia*,[96] and her shrine was prominent in the Panathenaic ceremony.[97] The cult statue cannot be located and we do not know what it looked like, but we know of another statue of Athena Hygieia, which stood on a surviving base near the south side of the *Propylaia*. This statue was of bronze and made by the sculptor Pyrrhos, and was dedicated by the Athenian people. According to Robertson, it may have been an up-to-date, realistic rendering of the cult statue. The cuttings in the base reveal that its feet were positioned apart, the right foot forward, and the left only touching the ground with the toes; and that it carried a spear in the left hand, which was anchored to a hole in the marble surface. For this reason, Ridgway proposes an armed Athena,[98] while Robertson identifies it as the Athena Promachos type.[99] Robertson even suggests that the Athena statue depicted on the Panathenaic amphoras may have been the cult statue of Athena Hygeia, based on Pyrrhos' bronze statue and the fact that this statue (if it truly is a statue) is shown next to an altar, often between two free-standing columns.[100] There appear, then, to have been both a cult and a cult statue of Athena Hygieia, but again, we are left without much knowledge about its appearance.

Athena Ergane

Athena Ergane ('the Worker') was the patron deity of skilful craftsmanship and art (from *erga* as the product of *technē*, 'artful craft'). In myth, the goddess is credited with

Appendix 1. The peplos of Athena at Athens

inventing and teaching various crafts, including textile production.[101] Her cult is well attested in Greece, and she also had a cult on the Acropolis. There is uncertainty about when this cult was established, but Ridgway suggests a date in the Archaic Period.[102] Pausanias alludes to a statue of Athena Ergane on the Acropolis (1.24.3), but he does not provide details about the statue or the cult.

Though a statue of Athena in this capacity did exist, it is now lost, and it is impossible to determine its original appearance with any certainty.[103] That being so, the so-called Endoios Athena, which dates to 530–520 BC and was recovered on the north slope of the Acropolis, has been restored – suggestively – as a spinner and identified tentatively as the statue of Athena Ergane (Fig. 123).[104] The statue is seated, and is clad in a long *chitōn* and *aegis*.[105] This seated image has been equated with a dedicatory statue set up by Kallias in the vicinity of the Erechtheion, which is mentioned by Pausanias (1.26.4/5) and has been attributed to the sculptor Endoios, who worked in Athens around 530–500 BC.[106] This attribution has been questioned, on the basis of an assumption that the statue must have rolled down the slope when the upper retaining wall containing the debris gave way. In this interpretation, the statue was damaged during the Persian attack and, after being removed, ended up as part of a deposit – which means that it would not have been visible to Pausanias. If this was the case, the statue in question is probably not the Athena Ergane of Endoios mentioned by Pausanias.[107] In a recent article, however, Marx convincingly argues that the assumption that the statue was found where it had fallen from above is incorrect. She argues

Fig. 123. Endoios Athena, ca. 530–520 BC. © Acropolis Museum, Athens, inv. no. Acr. 625. Photo: Socratis Mavrommatis

further that the statue most probably remained on the Acropolis until it was mutilated in Late Antiquity, which strengthens the attribution to Endoios. Shortly after being mutilated, it was taken down to the North Slope, where it was built into a wall.[108] The statue in question is therefore not a cult image, but rather a rare representation of Athena Ergane.

A group of terracotta *pinakes* recovered on the Acropolis are of particular interest in connection with representations of Athena Ergane.[109] They can be dated to the period between the end of the 6th and the beginning of the 5th century BC.[110] The *pinakes* are all fragmented, but are sufficiently extant to show that they were of uniform technique, clay, and size (with measurements ca. 22 × 16 cm), and that they were originally painted in bright colours.[111] Hutton divides the *pinakes* into two main groups: a) representations of Athena and b) those of other individuals. He identifies three different Athenas: Athena Ergane, Athena Polias, and Athena Promachos.[112] The representations of Athena Ergane show the goddess as a young girl, dressed in a *chitōn* and wearing a *sakkos* or *polos*, and seated on a bench with her feet on a footstool (Fig. 124a, b). Hutton suggests that they represent her as one of the women who worked the wool for her *peplos*,[113] but Ridgway identifies the representations with a *sakkos* as human, while the ones with a *polos* as divine.[114]

A further possible representation of Athena Ergane from the Acropolis is a marble relief dated to the beginning of the 5th century BC (Fig. 125).[115] The relief depicts

Fig. 124. Athena Ergane. Pinakes. © Acropolis Museum, Athens, inv. nos. Acr. 13054, Acr. 13055, Acr. 13057. Photo: Vangelis Tsiamis

Appendix 1. The peplos of Athena at Athens

Fig. 125. Athena Ergane. Relief, 5th century BC. © Acropolis Museum, Athens, inv. no. 577. Photo: Socratis Mavrommatis

Fig. 126. Terracotta figurine from the sanctuary of Athena at Lindos. © National Museum, Denmark, inv. no. 10575.

a standing Athena dressed in *chitōn*, *himation*, and *aegis*; the goddess' head is missing. In front of her, behind a workbench, there is a smaller, male figure seated on a stool, who is offering a gift to the goddess. It has been suggested that the relief depicts Athena Ergane receiving a primal offering (*aparchē*) from an artisan in his office or workshop.[116] Besides these depictions, there are several votive inscriptions, primarily belonging to the 4th century BC, which refer to Athena or Thea Ergane from the Acropolis.[117]

Pictorial representation of Athena Ergane holding a spindle and/or distaff are generally quite rare, and, according to Villing, representations of a spinning Athena are virtually non-existent on the Greek mainland.[118] Many of the known examples have instead been recovered in Italy or western Anatolia.[119] The earliest example is a terracotta figurine from Scornavacche, Sicily, dating to the late 5th or the 4th century BC, which depicts the goddess wearing an Attic helmet and holding a distaff.[120] Other examples include two 5th century terracotta figurines of a seated, spinning woman, possibly identifiable as Athena, which were recovered at Lindos (Fig. 126).[121]

Athena Lemnia

The lost Athena Lemnia of Pheidias also stood on the Acropolis. The statue is mentioned by Pausanias (1.28.2), who writes only that the statue was the one most worth seeing on the Acropolis, and that it was named after its dedicants. The statue was thus a dedication, and not a cult statue.[122]

The receiving cult statue

Since there were several cult statues of Athena on the Acropolis, there were thus several possible recipients of the *peploi*.[123] Closer scrutiny, however, can establish that not all are actual candidates for this particular offering. First of all, it has been stated that only cult statues received offerings, not dedicatory statues or votive offerings, no matter how beautiful, expensive, or colossal.[124] This means that we can exclude the statue of Athena Lemnia, but perhaps also Pheidias' Athena Parthenos and the so-called Athena Promachos, although we should be very careful about making such clear cut distinctions between votive and cult statues. As Mylonopoulos observes, the idea that the Athena Parthenos was a 'votive' statue and that the Athena Polias was the *only* cult statue on the Athenian Acropolis is a phantom in the scholarship.[125] Donohue likewise argues that we are working from false premises when we assert that Athena Parthenos is not a 'true cult statue', but merely a display piece, in contrast to e.g. the statue of Athena Polias, since such an assertion is based on the notion that old was more important than new, as well as the view that works of art could not have religious meaning.[126]

The cultic rite of offering a *peplos* to Athena can be traced back to at least the 7th century BC, and possibly much earlier. Since the cult of Athena Hygieia dates back to the 6th century BC, and the cult of Athena Ergane is unattested before the Archaic Period, they seem less likely recipients for the *peploi*. That said, there is of course the possibility that the cult of Athena Ergane was older, but there is no evidence to confirm or disprove this. Athena Ergane's function as goddess and protector of arts and crafts, especially weaving, makes her involvement in the production and dedication of the *peplos* feasible and, at least to modern thinking, also a likely recipient.[127] At present, however, this remains mere speculation.

This leaves us with two viable candidates: the statues of Athena Polias and Athena Nike, which both appear to have been old, wooden *xoana*, smaller than life-size. The epigraphic material does not provide us with the epithet of Athena as receiver of the *peplos*, but the Praxiergidai Decree as well as a group of inscriptions/decrees in honour of 120 *parthenoi* who worked the wool for Athena were all set up on the Acropolis behind or south of the Old Athena Temple (Dörpfeld Foundations) and beside the Karyatid Temple. If this was indeed the location of the cult statue of Athena Polias, it might indicate that Athena Polias was the recipient.[128] Indeed, in Aristophanes' *Birds*, when Euelpides asks '*For whom shall we work the wool for the peplos?*', Peisthetaerus answers: '*Why not choose Athena Polias?*',[129] which implies that she was the recipient. Finally, Pausanias refers to this particular statue as the most holy, which makes it likely that the statue of Athena Polias was the receiver of the *peplos*. This conclusion is based on rather circumstantial evidence, however, and should therefore be accepted only with great caution. As previously stated, Mansfield has suggested that an annual and a quadrennial *peplos* were dedicated to Athena. In his view, Athena Polias received the annual *peplos*.[130] Yet, as Nick argues, the city goddess Athena Polias cannot be equated with a single defined domain (*wirkungsbereich*), since she also encompasses the cults of Athena Nike, Athena Ergane, and Athena Hygeia.[131] It can thus be extremely difficult to determine which aspect of the city goddess was being addressed in the dedication of the *peplos*.

But what about the quadrennial *peplos*? Some information can be derived from an Attic inscription dating to 270/69 BC, which honours a certain Kallias of Sphettos for his many services to Athens.[132] Lines 55–70 state that Kallias was sent to the court of Ptolemy II, and that he there persuaded the king to donate a gift of ropes for conveying the *peplos* at the Greater *Panathenaia* in the following year:

> 'and since the Demos was then about to [celebrate] the Panathenaia for Athena Archegetis [for the first time after] they had recovered the City, [Kallias] conversed with the king about the ropes which it was necessary to prepare for the peplos, and the king having donated them to the city, he endeavoured to see that they be as fine as possible for the Goddess and that the delegates elected with him bring [the ropes back here] at once.'[133]

The text (line 65) refers to the festival as the '*Panathenaia* for Athena Archēgetis'.[134] Archēgetis ('First Leader' or 'Founder') was one of Athena's many epithets at Athens. Kroll argues that Athena Archēgetis and Athena Polias were one and the same.[135] This is, however, somewhat unfounded and appears to be based on the assumption that Athena Polias received both *peploi* (the annual and the quadrennial).[136]

The quadrennial *peplos* was used as a sail on the Panathenaic ship, and is therefore thought to have been very large (on the size of the *peploi*, see further below). It therefore follows that the cult statue receiving the garment must also have been very large (large *peplos* = large statue, small *peplos* = small statue). There is, however, no indication that this was actually the case. Nevertheless, this appears to be the argument behind the assignment of the large *peplos* to the Athena Parthenos or the so-called Athena Promachos. For example, Lewis suggests that the large *peplos* dedicated every fourth year was draped around the colossal statue of Athena Parthenos.[137] This is, however, unlikely for several reasons: first, it would have been extremely difficult to dress such a large statue; second, the golden *peplos* made by the sculptor had important features and extraordinary decorative elements, and it seems strange that those would have been covered; and, third, if the *peplos* were draped on the statue, it would have been hard to discern the tapestry-woven scenes thereupon.

Another possibility is that the *peplos* was used to decorate the walls of the temple, perhaps as a backdrop to the cult statue in the Parthenon or the Temple of Athena Polias – or as Kardara has proposed, that the *peplos* was hung in front of the cult image in her temple.[138] Hurwit suggests that, after the celebration, the new *peplos* may have been hung on a wall, perhaps the southern wall of the Erechtheion.[139] But the idea that such a valuable and precious textile would have been hung outside, where it would be ruined by rain and wind as well as bleached by the scorching sun is highly improbable – at least to modern thinking. If the *peplos* was not used to dress

Appendix 1. The peplos of Athena at Athens

a statue, exposing it as a back-drop in the temple seems very likely.[140] Of course, if it was only presented outside on the wall for a shorter period, this option cannot be excluded. In this way, it would have been visible to people visiting the sanctuary, but not necessarily allowed to enter the temple. Alternatively, the *peplos* could have simply been deposited in either the Parthenon, the Temple of Athena Polias, or the *peplothēkē* (see below).[141]

The east frieze of the Parthenon temple

The east frieze of the Parthenon temple, dated to 438–432 BC, is especially important for any discussion of the *peploi* of Athena. The scene depicted in the centre of the east frieze is the culmination of the procession depicted on all four sides of the building, and placed above the porch columns at the entrance to the temple, and it is thus the high point of the narrative. The scene is carved on a very long block of stone, and depicts a ceremony involving five figures and a folded cloth (see Fig. 94). In the centre of the scene there is a bearded man, clad in a long unbelted garment, facing a child who is naked under a draped *himation*. Together, they hold a folded garment. To the left, a woman, turning her back to the man and child, is faced by two young women, who approach her carrying cushioned stools on their heads.

The generally accepted view is that the entire frieze represents the Greater Panathenaia,[142] mainly because of the representation of the *peplos* on the east frieze.[143] There are, however, several variations on this interpretation: Kardara suggests that the frieze depicts the first or inaugural Panathenaic procession during the reign of Kekrops, and thus the Athenian heroic past,[144] while Holloway argues that the frieze depicts a renewal of the dedications of the Archaic Acropolis that was destroyed by the Persians in 480 BC, and that the *peplos* scene is thus a symbolic re-creation of the perished *peploi* of the previous Panathenaic festivals.[145] An altogether different interpretation has been put forward by Connelly, who suggests that the frieze represents the story of king Erechtheus, his wife Praxithea, and their three daughters, who gave their lives to save Athens.[146] In Connelly's view, the cloth depicted on the east frieze is a shroud for the child, identified as a girl, the youngest Erechtheid, who is to be sacrificed to save the city.[147] To my mind, however, there is no doubt that the east frieze depicts the Panathenaic *peplos* of Athena, whether in the heroic past, as a symbolic re-creation, or in a contemporary event, since this particular festival was such an important event in the religious life of Athens. Furthermore, even though important elements are missing (e.g. the hoplites and the ship),[148] it is possible to identify several important features from the *Panathenaia*, and the east frieze especially seems to have no other fully convincing interpretation other than that of the Panathenaic *peplos*. Harrison's identification of the frieze as a depiction of the *Panathenaia* probably provides the correct interpretive framework, and it is plausible that all of the depicted elements were part of the *Panathenaia*, inasmuch as many elements of the festival are also missing.[149] We are thus most likely dealing with a 'pastiche' of important events in the history and mythological past of Athens. In keeping with this interpretation of the frieze as (primarily) a treatment of a Panathenaic festival, the east frieze thus depicts the *peplos* of Athena that was conveyed on a ship to the Acropolis in the great Panathenaic procession. The man and the woman in the centre of the scene can be interpreted as a priest and a priestess, but the roles of the younger figures are more difficult to understand.[150] Two shorter girls carry stools with cushions, and one also a footstool. The two girls are possibly *arrēphoroi*.[151] The man and child are folding a large piece of cloth – the *peplos* for Athena.[152] The man could be the *archōn basileus*, the chief religious officer of Athens. The child is probably a boy, since he is naked under his *himation*, and is perhaps a temple servant. It has thus been suggested that the scene illustrates the handing over of the *peplos* to the *Praxiergidai* (mentioned in the inscriptions discussed above), an Athenian clan that had the right and duty to maintain the statue's garments.[153] But which *peplos* is actually depicted on the frieze? On the basis of context, the *peplos* is most likely to be the one presented every 4th year. According to M. Robertson, it is the old *peplos*, which was taken down and folded after 4 years of use.[154] Harrison, on the other hand, is correct in viewing it as the new *peplos*, and the act of folding it as the official rite of acceptance of the *peplos* by Athena.[155]

The appearance of Athena's *peploi*

The *peplos* depicted on the frieze is folded, and it is impossible to tell what it looked like. We do, however, have some knowledge of its appearance, since a few ancient authors provide scattered information about its decoration. For example, a passage in Euripides' tragedy *Hecuba* states:

> 'Or in the city of Pallas shall I represent in my weaving on the saffron peplos (*krokeōi peplōi*) the horses of Athena with their beautiful chariot, with elaborate flower-dyed[156] wefts,[157] or (shall I represent) the race of Titans, which Zeus destroyed with his fiery blast?'[158]

This indicates that the *peplos* was saffron yellow – a colour associated with women in Greek myth and ritual.[159] Euripides also refers to the decoration of the *peplos* of Athena in his play *Iphigenia in Tauris*, when Iphigenia laments that she neither sings to Hera with Argive women nor weaves the likeness of Pallas Athena and the Titans at the loom[160]:

> 'And now as a stranger I dwell in a house that borders on the Hostile Sea, with no husband, children, city, or friend. I do not sing in honor of Hera at Argos or weave with my shuttle upon the sounding loom the likeness of Athenian Pallas and the Titans in colors various.'[161]

Another source of information about the decoration of Athena's *peplos* is Plutarch's *Life of Demetrius*, which states that the Athenians 'also decreed that the figures of Demetrius and Antigonos should be woven into the sacred robe (*peplos*), along with

Appendix 1. The peplos of Athena at Athens

those of the gods.' The text subsequently refers to 'the sacred robe (*peplos*), for instance, in which they had decreed that the figures of Demetrius and Antigonos should be woven along with those of Zeus and Athena.'[162] Plato also describes the *peplos* of Athena as featuring images of the battle of gods: in his dialogue *Euthyphro*, Socrates says:

> 'And so you believe that there was really war between the gods, and fearful enmities and battles and other things of the sort, such as are told of by the poets and represented in varied designs by the great artists in our sacred places and especially on the robe (*peplos*) which is carried up to the Acropolis at the great Panathenaea? For this is covered with such representations. Shall we agree that these things are true, Euthyphro?'[163]

Finally, there is a relevant passage in Aristophanes' *Knights*, which reads:

> 'Let us sing the glory of our forefathers; ever victors, both on land and sea, they merit that Athens, rendered famous by these, her worthy sons, should write their deeds upon the sacred peplus.'[164]

Scholars often state that the quadrennial *peplos* of Athena was decorated with the battle of the Gods and the Giants, implying that this was always the case.[165] As the texts quoted above demonstrate, however, the actual evidence for this is quite late – and, although the subject-matter is usually described as the Gigantomachy, there is in fact no solid evidence that the entire mythical battle was depicted on the *peplos*, and among the gods only Zeus and Athena were certainly represented. Nevertheless, the imagery and the basic composition of the *peplos* were probably fixed by tradition, and it is therefore reasonable to assume that the *peplos* was decorated with these images every fourth year.[166] Each *peplos* would probably have been unique, though, and it would be a mistake to conceive of them as identical.

As noted above, several inscriptions refer to the working of the wool for the *peplos*, which makes this the likely fibre for Athena's garment. Furthermore, wool is generally perceived as the common textile fibre for *peploi*.

With regard to the size of the *peploi*, the annual robe for the wooden statue of Athena would probably have been relatively small, since the statue seems to have been under life-size. Barber suggests a size of roughly 5 by 6 ft (approximately 1.5 × 2 m).[167] The *peplos* presented every 4th year, on the other hand, appears to have been used as a sail on the Panathenaic ship, which implies that is was very large – the size of a normal sail, which Mansfield calculates as 4 – 8 m square.[168] The main source for the exceptionally large size of the *peplos*, however, is a profoundly corrupt fragment of Strattis' comedy *The Macedonians*, which was produced ca. 400 AD.[169] In this play, it is said that countless men were needed to hoist the *peplos* as a sail on the ship's mast. Yet there is not mention of a ship or an allusion to a sail in any other source.[170] Plutarch's *Life of Demetrios* is sometimes taken as evidence for the use of the *peplos* as a sail.[171] In fact, Plutarch provides no record of either sail or ship, but only of the sacred *peplos*:

> 'The sacred robe (*peplos*), for instance, in which they had decreed that the figures of Demetrius and Antigonus should be woven along with those of Zeus and Athena, as it was being carried in procession through the midst of the Cerameicus, was rent by a hurricane which smote it'.[172]

According to Barber, this implies that the ship's (!) *peplos* was rather large in proportion to its thickness as compared to a normal *peplos*, and she concludes that we need to imagine a cloth far bigger than the one for Athena's statue.[173] But, as noted, the sources on which this conclusion is based are rather late, and only one specifically mentions a ship and sail. The exceptionally large size of the *peplos* is therefore somewhat doubtful, or at least undocumented.

On the basis of the way that the folded *peplos* on the Parthenon frieze is depicted, Harrison concludes that the *peplos* was probably a so-called Argive *peplos*, which had an overfold about 1/3 the length of the dress – the type worn by Hera and Athena on the Parthenon frieze. She further remarks that the overall size of the cloth depicted appears to be right for a grown woman, but too big for a child.[174] Jeppesen has also performed calculations based on the depiction on the Parthenon frieze, concluding: 'Judging from the number of foldings indicated the cloth should be understood to represent a garment a little longer than, and approximately as wide as the height of the frieze', which would match the size of the so-called high priest.[175] To my mind, these 'calculations' of size and type are highly speculative and thus problematic, in that the cloth is folded an unknown number of times, and we cannot be confident that the depiction is intended to be wholly realistic.

What about the manner in which the *peploi* were produced? According to Barber, the annual (and smaller) *peplos* was woven on a warp-weighted loom, but we have no information about the quadrennial 'large *peplos*'. Barber does, however, raise the possibility that it was made on a tapestry loom, since this was well-known elsewhere in the Mediterranean and could have been imported.[176] Tapestry is, in fact, the likely technique employed in producing these decorated textiles; after all, it is the most well-known method of weaving patterns into a cloth. An alternative technique is the so-called supplementary weft-float technique, which entails floating a coloured pattern-weft across the top of a ground weave: this means that one weaves a plain background cloth, but between each row of weft the weaver inserts an extra coloured weft-thread, bringing it to the top as needed to make the pattern, and otherwise leaving it to ride behind. If the pattern thread is thicker than the weft, it will cover the ground weft entirely, producing the same illusion as tapestry.[177]

As far as the designs are concerned, Barber has suggested that, in case of the 'large *peplos*', the producers may have employed the stacked friezes layout of textile designs that had already emerged in the Bronze Age. She argues, further, that substantially later story-cloths of large dimensions that have survived from areas of Hellenic influence were designed in this way.[178] There is, moreover, no evidence whatsoever in the ancient Mediterranean world for the weaving of a single large scene on a cloth, as in for example later gobelin tapestries.[179] In case of the annual *peplos*, Barber suggests two possible designs on the basis of depictions in Ancient Greek art: a) successive scenes in a series of horizontal friezes running the entire width of the cloth, or b) square panels in a ladder-like arrangement running down the front of the garment.[180] Both designs are plausible, but the panel-like decoration down the front is rarer in ancient art. Besides, the horizontal panel decoration is reminiscent of the decorated garments used to decorate, for example, the Anatolian cult images and the veil of Despoina from Lycosoura (described in Part III).[181]

The old *peploi*

It is difficult to determine precisely what was done with the old garments after a new *peplos* had been presented to the goddess. One thing is certain: the garments would remain in the sanctuary.[182] It is possible that they were folded and stored in one of the temples on the Acropolis. The inventory of the Treasurers of Athena and the Other Gods (ca. 329/8–322/1 BC) refers to a so-called *peplothēkē*, perhaps a closet or an area in one of the temples designated for the storage of the sacred *peploi*, or even a separate building on the Acropolis.[183] Nagy also offers a third interpretation: that the *peplothēkē* was a room or large closet in the *Khalkothēkē*, a building used for storing votive offerings which is described in the epigraphic sources, but has not yet been identified in the architectural remains on the Acropolis.[184] Similarly, an inscription from Eleusis, dating to 329/8 BC, records Pentelic stones for a *Himatiothēkē* (*lithos pentelēikos en tēi himatiothēkēi*).[185] For Nagy, the context suggests that the Pentelic stone was intended for the *himatiothēkē*, a building that stored *himatia*.[186] Other epigraphic sources confirm the storage of textiles on the Acropolis: in 330/29 BC, due to the theft of sails and hangings for the Athenian fleet, these materials were stored in the *Khalkothēkē* at the *Opistodomos*.[187]

Notes

1. *IG* I^3, 7.
2. *IG* I^3, 7, fragment a, 10–11. Robertson 2004, 111, 115; Mansfield 1985, 141.
3. Translation: Robertson 2004, 118, 126.
4. Translation: Mansfield 1985, 141.
5. The term occurs in literary sources such as Hesychios. It is considered a title of honour dating back to the Classical Period. Robertson 2004, 143.
6. Robertson 2004, 124.
7. *LSCGS* 10, side A. Stafford 2007, 77. Nicomachos was later accused of being a lawbreaker who had deliberately perverted the laws of the ancestors. Evans 2010, 220.
8. Aleshire & Lambert 2003, 65.

9. Aleshire & Lambert 2003, 65.
10. Robertson 2004, 140.
11. The *athlothetai* were a board of officials with administrative responsibilities for the *Panathenaia*, e.g. overseeing the making of the *peplos*. Aleshire & Lambert 2004, 71.
12. Aleshire & Lambert 2003, 70.
13. IG II/III², 1060 and 1036. Mansfield 1985, 141.
14. Robertson 2004, 139, 146.
15. Aleshire & Lambert 2004, 71. It is important to note that SEG 28: 90 restores the text: ἀ ντὶ πέ]πλου, 'instead of a *peplos*'.
16. Aleshire & Lambert 2003, 75. Lines 11–14.
17. The decree of 103/2 BC is surmounted by a relief depicting two objects that could be pieces of woven cloth with holes at the edges. Robertson 2004, 144; Aleshire & Lambert 2003, 65, n.1.
18. IG II² 1034 and 1943 (103/2 BC); IG II² 1942 (a mere fragment) (100 BC). Aleshire & Lambert 2003, 65–66.
19. Robertson 2004, 140.
20. Ar. Av. 827. Translation: Mansfield 1985, 142. Here the verb ξαίνω is used, which can be translated as 'comb' or 'card', *LSJ* s.v. ξαίνω.
21. Xen. Hell. 1.4.12. Translation: Mansfield 1985, 143. From the verb *katakalūpto*.
22. Plut. Alc. 34.1. Translation: Perrin 1916.
23. Mansfield 1985, 140. *Lex. Seg. Gloss. rhet.* s.v. *kallion*.
24. In fact, the inscription says *kata*, 'along'.
25. Mansfield 1985, 142. *Lex. Seg. Gloss. rhet.* s.v. *kataniptēs*.
26. Sourvinou-Inwood 2011, 139.
27. Sourvinou-Inwood 2011, 142.
28. Sourvinou-Inwood 2011, 149–150.
29. Mansfield 1985, 372; Simon 1983, 46.
30. Sourvinou-Inwood 2011, 149, 156; Robertson 2004, 136–137.
31. *LSJ* s.v. *kallynō*.
32. Sourvinou-Inwood 2011, 155; Parker 2007, 474–475.
33. Robertson 2004, 136–137; Sourvinou-Inwood 2011, 156.
34. Mansfield 1985, 139–140.
35. Wachsmann 2012, 237–238.
36. Neils 2001, 23.
37. Wachsmann 2012, 238.
38. Mansfield 1985, 5–7, 16–17, 51, 55.
39. Robertson 1983, 276.
40. The epithet Polias or the name Athena Polias does not appear in written sources before the 5th century BC. Nick 2002, 144.
41. Philostr. V. A. 3.14.
42. Plut. Fr. 158.
43. Plut. Them. 10.4.
44. Plut. Them. 10.4.
45. Ridgway 1992, 122.
46. For the evidence for a standing versus a seated statue, see Alroth 1989, 49.
47. Ridgway 1992, 122.
48. Kroll 1982, 68.
49. Athenag. Leg. 17.3.
50. Dawson 2002, 67.
51. Ridgway 1992, 124.
52. Ridgway 1992, 126; Robertson 1996, 31; Nick 2002, 142.

Appendix 1. The peplos of Athena at Athens

53. Ridgway 1992, 125.
54. Hurwit 1999, 23.
55. *IG* II², 1424, 11–16; 1425, 307–312; 1426, 4–8; 1428, 142–146; 1429, 42–47; 1424a, 362–366. Translation: Kroll 1982.
56. Ridgway 1992, 135.
57. Nicarchos, *Anth. Pal.* 3.576.
58. Heliod. Hist. *FGrH* 4.425f.; Romano 1980, 58.
59. Acropolis Museum, inv. no. Acr. 2605+. Robertson 1996, 44; Ridgway 1992, 136.
60. Robertson 1996, 44; Beschi 1967–68, 533–534.
61. Mark 1987, 288.
62. Acropolis Museum, inv. no. 2556.
63. Mark 1987, 288; Mark 1993, 24, fig. 3. Unfortunately, the base bears no inscription to help in its identification.
64. Romano 1980, 59–60.
65. Romano 1980, 59.
66. Romano 1980, 62.
67. *IG* II² 88. Romano 1980, 58.
68. For more on the statue, see Nick 2002, 158–176.
69. The original statue is dated to ca. 448–447 BC. Hurwit 1999, 25.
70. Hurwit 1999, 25.
71. National Archaeological Museum, Athens, inv. no. 129.
72. Paus. 1.24.7. Translation: Jones & Ormerod 1918.
73. Hurwit 1999, 25–26. Pliny gives the height of the statue as 26 cubits, Plin. *Nat.* 36.18.
74. Paus. 5.11.10. Blundell 1998, 68.
75. Hurwit 1999, 27.
76. For a discussion on whether the Parthenon was a temple or a treasury, see e.g. Nick 2002, 119–132.
77. Nick 2002, 175.
78. Ridgway 1992, 130; Raubitschek & Stevens 1946, 112.
79. Raubitschek & Stevens 1946, 107; Hurwit 1999, 152.
80. Lundgreen convincingly rejects the depictions on most coins, Roman lamps, and Byzantine miniatures as reliable representations of the statue. Lundgreen 1997.
81. Lundgreen 1997, 191.
82. Hurwit 1999, 24.
83. Robertson 1975, 294.
84. Ridgway 1992, 127. On the Panathenaic amphoras, see e.g. Neils 1992.
85. E.g. British Museum, inv. no. B130.
86. E.g. National Museum Denmark, inv. no. ChrVIII 797. According to Neils, this reflects a chronological development: Athena is rendered in a *peplos* on amphoras dating to the 6th century BC. At the end of the 6th century BC the sleeved *chitōn* appears, while the *ependytēs* appears in the early 5th century BC. Neils 1992, 33.
87. E.g. British Museum, inv. no. B605.
88. Hurwit 1999, 24.
89. Hölscher 2010, 109.
90. *IG* II² 4225, L4. Lundgreen 1997, 190.
91. *Ad Dem.* 22.45.
92. Ridgway 1992, 129. E.g. Paus. 1.28.2; 9.4.1.
93. Hurwit 1999, 24.
94. Hurwit 1999, 25.
95. Lundgreen 1997, 197.

96. Ridgway 1992, 137.
97. Robertson 1996, 47; Ridgway 1992, 138.
98. Ridgway 1992, 137.
99. Robertson 1996, 48.
100. Robertson 1996, 29, 47, 48.
101. Villing 1998, 154.
102. Ridgway 1992, 139.
103. Ridgway 1992, 137.
104. For the statue as a spinner, see Consoli 2004, 15.
105. Acropolis Museum, inv. no. Acr. 625.
106. Marx 2001, 222.
107. Ridgway 1992, 138; Bundgaard 1974, 16, 31, note 58; Keesling 1999, 524–525.
108. Marx 2001, 244.
109. Acropolis Museum, inv. nos. Acr. 13054, Acr. 13055, Acr.13057.
110. Consoli 2004, 18.
111. Hutton 1897, 306.
112. Hutton 1897, 309.
113. Hutton 1897, 310 specifically suggests that she is in the guise of one of the *ergastinai*.
114. Ridgway 1992, 139.
115. Acropolis Museum, inv. no. Acr. 577.
116. Consoli 2004, 14; Hurwit 1999, 16.
117. *IG* I^2, 561, *IG* II^2, 4318, 4328, 4329, 4334, 4338, 4339. Consoli 2004, 24.
118. Villing 1998, 156.
119. Villing 1998, 154.
120. Museo Archaeologico, Ragusa, inv. no. 1851 RG.
121. Blinkenberg 1931, 535–536, no. 2210, pl. 102.
122. Ridgway 1992, 140.
123. For the production of the *peploi*, see Brøns forthcoming b.
124. Romano 1980, 412.
125. Mylonopoulos 2010, 4.
126. Donohue 1997, 37.
127. For depictions of Athena Ergane, see Consoli 2004; Consoli 2010.
128. In the 3rd century BC, the Erechtheion was termed '*tēs Poliados neōs*', Philoch. *FGrH* 328 F 67.
129. Ar. Av. 826.
130. Mansfield 1985, 2.
131. Nick 2002, 156. According to *IG* II^2 334, Athena Hygieia received a sacrifice at the small Panathenaia, which, according to Nick, indicates that she was honoured as an aspect of Athena Polias. Nick 2002,142. With regard to Athena Nike, Nick bases his argument on Soph. *Phil*. 134: '*Nikē t' Athēna Polias*'. Nick 2002, 140.
132. The stele was recovered during excavations in the Athenian Agora in 1971. Published by Shear 1978.
133. Translation: Shear 1978.
134. Kroll 1982, 69.
135. Kroll 1982, 69.
136. Kroll 1982, 69–72, includes the scholiast on Ar, Av. 516, who states that the *agalma* of Athena Archēgetis had an owl in her hand.
137. Lewis 1979/80, 28–29.
138. Kardara 1969, 185.
139. Wachsmann 2012, 239; Hurwit 1999, 45.

Appendix 1. The peplos of Athena at Athens 391

140. It is also a possibility that the wear and tear of the *peplos* was intentional. Perhaps this was not considered to be destruction of the garment, but rather similar to e.g. burning a sacrifice on the altar. I thank Hedvig Landenius Enegren and Ellen Harlizius Klück for this suggestion.
141. Mansfield 1985, 55; Sourvinou-Inwood 2011, 268.
142. Neils 2001, 173; Sourvinou-Inwood 2011, 284; Blundell 1998, 59. There have been several objections to this interpretation, primarily because the frieze does not depict e.g. the Panathenaic ship or the Athenian armed hoplites, e.g. Holloway 1966, 223.
143. Some scholars such as Jeppesen, however, refute the interpretation of the garment as being the *peplos* for Athena, which was presented to her at the Panathenaic festival. Instead, Jeppesen believes that the garment belonged to the person – usually interpreted as the high priest - handing it over to the child. Jeppesen 2007, 114, 164.
144. Kardara 1964; Neils 2001, 175.
145. Holloway 1966. For further on these and other interpretations of the Parthenon frieze, see Neils 2001, 173–201.
146. Connelly 1996; 2014.
147. Connelly 1993b; 2014.
148. The absence of the ship might be explained by the fact that it was a later addition to the Panathenaic festival. The earliest attestation of the ship dates to the Hellenistic Period. The earliest attestation is in an honorary decree for Philippides, *CIA* II 314, 14, dated to 284 BC. The ship and sail are not recorded in ancient sources again, however, until the 2nd century AD. Reuthner 2006, 313. Nevertheless, Barber suggests that the use of a ship in the procession possibly began shortly after the Persian Wars, in the form of one of the boats from the Battle of Salamis. Barber 1992, 114.
149. Harrison 1996, 209.
150. Neils 2001, 167.
151. Sourvinou-Inwood 2011, 298; the two girls have also been interpreted as '*Diphrophoroi*'. Reuthner 2006, 318.
152. Neils 2001, 170.
153. Barber 1992, 113.
154. Robertson 1975, 11.
155. Harrison 1996, 203.
156. *LSJ s.v. anthokrokos*: 'worked with flowers'.
157. *LSJ s.v. pēnē*: 'bobbin, spool, woof, web'. Tuck 2009, 153 translates *pēnē* as 'weft', 'web', or 'weaving'. According to Tuck 2009, 155, *pēnais* here simply refers to the *peplos* of Athena.
158. Eur. *Hec.* 466–474. Translation: Tuck 2009.
159. Barber 1992, 116. Eur. *Hec.* 466–474.
160. See also Tuck 2009, 156.
161. Eur. *IT* 222–224. Translation: Kovacs 1999.
162. Plut. *Demetr.* 10.4, 12.2. Translation: Perrin 1920.
163. Pl. *Euthphr.* 6b–c.
164. Ar. *Eq.* 565–566. Translation: O'Neill 1938.
165. Barber 1992, 103; Carpenter 2007, 402; Deacy 2007, 230–231; Robertson 1996, 63. According to Sourvinou-Inwood, the annual *peplos* also featured a depiction of the Gigantomachy. Sourvinou-Inwood 2011, 267.
166. Mansfield 1985, 58.
167. Barber 1992, 113.
168. Mansfield 1985, 6–7.
169. Stratt. fragment 31.

170. Reuthner 2006, 312.
171. E.g. Barber 1992, 114.
172. Plut. *Demetr.* 12.2/3. Translation: Perrin 1920.
173. Barber 1992, 114.
174. Harrison 1996, 202.
175. Jeppesen 2007, 109, fig. 9.
176. Barber 1992, 114.
177. Barber 1992, 111.
178. Barber 1992, 115.
179. Barber 1992, 115.
180. Barber 1992, 115.
181. For the storage of the old *peploi*, see Nagy 1984; Brøns forthcoming b. For the storage of other textiles (sails) on the Athenian Acropolis, see Nosch 2014, 37; Gabrielsen 1994, 148.
182. Yet one more option should be mentioned, even though there is no evidence to substantiate it: perhaps the old *peploi* were not stored in the temple, but instead cut up and given to devotees – perhaps important participants in the Panathenaic procession and/or ritual personnel. Such practices are attested in ethnographic sources: for example, the *kiswa*, a large decorated textile used to decorate the ka'ba in Mekka, was cut into pieces and given to devotees. This practice of dressing the ka'ba in cloth dates back to the 5th century AD. McGregor 2010, 260.
183. Mansfield 1985, 55. *IG* II/III^2 1462, 11; Nagy 1984, 231.
184. Nagy 1984, 231.
185. *IG* II^2 1672.309.
186. Nagy 1984, 231.
187. Nosch 2014, 37; Gabrielsen 1994, 148.

Appendix 2

Temple inventories. Greek texts and translations

No. 1.
Boeotia. Temple inventory from Tanagra, side B
SEG 43: 212 **(B)**
Ca. 260–250 BC
Translation: Roller 1989 (with modifications)

[Ε]ὐαέθλω ἄρχοντος, ἱαραρχι-
όντων Ἀσωποκρί τω Κλιωνίω,
Μού[ρτ]ωνος {²⁷Μού[ρσ]ωνος}²⁷ Καφισοδοτίω, γραμ-
ματίδδοντος Διοδώρω [Π]εδαγε-
(5) νείω, ἀνέθιαν *vacat*
Φιλοτίμα κιθῶνα πορφούριον παιδικόν·
Δαμονίκα ἀμόργινον κιθῶνα κορικόν·
Φιλλὼ κροκωτόν· *vacat*
Ὁμολωΐς λίνινον κιθῶνα πορφ[ού]ριον·
(10) Νικασία σουμμετρίαν γουνηκίαν·
Φιλονίκα ἀμπεχόνιον γουνηκῖον λευ[κό]ν·
Φιλόκλια [χλ]ανίδα· *vacat*
Γαναξὶς [χλ]ανίδα· *vacat*
Γοργὶς κιθ[ῶ]να γουν[ηκῖ]ον [παρ]-
(15) [ο]υφὰς ἔ[χοντ]α [— — — — — — —]
Ἑρμαΐς [— — — — — — — — — —]
Ξεννὼ μάλινον χιτ[ῶ]να [π]ουρείνια [ἔ]χο[ν]-
τα ἑπτά : Διοκκὼ γαδὰν παΐλλω ἀργουρί[αν]·
Ἀμούνταο ἄρχοντος, ἱαραρχιόντων Θωρα[κί]-
(20) δαο Ματρωνίω, Σαμίαο Δαμαρετίω, ἐπάνθετα χιτώνια· Πουθὶς χιτῶνα κορικὸν κνώσι-
ον τέλειον· Εὐτύχα ἀμόργινον· Καλλίχα ἀν[δρ]-
[ε]ῖο[ν· Φι]λοτίμα [μ]άλινον· Δαμοτίμα λευκό[ν· ἔ]-
νωτίδια Ἰάρων [ο]ὑπὲρ Δαμῶς· ἐρ[ωτ]ίσκ[ως Ἀρ]-
(25) χελάα· *vacat*

Νίκωνος ἄρχοντος, ἰαραρχιόντων Εὐγίτου[ος]
Τυχωνίω, Τιμίναο Φρουνωνίω, γραμματίδδον[τος]
Φρούνωνος Τιμίναο, ἐπάνθετα χιτώνια· Ἐμπεδί[α]
χιτῶνα κορικὸν γευματικὸν ἐπισανδαλίδας ἔχον-
(30) τα ἕξ : Πτωιοδώρα σχιστὸμ μάλινον πουρεινί-
δας ἔχοντα ἕξ· vacat
Φιλοξένα τρίβωνα ἀνδρῖον· vacat
Ἀνδροκκὼ χλανιδίσκαν λευκάν· vacat
Εὐφανία χιτῶνα μάλινον πουρείνια vacat
(35) σάρδια ἔχοντα ἕξ· vacat
Λιουσὶς χλαμουδίσκαν· vacat
Φιλοκκὼ ταραντῖνον ῥάκινον· vacat
Ξενοκκὼ χλανίδας δύο, τεγίδιον λευκόν,
λίνινος παρπόρφουρος· οὖτα ἔχι ἁ ἰάρεια·
(40) Φουσὶς χιτῶνα κορικὸν παρορφνιδωτόν·
Δαμοτίμα χιτωνίσκον παρορφνιδωτὸγ κοριδί[ω]·
Ἀριστόκλια χιτῶνα φάρινογ κοριδίω παρορφνιδω-
τὸν ἀνεπίγραφον, χιτῶνα παΐλλω παρουφὰς ἔχοντα·
Δαμοκρίτα χιτῶνα λίνινομ παρπόρφυρον, χιτῶνα
(45) κοριδίω παρορφνιδωτόν, ἀμόργινον κοριδίω χιτῶν[α]
ταραντινίναν παρορφνιδωτὸν κοριδίω ἀνεπίγρα[φον]
Θιοδώρα χιτῶνα παρορφνι[δ]ωτὸν κοριδίω· vacat
Χηρίππα χιτῶνα κοριδίω παρορφνιδωτόν· vacat
Ἰαρόκλια χιτῶνα κοριδίω ταραντινίναν παρορφνιδ[ω]-
(50) τόν, χιτῶνα παΐλλω παρπόρφυρον, ἀνεπίγραφα·
Χρουσία : δακκύλιος ὁλκὰ χρούσιος, ἁλύσιον
Πασίκλια ὁλκὰ χρούσιος πέτταρες ὀβολύ· Ὁμολωΐ[ς],
Τιμηνέτα ἐνωτίδια ὁλκὰ εἰμίχρουσον, ἐνωτίδια
περίδδυγα ὁλκὰ πέτταρες ὀβολύ, ἐρ[ω]τίσκυ ὁλκὰ
(55) εἰμίχρουσον : ὁλκὰ [π]άντων χρούσιοι τρῖς εἰμίχρουσον
vacat δύ' ὀβολύ.

With Euaethlos as archon, the high priests were Asōpokritos son of Kliōn, Mourtōn son of Kaphidotos; the recorder was Diodōrōs son of Pedagenēs, These [women] dedicated:
Philotīmā, a boy's purple *chitōn*
Dāmonīkā, a girl's amorgine *chitōn*
Phillō, a saffron-coloured garment (*krokōtos*)
Homolōis, a purple linen *chitōn*
Nīkasiā, a woman's robe (*soummetria*)
Philonīkā, a woman's *ampechonon*, white
Philoklia, a woman's *chlanida*
Wanaxis, a woman's *chlanida*

Gorgis, a woman's *chitōn*, having a border
Hermais [....]
Xennō, a yellow *chitōn* with seven buttons
Diokkō, a child's silver garment
With Amountas as arkhon, the high priests were Thōrakidas son of Matrōnios and Samias son of Dāmaretos;
Dedicated clothing:
Pouthis, a girl's *chitōn*, finely worked
Eutychā, an Amorgine
Kallichā, a man's
Philotimā, a yellow
Dāmotimā, a white
Earrings, Hiaron on behalf of Damōs
Little erotes, Archelaa
With Nikon as archon, the high priests were Eugiton son of Tykhon and Timinas son of Phrounōnos; the recorder was Phrounōnos son of Timinas;
Dedicated *chitōnes*/clothing:
Empedia, a girl's chitōn *geumatikos*[1] having six sandal straps (?)
Ptōiodōra, a yellow open garment with six buttons
Philoxenā, a man's worn *tribōn*
Androkkō, a white *chlanidiska*
Euphania, a yellow *chitōn* with six Sardian buttons
Liousis, a small *chlamys*
Philokko, a ragged *tarantinon*
Xenokkō, two *chlanides*, a white *tegidion*,
and a linen garment with purple borders. These the priestess has
Phousis, a girl's *chitōn* with a dark border
Dāmotimā, a girl's *chitōniskos* with a dark border
Aristoklia, a girl's cloth *chitōn* with a dark border, uninscribed, and a boy's *chitōn*, with a border
Dāmokritā, a linen *chitōn* with a purple border,
a girl's *chitōn* with a dark border,
a girl's 'amorgine' *chitōn*,
a girl's *tarantinon* with a dark border, uninscribed
Thiodōra, a girl's *chitōn* with a dark border
Chērippa, a girl's *chitōn* with a dark border
Hiāroklia, a girl's Tarantine *chitōn* with a dark border,
a boy's *chitōn* with a purple border, uninscribed
Golden objects: A ring weighing one stater, a chain
Pasiklia weighing one stater and four obols. Homolois,
Timēneta earrings weighing one-half stater, round earrings weighing four obols,
Erotes weighing one-half stater
The weight of all the golden objects: 3,5 staters 2 obols

No. 2.
Boeotia. Inventory of the possessions of an unknown deity at Thebes
IG VII, 2421
Mid-3rd century BC

[Διω]νυσίχω ἄρχοντος, ἰαραρχίοντος Ἀγελ[— — — — — — — — — —],
γραμματίδδοντος Φιλοξένω Γλαύκω [— — — — — — — — — — —]
[Ἀρ]τεμισία Ταραντῖνα σιΟΚλάδος Θιοδώρα Θ[— — — — — — —]
[Tαρα]ντῖνον [ῥ]<ά>μματ' ἔχον· Ἀριστὼ Ταραντῖνον πα[ρπόρφυρον],
(5) [ῥά]μματ' ἔχον· Λυσιμάχα [χι]τῶνα παρπόρφυ[ρ]ον αε[— —· — — —]
[—]ππὶς χιτῶνα μάλινον <κ>οριδίω παρπόρφυρον, πο[υρεί]-
[νι]α πέτταρα · Ἰράνα σινδόνα παρραπτῶς πορφ[ύρας ἔχω]-
[σαν], πουρείνια ὀκτό· Θιοζότα σχιστὸν περιπόρφυρον, [— — ἔ]-
[χον]τα ἕξ, κὴ λεῖρον · Τελεσίππα Ἀριστοδά[μω σινδόνα παρ]-
(10) [ραπ]τῶς πορφύρας ἔχοντα κὴ ἱ[μά]τια πέ[ντε· — — — — — —]
[— —]αν βαπτ[ὰν] παρρ[α]πτ[ὼς πορφύρας ἔχωσαν].

When Dionysios was arkhon, high-priest was Agel[------]
Philoxenos son of Glaukos was secretary [------]
Artemisia 'Tarentine' garments *sioklados*,
Thiodōra [------] a *tarantinon* with seams.
Aristo a *tarantinon* ed[ged with purple], with seams
Lysimachā a *chitōn* edged with purple [------],
[-]-ppis (name) an apple/quince coloured girl's *chitōn* edged with purple, four small knobs
Irana a *sindōn* with purple seams, eight small knobs
Thiozotā a garment with a purple border open at the side, having six [????] and a gold ornament
Telesippa daughter of Aristodamos, a *sindōn* with purple seams and five *himatia* [------]
[---] a dyed [garment] [with purple] seams.

No. 3.
Greek islands. Account of the treasurers of the Heraion, Samos
IG XII 6, 1, 261
346/5 BC

(12) κόσμος τῆς θεοῦ· κιθ[ὼ]-
ν Λύδιος ἔξαστιν ἔχων ἰσάτιδος, Διογένης ἀνέθηκε· κιθὼν Λύδιος ἔ-
ξαστιν ὑακινθίνην ἔχων· κιθὼν Λύδιος ἔξαστιν ὑακινθίνην ἔχω-
(15) ν· κιθὼν Λύδιος ἔξαστιν ἀλοργῆν ἔχων· κιθωνίσκος λινοῦς ἔξαστιν
ἀλοργῆν ἔχων· κιθὼν κατάστικτος : κιθὼν Λύδιος ἔξαστιν λευκὴν ἔ-
χων· μίτρη λιτὴ στυππείου· κιθωνίσκος χρυσῶι πεποικιλμένος μύρ-
τον χρύσεον ἔχων· περίβλημα λίνου ῥάκινον· μίτρη πάραυλος, ταύτην
ἡ θεὸς ἔχει· παράλασσις, Ἶριν ἐμ μέσωι ἔχει ἀλοργῆν· σινδὼν λίς, ἥντινα
(20) τῆι θεῶι παραπιτνῶσι· κιθῶνος στυππίνου τόμος· πρόσλημμα τῆς θε-
οῦ παραλοργὲς ἀμφιθύσανον· σφενδόναι λιναῖ δύο· κρήδεμνα ἑπτά, τούτων ἓν ἡ Εὐαγγελὶς ἔχει : περίζωμα ἀλοργοῦν ῥάκινον ποικίλον· κεκ[...]
πλεκτὸς ἀλοργοῦς· ὑποκεφάλαια δύο ἡμιτυβίου λιτά· ὑποκεφάλαιον ὑπογ[ε]-
γραμμένον· σπληνίσκον ὑπογεγραμμένον ἱππέα· σινδονίσκη ὑπογεγ-
(25) ραμμένη : σπληνίσκον λινοῦν ἐρίνεον· καταπέτασμα τῆς τραπέζης ῥ-
άκινον : παραπετάσματα δύο βαρβαρικὰ ποικίλα : αὐλαῖαι δύο· πρόσλημ[μ]-
[α] λινοῦν : ἱμάτιον λευκόν, ἡ ὄπισθε θεὸς ἔχει : κιθῶνες Λύδιοι ἐξάστεις
ἀλοργᾶς ἔχοντες : κιθῶνες ἐπὶ Θρασυάνακτος, τούτους ἡ θεὸς ἔχει· κιθῶνες ἐπὶ Ἱπποδάμαντος δύο, τούτους ἔχει ἡ θεός· ἐπὶ δημιοργοῦ Δαμασικλ-
(30) έους χλάνδιον ἀλοργοῦν, τοῦτο ἐπὶ τοῦ ὁδοῦ· ἐπὶ Δημητρίου ἄρχοντο[ς]
κιθῶνες δύο, τούτους ἡ θεὸς ἔχει : ἱμάτια Ἑρμέω : κιθῶνες ΔΔΔΓΙΙΙ, τ[ο]-
ύτων ὁ Ἑρμῆς ἕνα ἔχει : ἱμάτια : ΔΔΔΓΙΙΙΙ· τούτων ὁ Ἑρμῆς ἔχει ἕν· ἀπὸ
τούτων τῶν ἱματίων ὁ Ἑρμῆς ὁ ἐν Ἀφροδίτης ἔχει δύο· στρουθοὶ ὑπὸ τῆ[ι]
τραπέζηι : στρουθοὶ ἐπίχρυσοι δύο, στρουθοὶ ὑπάργυροι δύο, τῶν στρουθῶν
(35) τῶν ἐπιχρύσων ἐγλείπει τὰ ὀρσοπύγια : Φιλόστρατος ἀπέγραψε· σπληνίσκον, μίτρη, κρήδεμνον, χλάνδια δύο ἀλοργᾶ ἐπὶ τοῦ ὁδοῦ τῆς Ἥρας, μίτραι δύ[ο]
στύππιναι· κιθῶνες δύο, ἐνδυτὰ τῆς Εὐαγγελίδος· τρίχαπτον παλαιόν·

Adornment of the goddess
Lydian *chitōn* with edges of dark blue. Dedicated by Diogenēs.
Lydian *chitōn* with purple fringe
Lydian *chitōn* with purple fringe
Lydian *chitōn* with purple fringe
Linen *chitōniskos* with purple fringe
"Embroidered" *chitōn*
Lydian *chitōn* with white edges
Simple *mitrē* of coarse [material]
Chitōniskos interwoven with gold [and] which has a golden myrtle

Ragged linen *periblēma*
the Goddess has a *mitrē paraulos*
Paralassis[2] with a purple iris in the middle
Smooth *sindōn*, which they spread before the goddess
Piece of coarse *chitōn*
a *proslēmma* of the goddess with purple edges and fringes/tassels on either side
Two linen headbands (*sphendonai*)
Seven veils, of which Euangelis has one
Purple ragged patterned loincloth
Plaited purple hair veil (*kekryphalos*)[3]
Two simple cushions of linen cloth/towel[4]
Cushion, inscribed
Inscribed ribbon/strip (*spleniskos*) for a girl's ornament(?)[5]
Sindoniskē inscribed
Linen woollen strip (*spleniskos*)
Ragged table cover
Two foreign, patterned curtains
Two curtains
Linen *proslēmma*
The goddess at the back has a white *himation*[6]
Lydian *chitōnes* with purple edges:
Chitōnes, under Thrasyanax, these the goddess has
Two *chitōnes*, under Hippodamas, these the goddess has
At the time when Damasikles was demiourgos a purple *chlandion* [was] at the entrance
When Dēmētrios was arkhon, two *chitōnes*, these the goddess has
Himatia of Hermes:
38 *chitōnes*, of which Hermes has one
48 *himatia*, of which Hermes has one
Of these *himatia* the Hermes in [the temple] of Aphrodite has two
Birds under the altar-table:
two gilded birds, two birds of silver, the gilded birds lack the tail feathers
Philostratos inventoried
Ribbon/strip (*spleniskos*), *mitrē*, veil (*kredemnon*)
Two purple *chlanides* at the entrance of [the temple] of Hera
Two *mitrai* of coarse [material]
Two *chitōnes* garments of/from Euangelis
An old fine veil (*trichapton*)

No. 4.
Anatolia. Inventory of the Temple of Artemis *Kithōnē* (?) at Miletos
Milet VI, 3, 1357
2nd century BC
Translation: Herrmann 2006 (German); Cole 1998.

κα<λά>ρειρις μεσογλαύκινος περίχρ[υ]-
[σ]ος παλαιὸς ἠχρηωμένος, ἱμάτιον σελ<ά>γινον ? περιπόρφυρ[ον]
παλαιὸν ἠχρειωμένον, ἁλουργέα παλαιὰ κατακεκομμένα
ἀχρεῖα ὀκτώ, χλανίδες παλαιαὶ ἀχρεῖαι κατακεκομμέναι τ-
[ρ]εῖς, ἱμάτια πορφυρᾶ βαπτὰ ἀχρεῖα κατακεκομμένα τρία, κά[ρ]-
πασος παλαιός, σινδονίτης παλα[ι]ὸς ἀχρεῖος, ὀθόναι λιναῖ π-
[α]λαιαὶ ἀχρεῖαι τρεῖς, ἄλλαι ἡ[μ]ιτριβεῖς κεκομμέναι δύο, χλαμύδ[ες]
ν. ἐφηβικαὶ παλαιαὶ ἀχρεῖαι τέσσαρες, προ[σ]ωπίδια βομβύκινα πα-
[λ]αιὰ ἀχρεῖα τέσσαρα, ἄλλα ἐρεᾶ παλαιὰ ἀχρεῖα δύο, λινᾶ πα-
[λ]αιὰ ἀχρεῖα δεκαδύο, ἐπίκρηνον λ[ι]νοῦν παλαιόν, ἄλλα [ἀ]-
χρεῖα δύο, ἄλλο ἡμιτριβὲς κεκομμένον, ἄλλο βομβύκινον ἀχ-
ρεῖον κατατετιλμένον ἄλλο βομβύκινον ἡμιτριβὲς κεκομμέν-
[ο]ν, λημνίσκοι ξυστοὶ πράσινοι κατακεκομμένοι δύο, ἄλλος κόκκ[ι]-
[ν]ος παλαιὸς κατακεκομμένος, στρόφοι παλαιοὶ <ἐ>πίχρυσοι δύο, [ἄ]-
λλος σπα{ν}δίκινος παλαιὸς ἔχων κεραύνιον χρυσοποίκιλον, διά[ζω]-
μα ἐρεοῦν ἐπίχρυσον παλαιὸν κατακεκομμένον, ἄλλο λινο[ῦν]
καὶ ὑποκλείδιον ἡμιτριβές, ἃ [ἔφη]σεν ἀνατεθεικέναι Ἀνα<ῖ>ος, ζω[ν]α[ι]
παλαια<ὶ> δύο, ἄλλαι μείζονες παλαιαὶ δ[ύ]ο, χλάνδιον καὶ εὐπάρυ[φ]ον
[π]αιδικὰ κατακεκομμένα ἁλουργεα, παιδικ[ὰ ἄλλα] κατακεκομμέν[α..]
‐ INΛIEI ‐ ‐ 7 ‐ 8 ‐ ‐

an old useless *kalaseīris*, bluish grey in the middle, with gold border
an old useless *himation*, bright in colour, with purple border
eight old useless purple garments, frayed
three old useless *chlanides*, frayed
three purple-dyed *himatia*, useless and frayed
an old Karpasian garment
an old useless *sindonītēs*
three old useless linen *othonai*
two other half-worn [*othonai*], frayed
four old useless ephebic *chlamydes*
four old useless silken *prosōpidia*
two other old useless of wool
twelve old useless pieces of linen
an old linen *epikrēnon*.
two other ones, useless.

another one, half worn out, frayed.
another useless silken one, frayed.
another silken one, half worn to pieces, frayed,
two green cut woollen ribbons (*lēmniskoi*), frayed.
another purple (scarlet) one, frayed.
two old girdles (*strophoi*) overlaid with gold.
another old bright red having a gold-"embroidered" thunder bolt motif.
an old woollen belt/girdle with gold overlaid, old and frayed.
another of linen with a little clasp, half worn out.
Anaios says he has dedicated.
two old belts/girdles.
two other old ones, larger.
a small purple *chlandion* and one with a fine purple border, for children, frayed.
and other children's clothing, frayed.

Notes

1. Schachter suggests that *geumatikos* might refer to a girl's dining dress, Schachter 1997, 278.
2. Type of garment.
3. Only the beginning of the word is preserved: *kek*[????]. It can thus tentatively be restored as *kekryphalos*.
4. *LSJ s.v. ēmitybion* – 'linen cloth, towel'.
5. *Hippea* might be translated as a girl's ornament, Hesychius (2nd century BC), cf. Wilken, *Griechische Ostraka aus Ägypten und Nubien* 1899, *323*.
6. Grammatically there is no link between the *himation* and the goddess. It is possible that the goddess has something else, but it seems likely that the *himation* referred to belongs to the goddess.

Appendix 3

Clothing regulations. Greek texts and translations

No. 1
The sacred law of Andania
IG V, 1, 1390; *LSCG* 65
92/91 BC
Translation: Gawlinski 2011

Lines 13-15:
στεφάνων. στεφάνους δὲ ἐχόντω οἱ μὲν ἱεροὶ καὶ αἱ ἱεραὶ πῖλον λευκόν,
τῶν δὲ τελουμένων οἱ πρωτομύσται στλεγγίδα. ὅταν δὲ οἱ ἱεροὶ παραγγείλωντι,
 τὰμ μὲν στλεγγίδα ἀποθέσθωσαν, στεφανούσθωσαν δὲ πάντες δάφναι.

Concerning wreaths
The sacred men are to wear wreaths, the sacred women a white felt cap,
and the recently initiated among the initiated a tiara.
But when the sacred men give the order, they are to take off their tiara,
and they are all to be wreathed with laurel.

Lines 15-26:
εἱματισμοῦ. οἱ τελούμενοι τὰ μυστήρια ἀνυπόδετοι ἔστωσαν καὶ ἐχόντω τὸν
εἱματισμὸν λευκόν, αἱ δὲ γυναῖκες μὴ διαφανῆ μηδὲ τὰ σαμεῖα ἐν τοῖς εἱματίοις
 πλατύτερα ἡμιδακτυλίου, καὶ αἱ
μὲν ἰδιώτιες ἐχόντω χιτῶνα λίνεον καὶ εἱμάτιον μὴ πλείονος ἄξια δραχμᾶν ἑκατόν,
 αἱ δὲ παῖδες καλάσηριν ἢ σιν-
δονίταν καὶ εἱμάτιον μὴ πλείονος ἄξια μνᾶς, αἱ δὲ δοῦλαι καλάσηριν ἢ σινδονίταν
 καὶ εἱμάτιον μὴ πλείονος ἄξια δρα-
χμᾶν πεντήκοντα. ——————— αἱ δὲ ἱεραί, αἱ μὲν γυναῖκες καλάσηριν ἢ ὑπόδυμα
 μὴ ἔχον σκιὰς καὶ εἱμάτιον μὴ πλέονος ἄξια δύο
(20) μνᾶν, αἱ δὲ [παῖδε]ς καλάσηριν ἢ εἱμάτιον μὴ πλέονος ἄξια δραχμᾶν ἑκατόν.
——————— ἐν δὲ τᾶι πομπᾶι αἱ μὲν ἱεραὶ γυναῖκες ὑποδύ-
ταν καὶ εἱμάτιον γυναικεῖον οὖλον, σαμεῖα ἔχον μὴ πλατύτερα ἡμιδακτυλίου, αἱ
 δὲ παῖδες καλάσηριν καὶ εἱμάτιον μὴ δια-
φανές· μὴ ἐχέτω δὲ μηδεμία χρυσία μηδὲ φῦκος μηδὲ ψιμίθιον μηδὲ ἀνάδεμα μηδὲ
 τὰς τρίχας ἀνπεπλεγμένας μηδὲ ὑπο-

δήματα εἰ μὴ πίλινα ἢ δερμάτινα ἱερόθυτα. δίφρους δὲ ἐχόντω αἱ ἱεραὶ εὐσυΐνους στρογγύλους καὶ ἐπ' αὐτῶν ποτικεφάλαια
ἢ σπῖραν λευκά, μὴ ἔχοντα μήτε σκιὰν μητὲ πορφύραν. ὅσα<ς> δὲ δεῖ διασκευάζεσθαι εἰς θεῶν διάθεσιν, ἐχόντω τὸν εἱματισμόν,
καθ' ὃ ἂν οἱ ἱεροὶ διατάξωντι. ──────── ἂν δέ τις ἄλλως ἔχει τὸν εἱματισμὸν παρὰ τὸ διάγραμμα, ἢ ἄλλο τι τῶν κεκωλυμένων, μὴ ἐπιτρπέ-
τω ὁ γυναικονόμος καὶ ἐξουσίαν ἐχέτω λυμαίνεσθαι, καὶ ἔστω ἱερὰ τῶν θεῶν.

Concerning clothing.
The men who are initiated into the mysteries must be barefoot and wear white clothing,
the women wearing neither transparent clothes nor stripes on their *himatia* more than half a *daktylos* wide.
And the free adult women must wear a linen *chitōn* and *himation* worth in total no more than 100 drachmas,
the girls a *kalasēris* or a *sindonitēs* and a *himation* worth in total no more than one mina,
and the female slaves a *kalasēris* or a *sindonitēs* and a *himation* worth in total no more than 50 drachmas.
Of the sacred women, the adults must wear a *kalasēris* or an undergarment without a coloured border and a *himation* worth in total no more than two minas,
and the girls a *kalasēris* and *himation* worth in total no more than 100 drachmas.
In the procession the sacred women must wear an undergarment and a woolen woman's himation with stripes no more than half a *daktylos* wide,
and the girls must wear a *kalasēris* and a *himation* that is not transparent.
No woman is to have gold, rouge, white lead make-up, a hair band, plaited hair, or shoes unless of felt or sacrificial leather.
The sacred women must have round wicker stools with white pillows or a round cushion on them, having neither a coloured border nor purple colour.
Whichever women are to dress themselves in representation of the goddesses must wear the clothing which the sacred men order.
If anyone otherwise has clothing contrary to the *diagramma*, or if anyone has something else that is prohibited, the *gynaikonomos* must not allow the item and is to have the right to have it mutilated, and it must become the property of the gods.

No. 2
Hymn of Isyllos
IG IV², 1, 128; *SEG* 46:375, lines 17–19.
280 BC
Translation: Austino 2012

τοῖσιν ἐπαγγέλλεν καὶ πομπεύεν σφε κομῶντας Φοίβωι ἄνακτι υἱῶι τε Ἀσκλαπιῶι ἰατῆρι εἵμασιν ἐν λευκοῖσι, δάφνας στεφάνοις ποτ' Ἀπόλλω,

(...) proclaim to these men that they are to march in a processions for lord Apollo and his son Asclepios the healer with their hair down (or long) and in white garments and wearing crowns of laurel march purified to the temple of Apollo,

No. 3
Alexandreion, Priene
LSAM 35; *IPr* no. 205
3rd century BC
Translation: Cleland 2003

ἔλαχε τὴν ἱερωσύν[ην]
Ἀναξίδημος Ἀπολλων[ίου].
Εἰσίναι εἰς [τὸ]
ἱερὸν ἁγνὸν ἐ[ν]
ἐσθῆτι λευκ[ῆι].

Anaxideros son of Apollonios
obtained the priesthood/sacrifice as his portion.
Go pure into the sanctuary
in white clothing.

No. 4
Zeus Kynthios & Athena Kynthia, Delos
LSCGS 59; *ID* 2529
116/115 BC
Translation: Rostad 2006

― ― ― ― ― ― ― ― ― ― ― ―

[ἱερεὺς γενόμενος] Δ[ιὸς]
[Κυνθίου καὶ Ἀ]θηνᾶς
[Κυνθίας] ἐν τῶι ἐπ[ὶ]
[Σαραπί?]ωνος ἄρχο[ν]-
[τος] ἐνιαυτῶι, v
[ζακορεύον]τος Νικηφόρου
[τὸ .. ἀ]ντὶ τῆς καταγεί-
[σης στ]ήλης κατὰ πρόστα-
[γμα ἀν]έγραψεν τὴν προγ-
[ραφήν]· v ἰέναι εἰς τὸ ἱερ-
[ὸν τοῦ] Διὸς τοῦ Κυνθίου
[καὶ τῆ]ς Ἀθηνᾶς τῆς Κυνθί-
[ας χερ]σὶν καὶ ψυχῇ καθα-
[ρᾷ, ἔ]χοντας ἐσθῆτα λευ-
[κήν, ἀνυ]ποδέτους, ἁγνεύοντα[ς]
[ἀπὸ γυν]αικὸς καὶ κρέως
[καὶ μηθὲ]ν εἰσ[φ]έρειν . . . .
[― ― ― ― ― ― ― μη]δὲ κλειδίον μηδὲ
δακτύλιον σιδηροῦν μηδὲ
ζώνην μηδὲ βαλλάντιον
μηδὲ ὅπλα πολέμια μηδ'
ἄλλο πράττειν τῶν ἀπηγο-
ρευμένων μηθέν, v τὰς δὲ
θυσίας ἐπιτελεῖν καὶ καλ-
λιερεῖν κατὰ τὰ πάτρια.

Having become priest of Zeus Kynthios and Athena Kynthia in the year when [Sarapi?]_n was archon, when Nikephoros was attendant of the temple ..., instead of the damaged stele he wrote down the edict according to the command: Enter the sanctuary of Zeus Kynthios and Athena Kynthia with pure hands and soul, wearing a white garment, barefooted, pure from women and meat, and do not carry anything ..... nor a key, nor a ring of iron, nor a belt, nor a purse, nor weapons of war, and do not do anything else that is forbidden, but perform the sacrifices and sacrifice with good omens according to ancient traditions.

No. 5
Athena Lindia, Rhodes
LSCGS 91; Blinkenberg 1941, no. 487
3rd century BC
Translation: Rostad 2006; Cleland 2003

[ὅ]πλα ἀρήια μὴ φέροντας·
αἰσθῆτας καθαρὰς ἔχοντας χωρὶς ἐπικρανίων·
ἀνυποδέτους ἢ ἐν λευκοῖς μὴ αἰγείοις ὑποδήμασι·
μηδέ τι αἴγιον ἔχοντας·
μηδὲ ἐν ζώναις ἄμματα·

It is religiously permitted to enter cleansed and purified inside the lustral basin and the [gates] of the temple, refraining from looking (?), children –, purified not only with regard to the body, but also to the soul from everything that is polluted, impure and unlawful, without carrying martial weapons, with clean clothing,[1] without headdress, barefooted or wearing white shoes not made of goatskin, carrying nothing of goatskin, nor knots in the belts.

No. 6
Demeter, Chios
LSAM 6
1st century AD
Translation: Cleland 2003

[- - - - - - ταῖς]
[δ' ἱ]λασσομέν[αις οἰκῖος]
λα<ι>τρ(ι)έτω ἀνήρ · πᾶσαι ἀ-
νιλίποδές τε [καὶ] ἵμασι
φαιδρυνθῖσαι τῷ καλά-
θῳ συνέπεσθε, τὰ δὲ
χρυσῖα θέτ' οἴκοις · λῆρ-
οι γάρ, τὰ μὲν εχθραίνει το[ῖ]-
σιν δὲ προσα[υ]δᾷ.

and for those who appease
a man must serve the goddess in the temple:
all who take part with the *kalathos*
[shall be] barefoot and in clean clothes
and the gold placed in the temple:
for frippery is hateful
from those you are addressed by.

No. 7
Asklepios, Pergamon
LSAM 14; *SEG* 4:681
3rd century AD
Translation: Cleland 2003

[- - εἰσπορευέσ]θω εἰς [τὸ ἱερὸν τοῦ Ἀσκληπιοῦ ἀπὸ μὲν γυ-]
[ναικὸς ἡμέ]ρας δέκα, ἀπὸ δὲ [τ]ετοκ[υίας ἡμέρας.... ἀπὸ κή-]
[δους ἡμέρας.......] εἰσιων λουσάμενος · ἐὰ[ν δέ τις θέληι τῶν πό-]
[νων ἀπαλ]λάσσεσθαι, περικαθαιρέ[τω- - - - - - -]
[ἀλεκτρυό]νι λευκῶι καὶ [θ]είωι καὶ δ[αδὶ- - - - - -]
[... σινδο]νιάσας περικ[α]θαιρέτω ω[- - - - - - -]
[.. εἰ]σπορευέσθω πρὸς τὸν θεὸν τ[ὸν Σωτῆρα Ἀσκληπιόν- -]
[εἰς τὸ] μέγα ἐνκοιμητήριον ὁ ἐγκο[ιμᾶσθαι βουλόμενος- - -]
[ἐν ἱμα]τίοις λευκοῖς, ἁγνοῖς ἐλάας ἔ[ρνεσιν ἐστεμμένος,]
[ἔχων μήτε δακτ]ύλιον, μήτε ζώνην, μ[ήτε χρυσίον, μήτε τὰς]
[τρίχας πεπλεγμένα]ς, [ἀν]υπόδητος- - - -

with a white cockerel, and brimstone and torch . . .
. . . having wrapped in muslin let him purify completely . . .
let him proceed to the god, the Saviour Asklepios
into the big *enkoimētērion* anyone who wishes to sleep in the temple,
in white clothing, crowned with pure wreaths of olive,
having neither ring, nor girdle, nor gold
nor bound up hair, barefoot - - - -

No. 8
Demeter Thesmophoros, Arcadia
LSCGS 32; *SEG* 11:1112
6th century BC
Translation: Mills 1984

[Εἰκὰν γυ]νὰ ϝέσετοι ζτεραῖον λῶπος,
[ἱερὸ]ν ἔναι τᾶι Δάματρι τᾶι Θεσμοφόροι·
[εἰ δὲ] μὲ ὑνιερόσει, δυ(σ)μενὲς ἔασα ἐπεϝέργο
[κακο͂]ς ζ' ἐξόλοιτυ, κὰ ὄζις τότε δαμιοϝοργε͂
[ἀφάε]σται δραχμὰς τριάκοντα · εἰ δὲ μὲ ἀφάετοι,
[ὀφλε͂ν] τὰν ἀσέβειαν · ἔχε ὅδε κῦρος δέκο ϝέτεα · ἔνα[ι]
[δ' ἱερὸν] τόδε.

If a woman wears a brightly-colored robe,
it is to be consecrated to Demeter Thesmophoros.
If she does not dedicate it, being ill-disposed toward the rite,
let her 'perish' in a gruesome way, and whoever is then demiourgos,
let him [pay/exact] thirty drachmas.
Let this [law?] have authority for ten years.

No. 9
Despoina, Arcadia
IG V, 2, 514; *LSCG* 68; *SEG* 35:354
3rd century BC
Translation: Loucas & Loucas 1994 (adapted from Horsley 1979)

Δεσποίνας.
⟦μὴ παρέρπην ἔχοντας⟧ μὴ ἐξέστω
παρέρπην ἔχοντας ἐν τὸ ἱερὸν τᾶς
Δεσποίνας μὴ χρ[υ]σία ὅσα <μ>ὴ ἰν ἀνά-
θεμα μηδὲ πορφύρεον εἱματισμὸν
μηδὲ ἀνθινὸν μηδὲ [μέλ]ανα μηδὲ ὑπο-
δήματα μηδὲ δακτύλιον· εἰ δ' ἄν τις
παρένθῃ ἔχων τι τῶν ἁ στάλα [κ]ωλύει,
ἀναθέτω ἐν τὸ ἱερόν. μηδὲ τὰς τ[ρί]-
χας ἀμπεπλεγμένας μηδὲ κεκαλυμ-
μένος, μηδὲ ἄνθεα παρφέρην μηδὲ
μύεσθαι ⟦μύεσθαι⟧ κύενσαν μηδὲ θη-
λαζομέναν· τὸς δὲ θύοντας πὸς θύ[η]-
σιν χρέεσθαι ἐλαίαι, μύρτοι, κηρίο[ι],
ὁλοαῖς αἰρολογημέναις, ἀγάλματ[ι],
μάκωνσι λευκαῖς, λυχνίοις, θυμιά-
μασιν, ζμύρναι, ἀρώμασιν· τὸς δὲ θ[ύ]-
οντας τᾶι Δεσποίναι θύματα θύ[ην]
θήλεα λευκ[ὰ . .]ο . . . ος καὶ κ . . . . .

Belonging to Despoina.
Let it not be permissible for those to pass in who are bringing into the sanctuary of Despoina
Any gold objects which are not intended for dedication
Neither purple, nor flower-coloured or black clothing, nor sandals, nor a ring.
If anyone does enter with any of these things which the stele prohibits,
Let him dedicate it in the sanctuary.
Neither (let it be permissible to enter) with the hair braided, nor with the head covered.
Neither (let it be permissible) to bring in flowers, nor for a woman who is pregnant or breastfeeding to become an initiate.
And let those making sacrifices use for sacrifice olive,
Myrtle, honeycomb
Barley-groats cleared of darnel, a figurine,
White poppies, lamps, (various kinds of) incense, myrrh, and aromatics.
And let those making sacrifice to Despoina white ...

No. 10
Demeter, Patras
LSCGS 33
3rd century BC
Translation: Cleland 2003

Side A:
.... [Δα]-
ματρίοις τὰς γ[υ]ν[αῖ]-
κες μήτε χρυσίον ἔ-
χεν πλέον ὀδελοῦ ὀλ-
κάν, μηδὲ λωπίον ποικί-
λον, μήτε πορφυρέαν,
μήτε αὐλῆν. εἰ δέ κα
παρβάλληται, τὸ ἱ-
ερὸν καθαράσθω
ὡς παρσεβέουσα.

.... at the Damatrieia,
Women may have neither gold
more than one obol in weight,
and not multi-coloured clothing, nor purple,
nor be painted with white lead, nor a flute.
If this is transgressed (any of these are brought in), the sanctuary shall be purified
on the grounds that she is committing sacrilege.

No. 11
Alektrona, Ialyssos, Rhodes
IG XII, 1 677; *LSCG* 136, lines 19–35
Before 300 BC
Translation: Rostad 2006

νόμος ἃ οὐχ ὅσιον ἐσίμειν οὐδὲ
ἐσφέρειν ἐς τὸ ἱερὸν καὶ τὸ τέμενος τᾶς Ἀλεκτρώνας· μὴ ἐσί-
τω ἵππος ὄνος ἡμίονος γῖνος
μηδὲ ἄλλο λόφουρον μηθὲν μηδὲ ἐσαγέτω ἐς τὸ τέμενος μη-
θεὶς τούτων μηθὲν μηδὲ ὑποδή-
ματα ἐσφερέτω μηδὲ ὕειον μηθέν, ὅ,τι δέ κά τις παρὰ τὸν νόμον
ποιήσηι, τό τε ἱερὸν καὶ τὸ τέμενος
καθαιρέτω καὶ ἐπιρεζέτω, ἢ ἔνο-
χος ἔστω τᾶι ἀσεβείαι· εἰ δέ κα
πρόβατα ἐσβάληι, ἀποτεισάτω ὑ-
πὲρ ἑκάστου προβάτου ὀβολὸν
ὁ ἐσβαλών· ποταγγελλέτω δὲ
τὸν τούτων τι ποιεῦντα ὁ χρήι-
ζων ἐς τοὺς μαστρούς.

Law regarding what is not permitted to enter or bring into the shrine and *temenos* of Alektrone: A horse, donkey, mule, hinny, or any other pack animal must not enter. Nor is anyone to bring any of these into the *temenos*. No one is to bring in shoes or anything made from pig. The person, who does anything contrary to the law, is to purify the shrine and the *temenos*, and offer a sacrifice afterwards, or be liable to impiety. If someone brings in cattle, he who brings them in, is to pay an obol for each animal. Let the one who so desires rapport him who does any of these things to the treasurers.

No. 12
Leto, Xanthos
SEG 36:1221
Late 3rd–early 2nd century BC
Translation: Lupu 2006

Ἄ μὴ νομίζεται εἰς τὸ
ἱερὸν καὶ τὸ τέμενος
εἰσφέρειν, ὅπλον μη-
θέν, πέτασον, καυσί-
αν, πόρπην, χαλκόν,
χρυσόν, μηδὲ δακτύ-
λιον ὑπόχρυσον, μηδὲ
σκεῦος μηθέν, ἔξω
ἱματισμοῦ καὶ ὑπο-
δέσεως, τοῦ περὶ τὸ
σῶμα, μηδ' ἐν ταῖς
στοιαῖς καταλύειν
μηθένα ἀλλ' ἢ τοὺς
θύοντας

Things which is not customary to carry into the sanctuary and precinct: no weapon, *petasos*, *kausia*, dress-fastener, copper (objects), gold (objects), nor gold-plated rings and any equipment at all except for clothes and footwear (worn) around one's body; nor shall anyone camp in the stoas except those offering sacrifice.

No. 13
Egyptian divinities, Delos
ID 2180; *LSCGS* 56, lines 1–8.
2nd century BC

Θεῶι Μεγάλωι καὶ Διὶ Κασίωι καὶ Ταχνήψει,
Ὧρος Ὥρου Κασιώτης
ὑπὲρ Λευκίου Γρανίου
τοῦ Ποπλίου Ῥωμαίου.
γυναῖκα μὴ προσάγειν
μηδὲ ἐν ἐρεοῖς ἄνδρα·
κατὰ πρόσταγμα.

Theos Megalos and Zeus Kasios and Tachnépsis (Isis)
Horos son of Horos Kasiōtēs
Under Leukios Granios
Of Roman Publius.
A woman shall not come near [i.e. enter the temple/*temenos*]
Nor shall a man [dressed] in wool
According to ordinance

No. 14
Egyptian divinities, Delos
IG XI, 4 1300, *LSCG* 94, lines 1–2
2nd Century BC

ἀπ' οἴνου μὴ προσιέναι μηδὲ ἐν ἀνθινοῖς.

One shall not accept/proceed from wine nor in flowered [clothing].

No. 15
Unknown goddess, Eresos, Lesbos
LSCG 124 (line 17); *IG* XII suppl. no. 126
2nd century BC
Translation: Rostad 2006

[— — — — — — — — — — — — — — —]
[— — — —]ς εἰστείχην εὐσέβεας
vac. ἀπὸ μὲν κάδεος ἰδίω
[περιμένν]αντας ἀμέραις εἴκοσι. ἀπὸ δὲ
[. . . . .]ω ἀμέραις τρεῖς λοεσσάμενον·
[ἀπὸ δὲ . . .]άτω v ἀμέραις δέκα· v αὔταν δὲ
[τὰν τετό]κοισαν ἀμέραις τεσσαράκοντα·
[ἀπὸ δὲ . . .]τω ἀμέραις τρεῖς· v αὔταν δὲ τ[ὰν]
[τε]τόκοισαν v ἀμέραις δέκα·
[ἀπὸ δὲ γ]ύναικος αὐτάμερον λοεσσάμενον·
[. . . . .] δὲ μὴ εἰστείχην v μηδὲ προδόταις.
[μὴ εἰσ]τείχην δὲ μηδὲ γάλλοις v μηδὲ
[γύ]ναικες γαλλάζην ἐν τῶ τεμένει·
[μ]ὴ εἰσφέρην δὲ μηδὲ ὄπλα πολεμιστήρι[α]
μηδὲ θνασίδιον·
[μη]δὲ εἰς τὸν ναῦον εἰσφέρην v σίδαρον
μηδὲ χάλκον πλὰν νομίσματος
μηδὲ ὑπόδεσιν μηδὲ ἄλλο δέρμα
μῆδεν. vv μὴ εἰστείχην δὲ μηδὲ γύ[ναικ]α
εἰς τὸν ναῦον v πλὰν τᾶς ἰρέας
καὶ τᾶς προφήτιδος.
[μὴ πο]τίζην δὲ μηδὲ κτήνεα μηδὲ βοσκήματα
ἐν τῶ τεμένει.

.........enter pious from the funeral rites of relatives purified for twenty days, from (the funeral) of others cleansed for three days, from death purified for ten days, the woman who has given birth herself purified for forty days, from provoked abortion (?) purified for three days, the woman who has given birth herself for ten days, from intercourse with a woman cleansed on the same day. Murderers must not enter, nor must traitors enter, nor *galloi*, nor must women who practice the cult of Cybele enter the *temenos*. One must not bring in weapons of war, nor the carcass of an animal. Nor is one to bring iron into the temple, nor copper except money, nor shoes, nor any other skin. No one is to enter the temple, not even a woman, except the priestess and the prophetess. One must not water (?) herds or cattle inside the *temenos*.

No. 16
Pan and the nymphs, Attica
SEG 36:267
61/60 BC
Translation: Lupu 2006

ἀγαθῆ τύχῃ· ἐπὶ Θεο-
φήμου ἄρχοντος· vvv
Πυθαγόρας καὶ Σωσι-
κράτης καὶ Λύσανδρος
οἱ συνέφηβοι Πανὶ καὶ
Νύμφαις ἀνέθηκαν.
ἀπαγορεύει ὁ θεός· μὴ
[ε]ἰσφέρειν χρωμάτιν[ον]
[μ]ηδὲ βαπτὸν μηδὲ Λ . .

To good Luck.
In the archonship of Theophēmos,
the fellow ephebes Pythagorās, Sōsikratēs, and Lysandros
dedicated [this stele] to Pan and the Nymphs
The god forbids to carry in either coloured (garment) or dyed (garment) or [- -]

No. 17
Dionysos Bromios, Smyrna
LSAM 84; *AGRW* 195; *ISmyrna* 728
2nd century AD
Translation: Rostad 2006

[. . . .]της Μενάνδρου ὁ θεοφάντης ἀνέθηκεν. | [πάν]τες ὅσοι τέμενος Βρομίου ναούς τε περᾶτε, | τεσσαράκοντα μὲν ἤματα ἀπ' ἐχθέσεως πεφύλαχθε | νηπιάχοιο βρέφους, μὴ δὴ μήνειμα γένηται, || ἔκτρωσίν τε γυναικὸς ὁμοίως ἤματα τόσσα• | ἢν δέ τιν' οἰκείων θάνατος καὶ μοῖρα καλύψῃ, | εἴργεσθαι μηνὸς τρίτατον μέρος ἐκ προπύλοιο• | ἢν δ' ἄρ' ἀπ' ἀλλοτρίων οἴκων τι μίασμα γένηται, | ἠελίους τρισσοὺς μεῖναι νέκυος φθιμένοιο, || μηδὲ μελανφάρους προσίναι βωμοῖσι ἄνακτ[ος, - -] | μηδ' ἀθύτοις θυσίαις ἱερῶν ἐπὶ χῖρας ἰάλ[λειν, - - -] | μηδ' ἐν Βακχείοις ᾠὸν ποτὶ δαῖτα τ[ίθεσθαι ? - - -] | καὶ κραδίην καρποῦν ἱεροῖς βωμοῖς [- - -] | ἡδεόσμου τ' ἀπέχεσθαι, ὃν Δημ[ήτηρ ἀμάθυνεν ?•] || ἐχθροτάτην ῥίζαν κυάμων ἐκ σπέ[ρματος ? - - -] | Τειτάνων προλέγειν μύσταις [- - -] | καὶ καλάμοισι κροτεῖν οὐ θέ[μιον εἶναι - - -] | ἤμασιν, οἷς μύσται θυσί[ας - - -] | [μηδ]ὲ φορεῖν ΣΥ (?) [- - -]

The *theophantes* ... son of Menandros dedicated (this stele). All who enter the *temenos* and temples of Bromios: avoid for forty days after the exposure of a newborn child, so that (divine) wrath does not occur; after the miscarriage of a woman for the same amount of days. If he conceals the death and fate of a relative, keep away from the *propylon* for the third of a month. If impurity occurs from other houses, remain for three days after the departure of the dead. No one wearing black clothes may approach the altar of the king, nor lay hands on things not sacrificed from sacrificial animals, nor place an egg as food at the Bacchic feast, nor sacrifice a heart on the holy altars [...] keep away from the smell, which [...] the most hateful root of beans from seed (?) [...] proclaim to the *mystai* of the Titans [...] and it is improper to rattle with reeds [...] on the days when the *mystai* sacri[fice ...], nor bring [...].

No. 18
Unknown deity, Tlos, Lydia
LSAM 77; *CIG* Add. 435; *SEG* 6:775
2nd century BC
Translation: Cleland 2003

[- - -]ἔχων στολὴν ἀνθινήν, ἄλλος μὴ ἀγειρέτω μ[ήτε - - - - -]
[-μηδεὶ]ς τούτων ἐγ γυναικείαι στολῆι · ἐὰν δέ τις παρα[βαίνηι]
[καὶ εἰ]ς τὸ ἱερὸν ἔλθηι, ἀποτινέτω ἡμέρας ἑκάστης [- - - - -]

- - - having flower-coloured clothing, others must not gather together, nor - -
not one of these in women's clothes. And if anyone should transgress
and go into the sanctuary, he must pay each day - - - - - -

Note
1. Cleland 2003 translates this part as 'having clean clothing'.

Appendix 4

Dress-fasteners in sanctuaries

Fibulas and pins

Among the archaeological material providing direct evidence of Greek dress are the pins and fibulas used to fasten the garments. Although we cannot determine it with certainty, it is likely that the dress-fasteners recovered from sanctuaries were originally attached to garments, these being the main dedication, but the dress-fasteners are all that remains.[1] Dress-fasteners were often made of metal – especially bronze, but also of silver, gold, and ivory and therefore often preserved in the archaeological record.[2] They are recovered in large numbers all over Greece from different contexts: burials, settlements, and sanctuaries.[3] The earliest types appear in the 13th century BC and continue in use up through the Hellenistic Period. This indicates that dress-fasteners were an important and well-known part of ancient Greek attire, and thus an important source to our knowledge and interpretation of ancient Greek dress. For example, it has been suggested that the emergence of a new fibula type can be a result of changes in garments.[4]

Fibulas and pins were used to fasten together two edges of a garment, like the modern button, or to fasten an upper garment to an undergarment.[5] While there is still debate among archaeologists about the specific way fibulas were worn, they tend to agree that fibulas belonged to the outer layer of clothing, and were thus visible. Scholars of Classical antiquity consider the fibula a fastener for the *peplos*, used from sub-Mycenaean times on, or for the Ionian *chitōn*, which was worn in Athens, in the Argolis, on Crete and the Aegean islands.[6] Pins, like fibulas, are also thought to have served to fasten the *apoptygma* of the *peplos* on the right and the left shoulder. This was done by drawing a triangular lappet over the shoulder from the back and clasping it on the front part with either a fibula or a pin, point up or down.[7] However, this turns out to be a very simplified perception of the use of fibulas and pins, since dress-fasteners could be used for a wider range of garments.

A question that arises is whether all fibulas and pins were meant to fasten garments, or whether some were meant solely as decoration? There has been a tendency to term bronze specimens dress-fasteners, while gold and silver specimens are often considered jewellery, – a separation which appears unsubstantiated.

Fig. 127. Silver pin from the Argive Heraion. 7th or 6th century BC. © Trustees of the British Museum, inv. no. 1896,0617.1.

For one thing, the type of metal does not have any influence on the pin's or fibula's use and function as a dress-fastener. Second, we know from other contexts that bronze could be used for jewellery and decoration. This reflects a modern sense of gender difference in the metals where gold and silver is considered female, while bronze is considered more male, which does not necessarily reflect the situation in ancient Greece. There is thus an underlying assumption that gold and silver dress-fasteners were in the category of jewellery and reserved for women, which, however, is an interpretation based on modern perceptions of male and female. This distinction between jewellery and dress-fastener is reflected in modern terminology; for example, it makes a difference whether a fibula is termed fibula, clasp, or brooch, the latter implying jewellery rather than a dress-fastener. Furthermore, a distinction between jewellery and dress-fastener can 'disturb' the respective typologies. This can cause a skewed image on the interpretation of dress-fasteners and, as follows, the garments. Yet the dress-fasteners were likely to have been visible, creating an enticement to make them decorative, and in most cases pins and fibulas were probably both utilitarian and decorative.[8]

The study of especially fibulas, but also pins, has therefore been dominated by typology and consequently their use as dating-tools. This has caused the shape, in case of the fibulas the bow or the plate to which the needle is attached, and its decoration, to become the most important determining aspects, while size, function, use, and context tend to be considered less important. As follows, many fibula types

Appendix 4. Dress-fasteners in sanctuaries

Fig. 128. Gilded bronze pin from the temple of Aphrodite on Paphos. 3rd-2nd century BC. © Trustees of the British Museum, inv. no. 1888,1115.2.

are named after the shape of the bow or the needle plate, e.g. violin bow fibulas, bow fibulas, leaf bow fibulas, snake fibulas, plate fibulas, spectacle fibulas, spiral fibulas, disc fibulas, and hinge fibulas. Other types are named after their place of origin on e.g. Cyprus, the Near East, Phrygia, or Italy. The pins are named after the shape and decoration of the pinhead.[9]

Pins and fibulas with dedicatory inscriptions

When discussing the use of fibulas and pins as votive offerings, a few dress-fasteners are of special interest since they carry dedicatory inscriptions to specific deities. These are very rare, however, and to my knowledge consist of only four examples.[10] One is a broken pin dated to the end of the 7th century BC from the sanctuary of Artemis Orthia at Sparta. The head of the pin is of bronze and the shaft of which much is broken off, is of iron. On the upper side of the pin head is an inscription with the name Eileithyia (*ELEYTHIA*).[11]

Another example is a silver pin from the Argive Heraion, dated to the 7th or 6th century BC (Fig. 127). The pin has an elaborate head composed of a disc engraved with a flower, two moulded rings on either side of an inverted truncated cone, a large ribbed bead, and six moulded rings. But most interestingly, on the shaft is a dedicatory inscription to Hera (*TAS HERAS*) on the shaft. The length of the pin is 11.6 cm and it weighs 461 g.[12]

A third example is a gilded bronze pin from the temple of Aphrodite on Paphos, dated to the 3rd–2nd century BC (Fig. 128).[13] At the head of the pin are four goats' heads, separated by lotus flowers, springing from acanthus foliage; above are four doves leaning forward to drink from the flower cup; on top are two pearl beads. On the shaft is a dedicatory inscription for Aphrodite:

> 'To the Paphian Aphrodite Eubola vowed this, the wife of Aratas, the kinsman [i.e. a member of the Ptolemaic court] and Tamisa.'

The length of the pin is 17.8 cm and it weighs 538 g.[14]

A final example is an Archaic bronze fibula in the shape of a lion recovered at the sanctuary of Apollo Tyritas at Kynouria on the Peloponnese. It carries an inscription to Apollo: "Ἀπόλōνος ἐμ['.[15] These four examples illustrate that dress-fasteners were offered to several deities, primarily female: Eileithyia, Hera, and Aphrodite. They further indicate that they were made specifically for dedication for a specific deity and thus not items worn by the donor before dedication, nor random items given as a votive offering, but deliberate gifts. This is interesting in relation to the dedication of textile offerings, which have been claimed to be personal garments, perhaps worn by the donor before dedication.

Distribution of dress-fasteners in Greek sanctuaries (Table 34)

A survey of the distribution of dress-fasteners in Greek sanctuaries is not a straightforward task. In many cases the material is still unpublished, while others have been published a century ago, leaving out important information on context, typology, etc. The following examples are therefore not necessarily representative and the situation might change when more material becomes available. Nevertheless, by an investigation of the publications, it becomes clear that several sanctuaries have revealed a large number of dress-fasteners.

In Thessaly, fibulas are also popular votive gifts in some sanctuaries. Thus, in the sanctuary of Enodia at Pherai, ca. 1798 fibulas and 48 pins were recovered,[16] while at the sanctuary of Athena Itonia at Philia, ca. 566 fibulas and 166 pins were recovered.[17] The pins at Philia date from the early and Middle Geometric Periods to the Hellenistic Period. The majority are from late Geometric or early Archaic Periods (114 examples), while only 3 are from the Hellenistic Period.[18] The fibulas at Philia date from the Late Geometric to the Hellenistic Periods, with the vast majority belonging to the late Geometric to the early Archaic Periods (473 examples), while only one fibula can be dated to the Hellenistic Period. They comprise different types such as bow fibulas, plate fibulas and spectacle fibulas, while the 38 fibulas from the Classical Period are hinge fibulas (except one).

Table 34. Dress-fasteners in selected Greek sanctuaries

Region	Site	Deity	Fibulas	Pins
Thessaly	Pherai	Enodia	1798	48
Thessaly	Philia	Athena Itonia	566	166
Peloponnese	Argive Heraion	Hera	110	800
Peloponnese	Perachora	Hera Limenia	122	Several hundred
Peloponnese	Perachora	Hera Akreia	5	0
Peloponnese	Sparta	Artemis Orthia	104	Unknown no.
Peloponnese	Tegea	Athena Alea	32	Several hundred
Aegina		Aphaia	57	54
Rhodes	Lindos	Athena	1596	
Rhodes	Ialyssos	Athena	ca. 2000	
Rhodes	Kamiros	Athena	41	
Samos	Heraion	Hera	52	
Delos	Altar of Zeus		1	
Delos	Kabirion		1	
Delos	Artemision		1	
Delos	hieron of the peribolos		2	
Delos	Heraion	Hera	23	
Crete	Inatos	Eileithyia	Unknown no.	Unknown no.
Samothrace	Hall of the Votive Gifts		5	
Samothrace	Area of the stoa		Unknown no.	
Ephesos	Artemision		ca. 100	Several hundred

Several sanctuaries on the Peloponnese have also revealed dress-fasteners. From the Argive Heraion ca. 110 fibulas and ca. 800 pins have been recovered.[19] Many of the pins date to the second half of the 8th and the 7th centuries BC[20] and some of them show signs of usage, which might indicate that they have been worn and thus made exclusively for dedication.[21] The fibulas primarily consist of Greek types, dating to the late 8th to the 6th century BC.[22] At Perachora, in the sanctuary of Hera Limenia 122 fibulas and several hundred pins have been found, but only five fibulas in the sanctuary of Hera Akraia.[23] Most of the pins are of bronze and they primarily date from the 8th to the 5th century BC.[24] The fibulas primarily belong to four major types: Attico-Boeotian fibulas, East Greek fibulas, boat and leech fibulas, and spectacle fibulas. They are in bronze or ivory and date from the 8th to the 6th century BC.[25]

Besides the pins, a substantial number of other pin-like objects, sometimes termed 'spits',[26] have been recovered at the Argive Heraion and Perachora. They differ from the 'normal' pins by having their shaft covered by a series of globes, but they also differ in length, ranging from 30 to 80 cm.[27] At the Argive Heraion ca. 2000 of these objects have been recovered, while a Perachora the number is lower.[28] Like the pins they date to the second half of the 8th and the 7th centuries BC.[29] According to Kilian-Dirlmeier, they have also been recovered in a few burials and she therefore refuses that they are solely votive offerings. She therefore terms them 'representative pins'.[30] Yet these pins or spits almost exclusively occur in sanctuaries, which suggest that they were specially manufactured as votive offerings.[31] Because of the globes on the shaft and their immense size they can hardly have been worn as dress-fasteners and they must have had another, probably representative, function.[32] Jacobsthal considers them ritual pins, made for the goddess.[33] Foley has suggested that these large pins were employed in the clothing of the cult statue of Hera, which would perhaps have been very large and therefore in need of large garments and dress-pins,[34] but as argued by Baumbach, the early cult image at the Argive Heraion may have been quite small and furthermore, the immense numbers of pins can hardly all have been used for the dress for Hera, and he instead suggests that they may have been symbolic dress-pins.[35] It has also been suggested that they functioned as a *pars pro toto* without clothing.[36] The globes on these large pins seem to have made them unpractical for a use as dress-fasteners.

In the sanctuary of Artemis Orthia near Sparta ca. 104 fibulas have been recovered as well as an unknown number of bronze pins.[37] The pins and fibulas were found in Geometric contexts and with Laconian I and II pottery (700–beginning of the 6th century BC), but towards the end of the 7th century the use of pins and fibulas declined and eventually disappeared.[38]

In the sanctuary of Athena Alea at Tegea ca. 32 fibulas and several hundred pins have been recovered.[39] They are of bronze, and the fibulas include different types, e.g. bow fibulas, violin-bow fibulas, spectacle and spiral fibulas, plate fibulas, and disc fibulas.

Fibulas and pins have been recovered from two sanctuary sites on Aegina: the temple of Aphaia and the so-called Aphrodite temple (probably for Apollo). However, the finds from this temple still remain unpublished.[40] Approximately 54 pins and ca. 57 fibulas adhere from the temple of Aphaia.[41] The fibulas are of bronze and have been recovered in the Geometric layers of the eastern terrace and in the northern part of the *temenos*.[42]

The sanctuary of Athena at Ialyssos on Rhodes has revealed as a many as 2000 fibulas, thus even surpassing Lindos.[43] But in contrast, only 41 fibulas were recovered at the sanctuary of Athena at Kamiros,[44] which is surprisingly few in comparison to the two other Rhodian sanctuaries.

Approximately 52 bronze fibulas have been recovered from the Heraion on Samos. They represent different types such as Phrygian, which are the largest group with 17 examples, island fibulas with one globe on the bow, fibulas with globes on the bow, mainland fibulas with globes on the bow, three Italian fibulas, and two violin bow fibulas.[45]

Delos has generally revealed few fibulas in comparison with other Greek islands, but a few have been recovered from different sanctuaries on the island: one fibula south-west of the altar of Zeus, one in the Kabirion, one in the Artemision, and two in the hieron of the Peribolos.[46] In addition, Deonna records 23 ivory spectacle fibulas from the Heraion and ca. 6 ivory fragments from west of the Artemision, all primarily from the 7th century BC.[47] The fibulas from Delos are made from bronze, ivory, or bone and they primarily date from the 8th century BC to the Archaic Period.[48]

A still unpublished and therefore unknown number of fibulas and pins have been recovered from the sanctuary of Eileithyia at Inatos on Crete.[49] They are of bone or bronze and date to the Archaic Period.[50]

Around five bronze fibulas have been found beneath the floor of the 'hall of the votive gifts' in the southern part of the sanctuary on Samothrace. Loeffler has dated them to the Geometric to the early Archaic Period.[51] Newer excavations have also revealed further fibulas (unpublished) from the area of the stoa, dating to the Archaic Period.[52]

Finally, the Artemision in Ephesos yielded more pins than any Greek sanctuary, except the Argive Heraion, while only very few pins are recorded from other Ionian sites.[53] About 100 fibulas were also recovered from the site.[54] The Ephesian pins are of electrum, gold, silver, bronze, ivory, bone, amber and some have metal shanks and heads of crystal and semi-precious stones.[55] The Ephesian pins date chiefly from around the second half of the 7th century BC. Many of the electrum, gold, and silver pins were found in the basis of the cult statue. Jacobsthal therefore argues that they cannot have been dedicated with clothes, since there was not enough room in the basis.[56] Yet, this can be discussed, since this context might not be the original one of deposition and the pins may have been placed there intentionally after the garments had disintegrated. Furthermore, pieces of thin gold and silver foil cut into the form of fibulas were found in the Artemision. These imitations could not have been worn and were probably made exclusively for the purpose of dedication.[57]

With regard to chronology this is a complicated issue, since many of the dress-fasteners are without exact context and in the great majority of early excavated sanctuaries no strata were recorded, but generally, the majority of fibulas recovered in the sanctuaries appear to belong to the 8th to the 6th centuries BC.

A note should be made on the condition of especially the pins: pins with an intentionally bent shank, which has made them impossible to wear, have been recovered from sanctuaries and tombs.[58] It has been suggested that this was done to secure the pins or because of wear, which is, however, unlikely since they are solid enough to resist pull and pressure. Instead, it appears that these pins have been put out of shape on purpose, to make sure that they would never be used again and thus represent a case of ritual destruction for dedication.[59] More than half of the 'spits' from the Argive Heraion are much out of shape, and intentionally bent, but in contrast, this has not been attested at Perachora[60] or the sanctuary of Artemis Orthia.[61]

This short survey illustrates that there are clear tendencies in the use of dress-fasteners as votive offerings. First of all, some sanctuaries reveal a large number and thus that they are popular offerings, while at other sanctuaries, there are only very few. For example, no pins have been found on the Athenian Acropolis itself (excluding the slopes), and only one fibula.[62]

Second, in the sanctuaries with large numbers of dress-fasteners, there is a clear tendency for a preference for either fibulas or pins. For example, in some sanctuaries, especially on the Peloponnese, pins are very common, e.g. in the Hera sanctuary at Argos, the Athena Alea sanctuary at Tegea, and the Artemis Orthia at Sparta,[63] while fibulas are rare. The opposite situation is the case at other sanctuaries, e.g. at Pherai, Thessaly, and the Athena sanctuaries at Lindos and Ialyssos on Rhodes. This difference cannot be explained by chronology, but is a clear indication of differences in votive dedications, perhaps caused by differences in local fashions, tastes, and traditions.

Third, according to the archaeological findings, pins and fibulas are especially common in sanctuaries dedicated to female deities, e.g. the Argive Heraion, the sanctuary of Hera Limenia at Perachora, the Athena Alea sanctuary at Tegea, the Artemis Orthia sanctuary, the sanctuary of Enodia at Pherai, the temple of Aphaia on Aegina, the Heraion on Samos, and the sanctuaries of Athena at Lindos and Ialyssos and the Artemision at Ephesos. Dress-fasteners are also recovered in sanctuaries for male gods, e.g. at Amyklaion at Sparta, but this is quite rare. In other sanctuaries it is more difficult to determine which deity the dress-fasteners were for. This is for example the case in Olympia, where the pins could be dedicated to either Zeus or Hera.[64] From the available evidence, it appears that especially Athena and Hera, but also Artemis, were the recipients of dress-fasteners. It thus seems that this type of offering was almost exclusive to female deities – a fact, which might imply a stronger relation of dress-fasteners to female garments.

Fourth, the majority of dedicated fibulas appear to have been of local types, reflecting that either it was local people who dedicated them or that visitors from afar acquired local fibulas for dedication. However, since foreign/imported types are present, the former seems the most likely. These imported fibulas can furthermore reflect two scenarios: travelling individuals returning to home bringing fibulas acquired abroad, and/or visitors from abroad bringing fibulas from where they come from.

Fifth, dress-fasteners were common votive offerings in some Greek sanctuaries in the Geometric to the Archaic Periods. Afterwards, the fibulas and pins appear to slowly disappear from the archaeological record, even though occasional dedications of these objects still occur up through the Hellenistic Period. The custom thus continues, but on a much, much, smaller scale. Since fibulas and pins were used to fasten different types of garments such as *peploi, chitōnes,* and mantles, the presence of these objects in the sanctuaries could indicate that the custom of donating garments as votive offerings dates back to the Geometric Period. Of course, we cannot be certain that the dress-fasteners were always given with garments, but it definitely seems likely. Perhaps the disappearance of the fibulas from the votive record merely reflects a change in fashion – that people stopped wearing garments fastened by fibulas and therefore also stopped dedicating them. It can therefore not be excluded that visitors to the sanctuary after this period, still dedicated garments, but without dress-fasteners. This further indicates that votive offerings were influenced by fashion and what was 'modern' and in use at the time and not archaic, conservative tendencies.

A sixth and final point is that dress-fasteners are also recorded in the temple inventories, especially in the inventory from Oia and the inventories from Delos. These recordings are made in the 5th century BC to the Hellenistic Period, but we cannot tell whether they are old or new offerings. In regard of the archaeological material, it seems most likely that the lists, at least the one from Oia, record old offerings made perhaps in the Archaic Period. Another possibility is that dress-fasteners were still given to female deities, as an expression of an archaic, conservative dress of the gods. But if this was the case, it is likely that pins and fibulas would also be recorded in e.g. the Brauron catalogues.

Notes

1. Simon 1986, 202.
2. Pins could of course also be made from organic materials such as wood, bone etc.
3. E.g. Sapouna-Sakellarakis 1978; Kilian 1975.
4. Donder 1994, 7.
5. Muscarella 1964, 35–36.
6. Sapouna-Sakellarakis 1978, 8; Furtwängler 1906, 404.
7. Jacobsthal 1956, 109.
8. Muscarella 1964, 36.
9. Brøns 2013.
10. Pingiatoglu 1981, 54.
11. Pingiatoglu 1981, 53–54.
12. British Museum, inv. no. 1896,0617.1. Baumbach 2004, 92; Kilian-Dirlmeier 1984, 249, no. 4373; Jacobsthal 1956, 96; Marshall 1911, 106, no. 1250.
13. For the sanctuary and the votive offerings, see e.g. Leibundgut Wieland 2009.
14. British Museum, inv. no. 1888,1115.2. Jacobsthal 1956, 96; Pingiatoglu 1981, 54; Marshall 1911, 223 no. 1999.
15. *IG* V,1, 1519. Pingiatoglu 1981, 54; Cahn 1950, 189.
16. Kilian 1975, 168.
17. Kilian-Dirlmeier 2002.

Appendix 4. Dress-fasteners in sanctuaries

18. Kilian-Dirlmeier 2002.
19. Waldstein 1905, 207–239 (pins), 240–250 (fibulas).
20. Baumbach 2004, 92; Strøm 1995, 78–81; Kilian 1975, 168.
21. Baumbach 2004, 92; Kilian-Dirlmeier 1984, no. 880, 1088, 1029.
22. Baumbach 2004, 93; Strøm 1995, 71–76.
23. Payne 1940, 73.
24. Baumbach 2004, 35.
25. Baumbach 2004, 36; Payne 1940, 167–172; Stubbings 1962, 433–441.
26. E.g. Payne 1940, 71–79. See also Jacobsthal 1956, 13.
27. Baumbach 2004, 36, 92.
28. Baumbach 2004, 92; Payne 1940, 71–79.
29. Baumbach 2004, 92; Strøm 1995, 78–80; Kilian-Dirlmeier 1984, 109–122.
30. Kilian-Dirlmeier 1984, 162.
31. Baumbach 2004, 93.
32. Baumbach 2004, 36.
33. Jacobsthal 1956, 15.
34. Foley 1988, 85.
35. Baumbach 2004, 92.
36. Baumbach 2004, 36; Boehringer 2001, 196.
37. Droop 1929, 197–199, 200.
38. Droop 1929, 200.
39. Dugas 1921, 341, 375–381 (pins), 381–385 (fibulas); Kilian 1975, 168.
40. Sapouna-Sakellarakis 1978, 9.
41. Furtwängler 1906, 397–400 (pins), 400–404 (fibulas); Sapouna-Sakellarakis 1978, 10.
42. Sapouna-Sakellarakis 1978, 10.
43. Dietz & Trolle 1974, 48.
44. Jacobi 1932–33.
45. Sapouna-Sakellarakis 1978, 29.
46. Sapouna-Sakellarakis 1978, 13.
47. Deonna 1938, 285, pl. 86.728–729; Payne 1962, 434.
48. Sapouna-Sakellarakis 1978, 13.
49. Baumbach 2004, 37; Pingiatoglu 1981, 51; Sapouna-Sakellarakis 1978, 18.
50. Pingiatoglu 1981, 51.
51. Loeffler 1962, 151–153.
52. Sapouna-Sakellarakis 1978, 31.
53. Jacobsthal 1956, 33.
54. Hogarth 1908, 147.
55. Jacobsthal 1956, 34.
56. Jacobsthal 1956, 34.
57. Muscarella 1964, 38; Hogarth 1908.
58. Jacobsthal 1956, 114.
59. Jacobsthal 1956, 114.
60. Jacobsthal 1956, 15.
61. Droop 1929, 197.
62. Jacobsthal 1956, 96.
63. Baumbach 2004, 36.
64. Kilian-Dirlmeier 1984, 162.

Bibliography

Akurgal, E. 1970, *Ancient Civilizations and Ruins of Turkey*, Istanbul.
Alden, M. 2003, 'Ancient Greek Dress', *Costume* 37, 1, 2003, 1–16.
Aleshire, S. B. 1989, *The Athenian Asklepieion. The People, their Dedications, and the Inventories*, Amsterdam.
Aleshire, S. B. 1991, *Asklepios at Athens. Epigraphic and Prosopographic Essays on the Athenian Healing Cults*, Amsterdam.
Aleshire, S. B. & Lambert, S. D. 2003, 'Making the Peplos for Athena. A new edition of IG II^3 1060 + IG II^3 1036', *ZPE* 142, 2003, 65–86.
Allgrove-McDowell, J. 2003, 'Ancient Egypt, 5000–332 BC', in Jenkins, D. (ed.), *The Cambridge History Of Western Textiles, vol. I*, Cambridge 2003, 30–38.
Alroth, B. 1992, 'Changing Modes in the Representation of Cult Images', in Hägg, R. (ed.), *The Iconography of Greek Cult in the Archaic and Classical Periods. Proceedings of the First International Seminar on Ancient Greek Cult, organised by the Swedish Institute at Athens and the European Cultural Centre of Delphi (Delphi, 16-18 Novembre 1990)*, Athens, 9–46.
Alroth, B. 1989, *Greek Gods and Figurines. Aspects of the Anthropomorphic Dedications*, Uppsala.
Andronicos, M. 1984, *Vergina. The Royal Tombs*, Athens.
Arnaoutoglou, I. 1998, *Ancient Greek Laws: A Source Book*, London & New York.
Arthur, L. B. 2000, *Undressing Religion: Commitment and Conversion from a Cross-Cultural Perspective*, Oxford.
Arthur, L. B. 1999, *Religion, Dress and Body*, Oxford & New York.
Attyah Flower, M. 2008, *The Seer of Ancient Greece*, Berkeley.
Austino, C. E. 2012, *Adaptation and Tradition in Hellenistic Sacred Laws*. Dissertation, Department of Classical Studies, Duke University.
Badian, E. 1989, 'History from 'Square Brackets', *ZPE* 79, 1989, 59–70.
Baert, B. 2007, 'Mantle, Fur, Pallium: Veiling and Unveiling in the Martyrdom of Agnes of Rome', in Rudy, M. & Baert, B. (eds.), *Weaving, Veiling, and Dressing. Textiles and their Metaphors in the Late Middle Ages*, Turnhout, 215–238.
Baert, B., & Sidgwick, E. 2011, 'Touching the Hem: the thread between garment and blood in the story of the woman with the hemorrhage (Mark 5:24b-34parr)', *Textile: The Journal of Cloth and Culture* 9, 3, 2011, 308–351.
Balfour-Paul, J. 1997, *Indigo in the Arab World*, London.
Banck-Burgess, J. 1999, *Hochdorf IV: Die Textilfunde aus dem späthallstattzeitlichen Fürstengrab von Eberdingen-Hochdorf (Kreis Ludwigsburg)*, Stuttgart.
Barber, E. 1991, *Prehistoric Textiles: The Development of Cloth in the Neolithic and Bronze Ages with Special Reference to the Aegean*, Princeton.
Barber, E. 1992, 'The Peplos of Athena', in Neils, J. (ed.), *Goddess and Polis. The Panathenaic Festival in Ancient Athens*, Princeton, 103–118.
Barber, E. 1994, *Women's Work. The first 20.000 Years. Women, Cloth, and Society in Early Times*, New York & London.
Barber, E. 1999, 'Colour in Early Cloth and Clothing', *CAJ* 9, 1, 117–120.
Barthes, R. 1990, *The Fashion System*, Berkeley.

Batten, A. J. 2009, 'Neither Gold nor Braided Hair (1 Timothy 2.9; 1 Peter 3.3): Adornment, gender and honour in Antiquity', *New Testament Studies* 55, 484–501.

Baumbach, J. D. 2004, *The Significance of Votive Offerings in Selected Hera Sanctuaries in the Peloponnese, Ionia and Western Greece*, BAR International Series 1249, Oxford.

Baumbach, J. D. 2009, 'Speak, Votives, ...'. Dedicatory Practice in sanctuaries of Hera', in Prêtre, C. (ed.), *Le donateur, l'offrande et la déesse. Systèmes votifs dans les sanctuaires de déesses du monde grec. Actes du 31e colloque international organisé par l'UMR Halma-Ipel (Université Charles-de-Gaule, Lille 3, 13-15 décembre 2007)*, Kernos Supplement 23, Liege, 203–224.

Bean, G. 1968, *Turkey's Southern Shore*, New York.

Beard, M. 1991, 'Adopting an Approach II', in Rasmussen, T. & Spivey, N. (eds.), *Looking at Greek Vases*, Cambridge, 12–35.

Becker, M. J. 1993, *Human Skeletal Remains from Cremations Urns in the National Museum of Denmark*. Unpublished report.

Becq, J., Jeammet, V. & Mathieux, N. 2010, 'Divinities and Figurines in Boeotia', in Jeammet, V. (ed.), *Tanagra. Figurines for Life and Eternity*, Valencia, 161–237.

Bélis, M. 1998, 'The Use of Purple in Cooking, Medicine, and Magic. An Example of Interference by the Imaginary in Rational Discourse', in Buxton, R. (ed.), *From Myth to Reason? Studies in the Development of Greek Thought*, Oxford, 295–316.

Bell, C. 2007, 'Response: Defining the Need for a Definition', in Kyriakidis, E. (ed.), *The Archaeology of Ritual*, Los Angeles, 277–288.

Benda-Weber, I. 2014, 'Krokotos and crocota vestis: saffron-coloured clothes and muliebrity', in Alfaro, C., Tellenbach, M. & Ortiz, J. (eds.), *Purpurae Vestes IV. Production and Trade of Textiles and Dyes in the Roman Empire and Neighbouring Regions*, Valencia, 129–142.

Bender Jørgensen, L. 2013, 'The Question of Prehistoric Silks in Europe', *Antiquity* 87, 581–588.

Bennett, F. M. 1917, 'A Study of the Word Xoanon', *AIA* XXI, 8–21.

Benveniste, E. 1973, *Indo-European Language and Society*, London.

Bérard, C. & Durand, J. L. 1989, 'Entering the Imagery', in Bérard et al. (eds.), *A City of Images. Iconography and Society in ancient Greece*, Princeton, 23–38.

Bergemann, J. 1997, *Demos und Thanatos. Untersuchungen zum Wertsystem der Polis im Spiegel der attischen Grabreliefs des 4. Jahrhunderts v.Chr. und zur Funktion der gleichzeitigen Grabbauten*, Munich.

Beschi, L. 1967–1968, 'Contributi di Topografia Ateniese. Lo Xoanon di Atena Nike e il Culto delle Charites', *ASAtene* 29–30, 531–536.

Bettinetti, S. 2001, *La statua di culto nella pratica rituale greca*, Bari.

Bieber, M. 1928, *Griechische Kleidung*, Berlin & Leipzig.

Bielefeld, E. 1954–1955, 'Götterstatuen auf attischen Vasenbildern. Eine religionsgeschichtlich-archäologische Studie', *Wissenschaftliche Zeitschrift der Ernst Moritz Arndt-Universität Greifswald* 4, 379–400.

Bille, M. & Sørensen, T. F. 2012, *Materialitet – en indføring i kultur, identitet og teknologi*, Copenhagen.

Blinkenberg, C. 1926, *Fibules Greques et Orientales, Lindiaka V*, Copenhagen.

Blinkenberg, C. 1931, *Lindos. Fouilles de l'acropole. 1902-1914. I. Les petits objets*, Berlin.

Blinkenberg, C. 1941, *Lindos. Fouilles de l'acropole. 1902-1914. II. Inscriptions*, Berlin & Copenhagen.

Blum, H. 1998, *Purpur als Statussymbol in der Griechischen Welt*, Bonn.

Blundell, S. 2002, 'Clutching at Clothes', in Llewellyn-Jones, Ll. (ed.), *Women's Dress in the Ancient World*, London, 143–170.

Blundell, S. & Williamson, M. 1998, (eds.), *The Sacred and the Feminine in Ancient Greece*, New York.

Boardman, J. 1999, *The Greeks Overseas. Their Early Colonies and Trade*, London.

Boehringer, D. 2001, *Heroenkulte in Griechenland von der geometrischen bis zur klassischen Zeit. Attika, Argolis, Messenien*, Berlin.

Booth, N. B. 1979, 'Two Passages in Aeschylos Agamemnon', *Eranos* 77, 85–95.

Bostock, J. 1855, *The Natural History. Pliny the Elder*. Translated by John Bostock, London.

Boyer, P. 1996, 'What Makes Anthropomorphism Natural: Intuitive Ontology and Cultural Representations', *Journal of the Royal Anthropological Institute* 2, 83–92.

Bradley, M. 2009, *Colour and Meaning in Ancient Rome*, Cambridge.
Bremmer, J. 1998, "'Religion', 'Ritual' and the Opposition 'Sacred vs. Profane'. Notes on a Terminological 'Geneaology'", in Graf, F. (ed.), *Ansichten griechischer Rituale. Geburtstags-Symposium für Walter Burkert*, Stuttgart & Leipzig, 9–32.
Bremmer, J. 2008, 'Priestly Personnel of the Ephesian Artemisium: Anatolian, Persian, Greek, and Roman Aspects', in Dignas, B. & Trampedach, K. (eds.), *Practitioners of the Divine. Greek Priests and Religious Officials from Homer to Heliodorus*, Cambridge & London, 37–54.
Brinkmann, V. 2004a, 'Polykromien i arkaisk græsk skulptur -med strejftog ind i klassisk tid', in Nielsen, M. & Østergaard, J. S. (eds.), *Classicolor. Farven i Antik Skulptur. Meddelelser fra Ny Carlsberg Glyptotek* 6, 50–65.
Brinkmann, V. 2004b, 'Den gådefulde 'Peplos-kore' fra Athens Akropolis', in Nielsen, M. & Østergaard, J. S. (eds.), *Classicolor. Farven i Antik Skulptur. Meddelelser fra Ny Carlsberg Glyptotek* 6, 66–70.
Brinkmann, V. 2009, 'Mädchen oder Göttin? Das Rätsel der "Peploskore" von der Athener Akropolis,' in Brinkmann, V. (ed.), *Bunte Götter. Die Farbigkeit antiker Skulptur*, München, 70–79.
Brinkmann, U. 2014, 'Girls and Goddesses', in Østergaard, J. S. & Nielsen, A. M. (eds.), *Transformations. Classical Sculpture in Colour*, Copenhagen, 116–139.
Brock, R. 1994, 'The Labour of Women in Classical Athens', *CQ* 44, 336–346.
Bruit, L. 1989, 'Pausanias à Phigalie. Sacrifices non-sanglants et discours idéologique', *Metis* 1, 71–96.
Brulé, P. 1987, *La Fille d'Athenes. La religion des filles à Athènes à l'epoque classique : cultes, mythes et société*, Paris.
Brøns, C. 2012, 'Dress and Identity in Iron Age Italy. Fibulas as Indicators of Age and Biological Sex, and the Identification of Dress and Garments', *Babesch* 87, 45–68.
Brøns, C. 2013, 'Guddommelige garderober. Tekstiler i græske helligdomme i det første årtusind f.Kr.', *Nationalmuseets Arbejdsmark*, 96–109.
Brøns, C. 2014, 'Representation and realities: Fibulas and pins in Greek and Near Eastern Iconography,' in Harlow, M. & Nosch, M.-L. (eds.), *Greek and Roman Textiles and Dress: an Interdisciplinary Anthology*, Ancient Textiles Series 19, Oxford, 60–94.
Brøns, C. 2015, 'Textiles and Temple Inventories. Detecting an Invisible Votive Tradition in Greek Sanctuaries in the second half of the 1st Millennium BC', in Fejfer, J., Moltesen, M. & Rathje, A. (eds.), *Tradition. Transmission of Culture in the Ancient World*. Acta Hyperborea 14, Copenhagen, 43–84.
Brøns, C. forthcoming a, 'Sacred Colours: Purple Textiles in Greek Sanctuaries in the Second half of the 1st Millennium BC.', in Landenius Enegren, H. & Meo, F. (eds.), *Treasures of the Sea. Sea Silk and Purple Dye in Antiquity*. Ancient Textiles Series 29. Oxford.
Brøns, C. forthcoming b, 'The Peplos Paradigm. Textiles and Textile Production on the Athenian Acropolis', in Abbe, M. & Normann, N. (eds.), *The Parthenon: Color, Materiality and Aesthetics*, Cambridge.
Brøns, C., & Droß-Krüpe, K. forthcoming, 'From Epigraphy to Papyrology. Reconsidering the Greek term *halourgos* (ἁλουργός) and its Relation to Textiles and the Colour Purple', *Textile History*.
Bücher, K. 1922, 'Zur griechischen Wirtschaftsgeschichte', *Beiträge zur Wirtschaftsgeschichte*, Tübingen, 1–97.
Bülow-Jacobsen, A. 2014, 'Texts and textiles on Mons Claudianus', in Tallet, G. & Zivie-Coche, C. (eds.), *Le myrte & la rose. Mélanges offerts à François Dunand par ses élèves, collègues et amis*, Montpellier, 3–7.
Bundgaard, J. A. 1974, *The Excavation of the Athenian Acropolis 1882-1890 / The Original Drawings ed. from the Papers of Georg Kawerau*, Copenhagen.
Bundgaard-Rasmussen, B. 1998, 'Gold Ornaments from the Mausoleum at Halikarnassos', in Williams, D. (ed.), *The Art of the Greek Goldsmith*, London, 66–73.
Bundrick, S. D. 2008, 'The Fabric of the City: Imaging Textile Production in Classical Athens', *Hesperia* 77, 2, 283–334.
Burgess, J. S. 2001, *The Tradition of the Trojan War in Homer and the Epic Cycle*, Baltimore.
Burke, B. 2012, 'Looking for Sea-silk in the Bronze Age Aegean', in Nosch, M.-L. & Laffineur, R. (eds.), *Kosmos. Jewellery, Adornment and Textiles in the Aegean Bronze Age. Proceedings of the 13th International Aegean Conference/13e Rencontre égéenne internationale, University of Copenhagen, Danish National Research Foundation's Centre for Textile Research, 21-26 April 2010*, Aegaeum 33, Liège, 171–178.
Burkert, W. 1983, *Homo Necans. The Anthropology of Ancient Greek Sacrificial Ritual and Myth*, Berkeley.
Burkert, W. 1985, *Greek Religion: Archaic and Classical*, Oxford.

Burtt, J. O. 1954, *Minor Attic Orators, Volume II: Lycurgus. Dinarchus. Demades. Hyperides.* Translated by J. O. Burtt, Cambridge, MA & London.

Buschor, E. 1930, 'Heraion von Samos. Frühe Bauten', *AM* 55, 1–99.

Cage, J. 1999, 'What Meaning had Colour in Early Societies?', *CAJ* 9, 1, 109–126.

Cahn, H. A. 1950, 'Die Löwen des Apollon', *MusHelv* 7, 185–199.

Cahn, H. A. 1985, 'Tissaphernes in Astyra', *Archäologischer Anzeiger*, 587–594.

Callon, M. 1986, 'Some Elements of a Sociology of Translation: Domestication of the Scallops and the Fishermen of St Brieuc Bay', in Law, J. (ed.), *Power, Action and Belief: A New Sociology of Knowledge?*, London, 196–223.

Campobianco, L. S. 2016, 'Clothing the Self in Metaphor: the Spartan *Phoinikis* as Sign of Double Identity', in Fanfani, G., Harlow, M. & Nosch, M.-L. (eds.), *Spinning Fates and Songs of the Loom.* Ancient Textiles Series 24, Oxford, 179–194.

Carbon, J.-M. & Pirenne-Delforge, V. 2012, 'Beyond Greek "Sacred Laws"', *Kernos* 25, 163–181.

Cardon, D. 2007, *Natural Dyes. Sources, Tradition, Technology and Science*, London.

Carpenter, T. H. 2007, 'Greek Religion and Art', in Ogden, D. (ed.), *A Companion to Greek Religion*, Oxford, 398–411.

Carstens, A. M. 2012, 'Bringing Wool to Zeus Labraundos', in Schrenk, S., Vössing, K. & Tellenbach, M. (eds.), *Kleidung und Identität in religiösen Kontexten der römischen Kaiserzeit*, Regensburg, 141–148.

Caskey, M. E. 1976, 'Notes on Relief Pithoi of the Tenian-Boiotian Group', *AJA* 80, 1, 19–41.

Caskey, M. E. 1981, *Ayia Irini, Kea: the Terracotta Statues and the Cult in the Temple*, Stockholm.

Castor, A. Q. 2008, 'Grave Garb: Archaic and Classical Macedonian Funerary Costume', in Colburn, C. S. & Heyn, M. K. (eds.), *Reading a Dynamic Canvas: Adornment in the Ancient Mediterranean World*, Newcastle, 115–145.

Cesare, M. de, 1997, *Le statue in imagine. Studi sulle raffigurazioni di statue nella pittura vascolare Greca*, Rome.

Chaniotis, A. 2008, 'Priests as Ritual Experts in the Greek World', in Dignas, B. & Trampedach, K. (eds.), *Practitioners of the Divine. Greek Priests and Religious Officials from Homer to Heliodorus*, Cambridge, MA & London, 17–36.

Christensen, L. B. 2009, '"Cult" in the study of Religion and Archaeology', in Jensen, J. T. et al. (eds.), *Aspects of Ancient Greek Cult. Context, Ritual and Iconography*, Aarhus, 13–28.

Chrysostomou, P. & Chrysostomou, A. 2012, 'The Lady of Archontiko', in Stampolidis, N. C. & Giannopoulou, M. (eds.), *Princesses of the Mediterranean in the Dawn of History*, Athens, 366–387.

Clark, I. 1998, 'The Gamos of Hera. Myth and Ritual' in Blundell, S. & Williamson, M. (eds.), *The Sacred and the Feminine in Ancient Greece*, London, 13–26.

Clark, R. J. H, Cooksey, C. J., Daniels, M. A. M. & Withnall, R. 1993, 'Indigo, Woad, and Tyrian Purple: Important Vat Dyes from Antiquity to the Present', *Endeavour*, 17, 4, 191–199.

Cleland, L. 2002, *Colour in Ancient Greek Clothing: A Methodological Investigation*. PhD dissertation, University of Edinburgh.

Cleland, L. 2005, *The Brauron Clothing Catalogues*, Oxford.

Cleland, L. 2010, 'A Hierarchy of Women: Status, Dress and Social Construction at Andania'. Unpublished paper for the conference volume of Inaugural Celtic Classics Conference (Maynooth 2000).

Cleland, L., Davies, G. & Llewellyn-Jones, Ll. 2007, *Greek and Roman Dress A to Z*, Oxford.

Cleland, L., Harlow, M. & Llewellyn-Jones, L. (eds.) 2005, *The Clothed Body in the Ancient World*, Oxford, xi–xvi.

Clinton, K. 1974, *The Sacred Officials of the Eleusinian Mysteries*, Philadelphia.

Clinton, K. 2007, 'The Mysteries of Demeter and Kore', in Ogden, D. (ed.), *A Companion to Greek Religion*, Oxford, 342–356.

Colburn, C. S. & Heyn, M. K. (eds.) 2008, *Reading a Dynamic Canvas. Adornment in the Ancient Mediterranean World*, Newcastle.

Cole, S. G. 1998, 'Domesticating Artemis', in Blundell, S. & Williamson, M. (eds.), *The Sacred and the Feminine in Ancient Greece*, New York, 27–43.

Cole, W. 1685, 'A Letter from Mr. William Cole of Bristol, to the Phil. Society of Oxford; Containing His Observations on the Purple Fish', *Philosophical Transactions* 1685, 15, 1278–1286, published 1 January 1685. http://rstl.royalsocietypublishing.org/content/15/167-178/1278.full.pdf+html

Connelly, J. B. 1993a, 'Narrative and Image in the Attic Vase Painting: Ajax and Kassandra at the Trojan Palladion', in Holliday, P. J. (ed.), *Narrative and Event in Ancient Art*, Cambridge, 88-129.
Connelly, J. B. 1993b, 'The Parthenon Frieze and the Sacrifice of the Erechtheids: Reinterpreting the Peplos Scene', *AJA* 97, 2, 309-310.
Connelly, J. B. 1996, 'Parthenon and Parthenoi: A Mythological Interpretation of the Parthenon Frieze', *AJA* 100, 1, 53-80.
Connelly, J. B. 2007, *Portrait of a priestess*, Princeton.
Connelly, J. B. 2008, 'In Divine Affairs – the Greatest Part: Women and Priesthoods in Classical Athens', in Kaltsas, N. & Shapiro, A. (eds.), *Worshipping Women. Ritual and Reality in Classical Athens*, New York, 186-241.
Connelly, J. B. 2014, *The Parthenon Enigma. A new understanding of the world's most iconic building and the people who made it*, New York.
Connolly, P. 1981, *Greece and Rome at War*, London.
Consoli, V. 2004, 'Atena Ergane. Sorgere di un culto sull'Acropoli di Atene', *ASAtene* 82, 31-59.
Consoli, V. 2010, 'Elmo, fuso e conocchia. Per un'iconografia di Atena Ergane', *Eidola* 7, 9-28.
Corbett, P. E. 1970, 'Greek temples and Greek worshippers: the literary and archaeological evidence', *BICS* 17, 149-158.
Culham, P. 1986, 'Again, What Meaning Lies in Colour!', *ZPE* 64, 235-245.
Dakoronia, F. & Gounaropoulou, L. 1992, 'Artemiskult auf einem Weihrelief aus Achinos bei Lamia', *AM* 107, 217-227.
Davreux, J. 1942, *La légende de la prophétesse Cassandre*, Liege & Paris.
Dawson, S. 2002, *The Setting and Display of Cult Images in the Archaic and Classical Periods in Greece*. McMaster University, Open Access Dissertations and Theses.
Day, J. 2011a, 'Crocuses in Context. A Diachronic Survey of the Crocus Motif in the Aegean Bronze Age', *Hesperia* 80, 337-379.
Day, J. 2011b, 'Counting Threads. Saffron in Aegean Bronze Age Writing and Society', *Oxford Journal of Archaeology* 30, 4, 369-391.
De Moor, A. & Fluck, C. 2009, *Clothing the House. Furnishing Textiles of the 1st millennium AD from Egypt and Neighbouring Countries. Proceedings of the 5th Conference of the Research Group 'Textiles from the Nile Valley' Antwerp, 6-7 October 2007*, Tielt.
Deacy, S. 2007, 'The Religious System at Athens', in Ogden, D. (ed.), *A Companion to Greek Religion*, Oxford, 221-235.
Deacy, S. & Villing, A. 2009, 'What was the colour of Athena's aegis?', *JHS* 129, 111-129.
Delivorrias, A. 2008, 'The Worship of Aphrodite in Athens and Attica: Cult Places, Rites, Iconography', in Kaltsas, N. & Shapiro, A. (eds.), *Worshipping Women. Ritual and Reality in Classical Athens*, New York, 106-123.
Demakopoulou, K. & Konsola, D. 1981, *Archaeological Museum of Thebes; A Guide*, Athens.
Demand, N. 1994, *Birth, Death, and Motherhood in Classical Greece*, Baltimore & London.
Deonna, W. 1938, *Le mobilier délien. Exploration Archéologique de Délos XVIII*, Paris.
Deshours, N. 2006, *Les Mystéres d'Andania: étude d'épigraphie et d'histoire religieuse (Scripta Antiqua 16)*, Paris.
Despinis, G. I. 2010, Ἄρτεμις Βραυρωνία: Λατρευτικά ἀγάλματα καί ἀναθήματα ἀπό τά ἱερά τῆς θεᾶς στῆ Βραυρώνα καί τήν Ἀκρόπολη τῆς Ἀθήνας, Athens.
Despinis, G. I. 2004, 'Die Kultstatuen der Artemis in Brauron', *AM* 119, 261-315.
Despinis, G. I. 2007, 'Neues zu der spätarchaischen statue des Dionysos aus Ikaria', *AM* 122, 103-138.
Dindorf, W. 1853, *Harpocrationis Lexicon in decem oratores Atticos*, Oxford.
Dietz, S. & Trolle, S. 1974, *Arkæologiens Rhodos*, Aarhus.
Dignas, B. 2002, 'Inventories or Offering Lists? Assessing the Wealth of Apollo Didymaeus', *ZPE* 138, 235-244.
Dignas, B. 2003, *Economy of the Sacred in Hellenistic and Roman Asia Minor*, Oxford.
Dignas, B. & Trampedach, K. 2008, *Practitioners of the Divine. Greek Priests and Religious Officials from Homer to Heliodorus*, Washington.
Dillon, M. 1997, *Pilgrims and Pilgrimage in Ancient Greece*, London.

Dillon, M. 1999, 'Post-nuptial Sacrifices on Kos (Segre, *ED* 178) and Ancient Greek Marriage Rites', *ZPE* 124, 63–80.
Dillon, M. 2002, *Girls and Women in Classical Greek Religion*, London.
Dillon, M. 2012, 'Hera', in Bagnall, R. et al. (eds.), *Blackwell Encyclopedia of Ancient History* vol. 6, Oxford, 3135–3136.
Donder, H. 1994, *Die Fibeln. Katalog der Sammlung antiker Kleinkunst des archäologischen Instituts der Universität Heidelberg*, Mainz.
Donohue, A. A. 1988, *Xoana and the Origins of Greek Sculpture*, Atlanta.
Donohue, A. A. 1997, 'The Greek Images of the Gods: Considerations on Terminology and Methodology', *Hephaistos* 15, 31–45.
Douny, L. & Harris, S. 2014, 'Wrapping and Unwrapping, Concepts and Approaches' in Douny, L. & Harris, S. (eds.), *Wrapping and Unwrapping Material Culture, Archaeological and Anthropological Perspectives*, Walnut Creek, 15–40.
Drogou, S., Saatsoglu-Paliadeli, C., Faklaris, P., Kottaridou, A. & Tsigarida, E.B. 1994, *Vergina. The Great Tumulus. Archaeological Guide*, Thessaloniki.
Droogan, J. 2012, *Religion, Material Culture and Archaeology*, London.
Droop, J. P. 1929, 'Bronzes', in Dawkins, R. M. (ed.), *The Sanctuary of Artemis Orthia at Sparta*, London, 196–202.
Droß-Krüpe, K. (ed.) 2014, *Textile Trade and Textile Distribution in Antiquity*. Philippika 73, Wiesbaden.
Droß-Krüpe, K. & Paetz gen. Schieck, A. 2014, 'Unravelling the Tangled Threads of Ancient Embroidery – a Compilation of Written Sources and Archaeologically Preserved Textiles', in Harlow, M. & Nosch, M.-L. (eds.), *Greek and Roman Textiles and Dress: an Interdisciplinary Anthology*, Ancient Textiles Series 19, Oxford, 207–236.
Dugas, C. 1921, 'Le sanctuaire d'Aléa Athéna à Tégée avant le IVe siècle', *BCH* 45, 335–435.
Dumézil, G. 1958, *L'idéologie tripartie des Indo-Européens*, Brussels.
Dunbabin, K. M. D. 1990, 'Ipsa deae vestigia... Footprints divine and human on Graeco-Roman monuments', *Journal of Roman Archaeology* 3, 85–109.
Dunbabin, T. J. 1936–1937, '῎Εχθρη παλαίη', *ABSA* 37, 83–91.
Durkheim, E. 1995, *The Elementary Forms of Religious Life*, New York.
Dörpfeld, W. 1919, 'Das Hekatompedon in Athens', *Jahrbuch des Deutschen Archäelogischen Instituts* 34, 1–40.
Edelmann, M. 1999, *Menschen auf griechischen Weihreliefs. Quellen und Forschungen zur Antiken Welt*, Munich.
Edgeworth, R. J. 1988, 'Saffron-coloured terms in Aeschylos', *Glotta* 66, 179–182.
Edmonds, J. M. 1912, *The Greek Bucolic Poets*. Translated by J. M. Edmonds, Cambridge, MA.
Ehrman, B. D. 2003, *Lost Scriptures: Books that Did Not Make it into the New Testament*, Oxford.
Ekroth, G. 2003, 'Inventing Iphigeneia? On Euripides and the Cultic Construction of Brauron', *Kernos* 16, 59–118.
Ekroth, G. 2009a, 'Why (not) Paint an Altar? A Study of Where, When and Why Altars Appear on Attic Red-figure Vases', in Nørskov, V. et al. (eds.), *The World of Greek Vases*, Rome, 89–114.
Ekroth, G. 2009b, 'Thighs or Tails? The Osteological Evidence as a Source for Greek Ritual Norms', in Brulé, P. (ed.), *Le norme en matiere religieuse*. Kernos supplement 21, Liege, 125–151.
Ekroth, G. 2011, 'Ull, pengar och sex: Tolkningar av ett attiskt, rödfigurigt vasmotiv', *Medusa. Svensk tidsskrift för antiken, Stockholm: Föreningen för en svensk antiktidskrift* 32, 1–12.
Ekroth, G. & Wallensten, J. 2013, 'Introduction: bones of contention?', in Ekroth, G. & Wallensteen, J. (eds.), *Bones, Behaviour and Belief. The Zooarchaeological Evidence as a Source for Ritual Practice in Ancient Greece and Beyond*, Stockholm.
Elsner, J. 2012, 'Material Culture and Ritual: State of the Question', in Wescoat, B. D. & Ousterhout, R. G. (eds.), *Architecture of the Sacred. Space, Ritual, and Experience from Classical Greece to Byzantium*, Cambridge, 1–26.
Evans, N. 2010, *Civic Rights. Democracy and Religion in Ancient Athens*, Berkeley.
Falconer, W. 1903, *The Geography of Strabo*, London.
Ferrari, G. 2003, 'What Kind of Rite of Passage Was the Ancient Greek Wedding?', in Dodd, D. B. & Faraone, C. A. (eds.), *Initiation in Ancient Greek Rituals and Narratives: New Critical Perspectives*, London, 27–42.
Figueira, T. 1991, *Athens and Aigina in the Age of Imperial Colonization*, London.

Figueira, T. 1993, *Excursions in Epichoric History. Aiginetan Essays*, Boston.
Flashar, M. 1999, 'Zur Datierung der Kultbildgruppe von Klaros [Klaros-Studien I]', in Bol, P. C. (ed.), *Hellenistische Gruppen. Gedenkschrift für Andreas Linfert*, Mainz am Rhein, 53–94.
Fleischer, R. 1973, *Artemis von Ephesos und verwandte Kultstatuen aus Anatolien und Syrien*, Leiden.
Fleischer, R. 1983, 'Neues zu kleinasiatischen Kultstatuen', *Archäologischer Anzeiger* 98, 81–93.
Fleischer, R. 1999, 'Neues zum Kultbild der Artemis von Ephesos', in *100 Jahre österreichische Forschungen in Ephesos. Akten des Symposions, Wien 1995*, Vienna, 605–609.
Fleischer, R. 2000–2001, 'Eine silberne Hand der Artemis von Ephesos im Archäologischen Museum der Universität Münster', *Boreas* 23–24, 191–194.
Fleischer, R. 2002, 'Die Amazonen und das Asyl des Artemisions von Ephesos', *JdI* 117, 185–216.
Flemestad, P. 2014, 'Theophrastos of Eresos on Plants for Dyeing and Tanning', in Alfaro, C., Tellenbach, M. and Ortiz, J. (eds), *Purpureae Vestes IV: Textiles and Dyes in Antiquity. Production and Trade of Textiles and Dyes in the Roman Empire and Neighbouring Regions*, Valencia, 203–209.
Flemming, R. & Hanson, A. E. 1998, *Hippocrates' Peri parthenōn (Diseases of Young Girls): Text and Translation*, Leiden.
Fletcher, J. 2016, 'The Curse as a Garment in Greek tragedy', in Fanfani, G., Harlow, M. & Nosch, M.-L. (eds.), *Spinning Fates and the Song of the Loom. The Use of Textiles, Clothing and Cloth Production as Metaphor, Symbol and Narrative Device in Greek and Latin Literature*. Ancient Textiles Series 24, Oxford, 101–113.
Foley, A. 1988, *The Argolid 800-600 BC. An archaeological Survey*, Göteborg.
Forbes, R. 1964, *Studies in Ancient Technology* Volume IV, Leiden.
Francis, E. D. & Vickers, M. 1984, 'Green Goddess. A Gift to Lindos from Amasis of Egypt', *AJA* 88, 1, 68–69.
Franke, P. R. & Nollé, M. K. 1997, *Die Homonoia-Münzen Kleinasiens und der thrakischen Randgebiete I*, Saarbrucken.
Fränkel, F. 1890–1915, *Die Inschriften von Pergamon VIII*, 1–2, Berlin.
Freedberg, D. 1989, *The Power of Images: Studies in the History and Theory of Response*, Chicago.
Frickenhaus, A. 1912, *Lenäenvasen. Zweiundsiebzigstes Programm zum Winckelmannsfeste der Archäologischen Gesellschaft zu Berlin*, Berlin.
Furtwängler, A. 1906, *Aegina. Das Heiligtum der Aphaia*, Munich.
Gabrielsen, V. 1994, *Financing the Athenian Fleet. Public Taxation and Social Relations*, Baltimore & London.
Gagarin, M. (ed.) 2010, *Oxford Encyclopedia of Ancient Greece and Rome*, Oxford.
Gaifman, M. 2005, *Beyond Mimesis in Greek Religious Art. Aniconism in the Archiac and Classical Periods*. PhD dissertation, Princeton University.
Gaifman, M. 2008, 'Visualized rituals and dedicatory inscriptions on votive offerings to the nymphs', *OpAthRom* 1, 85–103.
Garland, R. 1989, 'The well-ordered corpse: An investigation into the motives behind Greek funerary legislation', *BICS* 36, 1–15.
Garland, R. 1990a, 'Priests and Power in Classical Athens', in Beard, M. & North, J. (eds.), *Pagan Priests*, New York, 73–91.
Garland, R. 1990b, *The Greek Way of Life from Conception to Old Age*, London.
Gauer, W. 1984, 'Was geschieht mit dem peplos?', in Berger, E. (ed.), *Parthenon-Kongress Basel: Referate und Berichte*, Mainz, 220–229.
Gauthier, P. 1985, 'Les chlamydes et l'entretien des éphèbes athéniens', *Chiron* 15, 149–163.
Gawlinski, L. 2008, '"Fashioning" Initiates: Dress at the Mysteries', in Colburn, C. S. & Heyn, M. K. (eds.), *Reading a Dynamic Canvas: Adornment in the Ancient Mediterranean World*, Cambridge, 146–169.
Gawlinski, L. 2011, *The Sacred Law of Andania. A New Text with Commentary*, Berlin.
Geddes, A. 1987, 'Rags and Riches: The Costume of Athenian Men in the Fifth Century', *CQ* 37, 2, 307–331.
Gell, A. 1993, *Wrapping in Images*, Oxford.
Gell, A. 1998, *Art and Agency. An Anthropological Theory*, Oxford.
Gerziger, D. 1975, 'Eine Decke aus dem sechsten Grab der "Sieben Brüder"', *Antike Kunst* 18, 2, 51–55.
Gibson, C. 2003, 'Libanius, Hypotheses to the Orations of Demosthenes,' in Blackwell, C. W. (ed.), *Dēmos: Classical Athenian Democracy* (A. Mahoney and R. Scaife, eds, The Stoa: a consortium for electronic publication in the humanities [www.stoa.org]) edition of April 30.

Gladigow, B. 1988, 'Anikonische Kulte', in Cancik et al. (eds.), *Handbuch religionswissenschaftlicher Grundbegriffe* vol. I, 472–473.
Gleba, M. 2008a, '*Auratae vestes*: Gold Textiles in the Ancient Mediterranean', in Alfaro, C. & Karali, L. (eds.), *Purpureae Vestes II*, Valencia, 61–77.
Gleba, M. 2008b, *Textile Production in Pre-Roman Italy*, Oxford.
Gleba, M. 2012, 'Linen-clad Etruscan Warriors', in Nosch, M.-L. (ed.), *Wearing the Cloak. Dressing the Soldier in Roman Times*, Oxford, 45–55.
Gleba, M. 2014, 'Wrapped Up for Safe Keeping: Wrapping Customs in Early Iron Age Europe, Approaches' in Douny, L. & Harris, S. (eds.), *Wrapping and Unwrapping Material Culture, Archaeological and Anthropological Perspectives*, Walnut Creek, 135–146.
Gleba, M. & Krupa, T. 2012, 'Ukraine', in Gleba, M. & Mannering, U. (eds.), *Textiles and Textile Production in Europe. From Prehistory to AD 400*. Ancient Textiles Series 11, Oxford, 399–425.
Gleba, M. & Mannering, U. 2012, 'Introduction', in Gleba, M. & Mannering, U. (eds.), *Textiles and Textile Production in Europe. From Prehistory to AD 400*. Ancient Textiles Series 11, Oxford, 1–26.
Godelier, M. 1999, *The Enigma of the Gift*, Chicago.
Godley, A. D. 1920, *Herodotus. The Histories*, Cambridge.
Goette, H. R. 2006, 'Further Thoughts on the Athenian Girls at Brauron (Arktoi)', in Mattusch, C. C. et al. (eds.), *Common Ground. Archaeology, Art, Science, and Humanities. Proceedings of the XVIth International Congress of Classical Archaeology, Boston, August 23-26, 2003*, Oxford, 605.
Goette, H. R. 2012, Zur Darstellung von religiöser Tracht in Griechenland und Rom, in Wiegand, H. & Wieczorek, A. (eds.), *Kleidung und Identität in religiösen Kontexten der römischen Kaiserzeit*, Regensburg, 21–34.
Goff, B. 2004, *Citizen Bacchae. Women's ritual practice in ancient Greece*, Berkeley & Los Angeles.
Goldman, H. 1942, 'The Origin of the Greek Herm', *AJA* 46, 1, 58–68.
Good, I. 2001, 'Archaeological textiles: a review of current research', *Annual Review of Anthropology* 30, 209–226.
Granger-Taylor, H. 1985, 'Appendix 2: A Fragment of Sprang in the British Museum', in Jenkins, I. & Williams, D., *AJA* 89, 3, 417–418.
Granger-Taylor, H. 2012, 'Fragments of Linen from Masada, Israel – the Remnants of Pteryges? – and Related Finds in Weft- and Warp-twining Including Several Slings', in Nosch, M.-L. (ed.), *Wearing the Cloak. Dressing the Soldier in Roman Times*. Ancient Textiles Series 10, Oxford, 56–84.
Guthrie, S. 1993, *Faces in the Clouds*, Oxford.
Günther, L. M. 2013, 'Concepts of Purity in Ancient Greece, with Particular Emphasis on Sacred Sites', in Frevel, C. & Nihan, C. (eds.), *Purity and the Forming of Religious Traditions in the Ancient Mediterranean World and Judaism*, Leiden, 245–260.
Günther, W. 1988, 'Vieux et inutilisable dans un inventaire inédit de Milet', in Knoepfler, D. (ed.), *Comptes et inventaires dans la cité grecque*, Geneva, 215–237.
Halvorsen, S. 2012, 'Norway', in Gleba, M. & Mannering, U. (eds.), *Textiles and Textile Production in Europe. From Prehistory to AD 400*. Ancient Textiles Series 11, Oxford, 275–292.
Hamilton, R. 2000, *Treasure Map. A Guide to the Delian Inventories*, Ann Arbor.
Hampe, R. 1936, *Frühe griechische Sagenbilder in Böotien*, Athens.
Hansen, M. H. 1999, *The Athenian Democracy in the Age of Demosthenes. Structure, Principles, and Ideology*, Norman.
Harlizius-Klück, E., *Gesponnen und Verworben. Textiles zu Zeiten von Römern und Germanen*, Tuchmacher Museum Bramsche.
Harlow, M. 2004, 'Female dress 3rd–6th Century: The Message in the Media', *Antiquité Tardive* 12, 203–216.
Harlow, M. & Nosch, M.-L. 2014, 'Weaving the Threads: Methodologies in Textile and Dress Research for the Greek and Roman World – the State of the Art and the Case for Cross-disciplinarity', in Harlow, M. & Nosch, M.-L. (eds.), *Greek and Roman Textiles and Dress: an Interdisciplinary Anthology*, Ancient Textiles Series 19, Oxford, 1–33.
Harris, D. 1995, *The Treasures of the Parthenon and the Erechteion*, Oxford.

Harrison, E. B. 1996, 'The Web of History. A Conservative Reading of the Parthenon Frieze', in Neils, J. (ed.), *Worshipping Athena. Panathenaia and Parthenon*, London, 198–214.
Hartog, F. 1988, *The Mirror of Herodotus. The Representation of the Other in the Writing of History*, Berkeley.
Harvey, G. 2005, *Animism. Respecting the living world*, London.
Haynes, D. E. L. 1960–1961, 'A Pin and Four Buttons from Greece', *British Museum Quarterly* XXIII, 2, 48–49.
Helmecke, G. 2009, 'Textiles for the Interiors. Some Remarks on Curtains in the Written Sources', in De Moor, A. & Fluck, C. (eds.), *Clothing the House. Furnishing Textiles of the 1st millennium AD from Egypt and Neighbouring Countries. Proceedings of the 5th Conference of the Research Group 'Textiles from the Nile Valley' Antwerp, 6-7 October 2007*, Tielt, 48–53.
Henrichs, A. 2008, 'What is a Greek Priest?', in Dignas, B. & Trampedach, K. (eds.), *Practitioners of the Divine: Greek Priests and Religious Officials from Homer to Heliodorus*, Cambridge, 1–14.
Herrmann, H. V. 1959, *Omphalos*, Münster.
Herrmann, P., Günther, W. & Ehrhardt, N. 2006, *Inschriften von Milet, Teil 3. Inschriften n. 1020-1580*, Berlin & New York.
Hewitt, J. W. 1909, 'Major Restrictions on Access to Greek Temples', *TAPA* 40, 83–91.
Hildebrandt, B. 2009, 'Seide als Prestigegut in der Antike', in Hildebrandt, B. & Veit, C. (eds.), *Der Wert der Dinge – Güter im Prestigediskurs*, Munich, 183–239.
Hildebrandt, B. 2012a, 'Some Thoughts on the Unravelling of Chinese Silks in the Roman Empire. A Reassessment of Lucan, Bellum Civile 10.141-143', in Tzachili, I. & Zimi, E. (eds.), *Textiles and Dress in Greece and the Roman East: A Technological and Social Approach. Proceedings held at the Department of History, Archaeology and Cultural Resources Management of the University of Peloponnese in Kalamata in collaboration with the Department of History and Archaeology of the University of Crete on March 18-19, 2011*, Athens, 107–115.
Hildebrandt, B. 2012b, 'Der Römer neue Kleider. Zur Einführung von Seide im kaiserzeitlichen Rom', in Lehmann, G. A., Engster, D. & Nuss, A. (eds.), *Von der bronzezeitlichen Geschichte zur modernen Antikenrezeption*, Göttingen, 11–53.
Hiller von Gaertringen, F. 1906, *Inschriften von Priene*, Berlin.
Hiller, S. 1983, 'Mycenaean Traditions in early Greek Cult Images' in Hägg, R. (ed.), *The Greek Renaissance of the eighth century B.C. Tradition and Innovation. Proceedings of the Second International Symposium at the Swedish Institute at Athens, 1-5 June, 1981*, Stockholm, 91–99.
Himmelmann-Wildschütz, N. 1959, *Zur Eigenart des klassischen Götterbildes*, Munich.
Hoff, R. v. 2008, 'Images of Cult Personnel in Athens between the Sixth and First Centuries BC', in Dignas, B. & Trampedach, K. (eds.), *Practitioners of the Divine. Greek Priests and Religious Officials from Homer to Heliodorus*, Cambridge & London, 107–143.
Hogarth, D. G. 1908, *Excavations at Ephesus, the Archaia Artemisia*, London.
Hollander, A. 1978, *Seeing Through Clothes*, London.
Holloway, R. 1966, 'The Archaic Acropolis and the Parthenon Frieze', *Art Bulletin* 48, 223–226.
Hooker, E. M. 1950, 'The Sanctuary and Altar of Chryse in Attic Red-Figure Vase-Paintings of the Late Fifth and Early Fourth Centuries B.C.', *JHS* 70, 35–41.
Horster, M. & Klöckner, A. 2012, (eds.), *Civic Priests. Cult Personnel in Athens from the Hellenistic Period to Late Antiquity*, Berlin.
Horster, M. & Schröder, T. 2013, 'Priests, Crowns and Priestly Headdresses in Imperial Athens', in Alroth, B. & Scheffer, C. (eds.), *Attitudes Towards the Past in Antiquity Creating Identities. Proceedings of an International Conference held at Stockholm University, 15-17 May 2009*, Stockholm, 233–239.
Houtman, D. & Meyer, B. 2012, 'Introduction', in Houtman, D. & Meyer, B. (eds.), *Things: Religion and the Question of Materiality*, New York, 1–26.
Humphreys, S. C. 1983, *Anthropology and the Greeks*, London.
Hundt, H. J. 1969, 'Über vorgeschichtliche Seidenfunde', *Jahrbuch des römisch-germanischen Zentralmuseums Mainz* 16, 59–71.
Hurwitt, J. 1999, *The Athenian Acropolis: History, Mythology, and Archaeology from the Neolithic Era to the Present*, Cambridge.

Hutton, C. A. 1897, 'Votive Reliefs in the Acropolis Museum', *JHS* 17, 306-318.
Huysecom-Haxhi, S. 2009, *Les Figurines en terre cuite de l'Artemision de Thasos. Etudes Thasiennes XXI*, Athens.
Hölscher, F. 2010, 'Gods and Statues – An Approach to Archaistic Images in the Fifth Century BCE', in Mylonopoulos, J. (ed.), *Divine Images and Human Imaginations in Ancient Greece and Rome*, Leiden, 105-120.
Hölscher, T. 2005, 'Kultbild', in *ThesCRA IV. Cult places. Representations of Cult Places*, 52-65.
Håland, E. J. 2004, 'Athena's Peplos: Weaving as a Core Female Activity in Ancient and Modern Greece', *Cosmos* 20, 155-182.
Ignatiadou, D. 2012, 'The Sindos Priestess', in Stampolidis, N. C. & Giannopoulou, M. (eds.), *Princesses of the Mediterranean in the Dawn of History*, Athens, 388-411.
Imhoof-Blumer, F. 1913, *Nomisma 8. Beiträge zur Erklärung griechischer Münztypen*, Berlin.
Imhoof-Blumer, F. W. & Gardner, P. 1964, *Ancient Coins Illustrating Lost Masterpieces of Greek Art: A Numismatic Commentary on Pausanias*, Chicago.
Jacobi, G. 1932-1933, *Esplorazioni Archeologici di Camiro II, Clara Rhodos* 6-7, Bergamo.
Jacobsthal, J. 1956, *Greek Pins and their Connexions with Europa and Asia*, Oxford.
Jenkins, I. & Williams, D. 1985, 'Sprang Hair Nets: Their Manufacture and Use in Ancient Greece', *AJA* 89, 3, 411-418.
Jeppesen, K. 2000, *The Maussolleion at Halikarnassos. Vol. 4. The Quadrangle. The Foundations of the Maussolleion and its Sepulchral Compartments*, Aarhus.
Jeppesen, K. 2007, 'A Fresh Approach to the Problems of the Parthenon Frieze', *Proceedings of the Danish Institute at Athens V*, Athens, 101-172.
Jim, T. S. F. 2012, 'Naming a Gift: the Vocabulary and Purposes of Greek Religious Offerings', *GRBS* 52, 310-337.
Johnson, M. 1964, *Ancient Greek Dress*, Chicago.
Jones, N. 1999, *The Associations of Classical Athens: The Response to Democracy*, New York & Oxford.
Jones, W. H. S. & Ormerod, H. A. 1918, *Pausanias, Description of Greece*, London.
Joseph, N. 1986, *Uniforms and nonuniforms: communication through clothing*, New York.
Jost, M. 1973, 'Pausanias en Megalopolitide', *Revue des Études Anciennes* 75, 245-267.
Jucker, I. 1967, 'Artemis Kindyas', in Ackermann, H. C. et al. (eds.), *Gestalt und Geschichte. Festschrift Karl Schefold zu seinem sechzigsten geburtstag am 26. Januar 1965*, Bern, 133-145.
Kaestner, D. W. 1976, 'The Coan Festival of Zeus Polieus', *Classical Journal* 71, 4, 344-348.
Kaltsas, N. 2002, *Sculpture in the National Archaeological Museum, Athens*, Los Angeles.
Kanold, I. & R. Haubrichs. 2008, 'Tyrian purple dying: an experimental approach with fresh murex trunculus', in Alfaro, C. & Karali, L. (eds.), *Purpureae Vestes II*, 253-255.
Karakasi, K. 2001, *Archaische Koren*, Munich.
Karali, L. & Megaloudi, F. 2008, 'Purple dyes in the environment and history of the Aegean a short review', in Alfaro, C. & Karali, L. (eds.), *Purpureae Vestes II*, 181-184.
Karanika, A. 2001, 'Memories of Poetic Discourse in Athena's Cult Practices', in Deacy, S. & Villing, A. (eds.), *Athena in the Classical World*, Leiden, 277-291.
Kardara, C. 1960, 'Problems of Hera's Cult Images', *AJA* 64, 4, 343-358.
Kardara, C. 1964, 'Γλαυκῶπις. Ο Αρχαίος Ναός και το θέμα της ζωφόρου του Παρθενώνος', *Archaiologike Ephemeris* 1961, 62-158.
Kardara, C. 1969, 'Πυκινός δόμος και Παναθηναϊκός πέπλος', *Archaiologike Ephemeris* 1960, 165-202.
Karousou, S. P. 1956, 'Hellenistika antigrapha kai epanalipseis archaion ergon', *Archaiologike Ephemeris*, 154-180.
Keesling, C. 1999, 'Endoios's Painting from the Themistoklean Wall: A Reconstruction', *Hesperia* 68, 4, 509-548.
Keesling, C. 2003, *The Votive Statues of the Athenian Acropolis*, Cambridge.
Keesling, C. 2010, 'Finding the Gods. Greek and Cypriot Korai Revisited', in Mylonopoulos, J. (ed.), *Divine Images and Human Imaginations in Ancient Greece and Rome*, 87-104.
Kendrick Pritchett, W. 1956, 'The Attic Stelai: Part II', *Hesperia* 25, 3, 178-328.
Keuls, E. C. 1993, *The Reign of the Phallus. Sexual Politics in Ancient Athens*, London.

Kilian, K. 1975, *Fibeln in Thessalien von der mykenischen bis zur archaischen Zeit*, Prähistorische Bronzefunde (PBF) XIV, 2, Munich.
Kilian-Dirlmeier, I. 1984, *Nadeln der frühhelladischen bis archaischen Zeit von der Peloponnes*, Prähistorische Bronzefunde (PBF) XIII, 8, Munich.
Kilian-Dirlmeier, I. 2002, *Kleinfunde aus dem Athena Itonia-Heiligtum bei Philia (Thessalien)*, Mainz.
Kindt, J. 2011, 'Ancient Greece', in Insoll, T. (ed.), *The Oxford Handbook of the Archaeology of Ritual and Religion*, Oxford, 696–709.
King, H. 1993, 'Bound to Bleed: Artemis and Greek Women', in Cameron, A. & Kuhrt, A. (eds.), *Images of Women in Antiquity*, London, 109–127.
Kleijwegt, M. 2002, 'Textile Manufacturing for a Religious Market. Artemis and Diana as Tycoons of Industry', in Jongman, W. & Kleijwegt, M. (eds.), *After the Past. Essays in Ancient History in Honour of H. W. Pleket*, Leiden, 81–134.
Kline, A. S. 2013, *Lucius Apuleius. The Golden Ass, Book VI.* Translated by A. S. Kline.
Knoepfler, D. 1977, 'Zur Datierung der grossen Inschrift aus Tanagra im Louvre', *Chiron* 7, 67–87.
Kolonia, R. 2012, *Delphi Archaeological Museum*, Athens.
Kontis, I. 1967, 'Ἄρτεμις Βραυρωνία', *Deltion* 22, 156–206.
Kosmetatou, E. 2001, 'The Delian Hieropoioi of 171 B.C.', *Epigraphica* 63, 256–258.
Kosmetatou, E. 2004, 'Persian Objects in Classical and Early Hellenistic Inventory Lists', *MusHelv* 61, 139–170.
Kosmopoulou, A. 2001, 'Working Women: Female Professionals on Classical Attic Gravestones', *ABSA* 96, 281–319.
Kottaridi, A. 2004, 'The Lady of Aegae', in Pandermalis, D. (ed.), *Alexander the Great. Treasures from an Epic Era of Hellenism*, New York, 139–147.
Kottaridi, A. 2012, 'The Lady of Aigai', in Stampolidis, N. C. & Giannopoulou, M. (eds.), *Princesses of the Mediterranean in the Dawn of History*, Athens, 412–433.
Kovacs, D. 1999, *Euripides. Trojan Women, Iphigenia among the Taurians, Ion*, Cambridge, MA.
Kroll, J. 1982, 'The Ancient Image of Athena Polias', *Hesperia Supplements, Vol. 20, Studies in Athenian Architecture, Sculpture and Topography. Presented to Homer A. Thompson*, 65–76.
Künze-Götte, E. 1992, *Der Kleophrades-Maler unter Malern Schwarzfiguriger Amphoren: Eine Werkstattstudie*, Mainz-am-Rhein.
Kyriakidis, E. 2007, 'Archaeologies of Ritual', in Kyriakidis, E. (ed.), *The Archaeology of Ritual*, Los Angeles, 289–308.
Lalleman, P. J. 1997, 'Healing by a Mere Touch as a Christian Concept', *Tyndale Bulletin* 48, 355–361.
Landenius Enegren, H. & Meo, F. (eds.) forthcoming, *Treasures of the Sea. Sea Silk and Purple Dye in Antiquity.* Ancient Textiles Series 29, Oxford.
Lange, K. 1881a, 'Tempelskulpturen von Sunion', *AM* 6, 233–237.
Lange, K. 1881b, 'Die Athena Parthenos', *AM* 6, 56–94.
Langlotz, E. 1964, *Aphrodite in den Gärten*, Heidelberg.
Lapatin, K. 2001, *Chryselephatine Statuary in the Ancient Mediterranean World*, Oxford.
Lapatin, K. 2005, 'The Statue of Athena and Other Treasures in the Parthenon', in Neils, J. (ed.), *The Parthenon. From Antiquity to the Present*, Cambridge, 261–292.
Larson, J. 2007, *Ancient Greek Cults. A Guide*, New York.
Larson Lovén, L. 2013, 'Female Work and Identity in Roman Textile Production and Trade: A Methodological Discussion', in Gleba, M. & Pázstókai-Szeöke, J. (eds.), *Making Textiles in Pre-Roman and Roman Times. People, Places, Identities*. Ancient Textiles Series 13, Oxford, 109–125.
Latour, B. 2005, *Reassembling the Social: an Introduction to Actor-network Theory*, Oxford.
Lee, M. M. 2004, 'Problems in Greek dress terminology. Kolpos and Apoptygma', *ZPE* 150, 221–224.
Lee, M. M. 2005, 'Constru(ct)ing Gender in the Feminine Greek *Peplos*', in Cleland, L., Harlow, M. & Llewellyn-Jones, Ll. (eds.), *The Clothed Body in the Ancient World*, Oxford, 55–64.
Lee, M. M. 2012a, 'Maternity and Miasma. Dress and the transition from *parthenos* to *gunē*', in Petersen, L. H. & Salzman-Mitchell, P. (eds.), *Mothering and Motherhood in Ancient Greece and Rome*, Austin, 23–40.

Lee, M. M. 2012b, 'Dress and Adornment in Archaic and Classical Greece', in James, J. L. & Dillon, S. (eds.), *A Companion to Women in the Ancient World*, Malden MA, 179–190.
Lee, M. M. 2015, *Body, Dress, and Identity in ancient Greece*, Cambridge.
Leibundgut Wieland, D. 2009, 'Dedicated to the Paphian Goddess. Votive Offerings from the Sanctuary of the Paphian Aphrodite at Palaipaphos', *Medelhavs Mus Focus* 5, 145–157.
Leitao, D. D. 1995, 'The Perils of Leukippos: Initiatory Transvestism and Male Gender Ideology in the Ekdusia at Phaistos', *ClAnt* 14, 130–163.
Levides, A. V. 2002, 'Why did Plato not suffer of color blindness?: An interpretation of the passage on color blending of Timaeus', in Tiverios, M. A. & Tsiafakis, D. S. (eds.), *Color in Ancient Greece: The Role of Color in Ancient Greek Art and Architecture 700-31 BC*, Thessaloniki, 9–21.
Lewis, D. M. 1979–1980, 'Athena's Robe', *Scripta Classica Israelica* 5, 28–29.
LiDonnici, L. R. 1992, 'The Images of Artemis Ephesia and Greco-Roman Worship: A Reconsideration', *Harvard Theological Review* 85, 4, 389–415.
Lindenlauf, A. 2006, 'Recyling of Votive Offerings In Greek Sanctuaries. Epigraphical and Archaeological Evidence', in Mattusch, C. C. et al. (eds.), *Common Ground. Archaeology, Art, Science, and Humanities. Proceedings of the XVIth International Congress of Classical Archaeology, Boston, August 23-26, 2003*, Oxford, 30–32.
Linders, T. 1972, *Studies in the Treasure Records of Artemis Brauronia Found in Athens*, Stockholm.
Linders, T. 1975, *The Treasurers of the Other Gods in Athens and their Functions*. Beiträge zur klassischen Philologie, 62, Meisenheim.
Linders, T. 1984, 'The Kandys in Greece and Persia', *OpAth* 15, 107–114.
Linders, T. 1988a, 'The purpose of Inventories. A Close Reading of the Delian Inventories of the Independence', in Knoepfler, D. (ed.), *Comptes et inventaires dans la cité grecque. Actes du colloque international d'épigraphie, Neuchâtel 23-26 septembre 1986 en l'honneur de Jacques Tréheux*, Neuchâtel, 37–47.
Linders, T. 1988b, 'The Delian Temple Accounts. Some Observations', *OpAth* 19, 69–73.
Linders, T. 1988c, 'Continuity in Change. The Evidence of the Temple Accounts of Delos. Prolegomena to a Study of the Economic and Social Life of Greek Sanctuaries', in Hägg, R. et al. (eds.), *Early Greek Cult Practice. Proceedings of the Fifth International Symposium at the Swedish Institute at Athens, 26-29 June, 1986*, Stockholm, 267–269.
Lipka, M. 2002, *Xenophon's Spartan Constitution: Introduction. Text. Commentary*, Berlin & New York.
Liu, J. 2012, 'Clothing Supply for the Military. A Look at the Inscriptional Evidence, in Nosch, M.-L. (ed.), *Wearing the Cloak. Dressing the Soldier in Roman Times*. Ancient Textiles Series 10, Oxford, 19–28.
Liverani, P. et al. (eds.) 2004, *I colori del bianco. Policromia nella scultura antica*, Rome.
Llewellyn-Jones, L. 2003, *Aphrodite's Tortoise. The Veiled Women of Ancient Greece*, Swansea.
Llewellyn-Jones, L. 2007, 'House and Veil in Ancient Greece', *ABSA* 15, 251–258.
Loeffler, E. P. 1962, 'Lamps and Minor Objects', in Lehmann, K. (ed.), *The Hall of Votive Gifts. Samothrace 4.1*, New York, 147–166.
Loraux, N. 1990, 'Herakles: The Super-male and the Feminine', in Halperin, M. et al. (eds.), *Before Sexuality. The Construction of Erotic Experience in the Ancient Greek World*, Princeton, 21–52.
Lorimer, H. L. 1950, *Homer and the Monuments*, London.
Losfeld, G. 1991, *Essai sur le costume Grec*, Paris.
Loucas, I. & Loucas, E. 1994, 'The Sacred Laws of Lycosoura', in Hägg, R. (ed.), *Ancient Greek Cult Practice from the Epigraphical Evidence*, Stockholm, 97–99.
Lullies, R. 1968, *Griechische Kunstwerke der Sammlung Ludwig, Aachen*. Aachener Kunstblätter des Museumsvereins 37.
Lundgreen, B. 1997, 'A Methodological Enquiry. The Great Bronze Athena by Pheidias', *JHS* 117, 190–197.
Lupu, E. 2006, *Greek Sacred Law: A Collection of New Documents*, Leiden.
MacLachlan, B. 1995, 'Love, War and the Goddess in Firth-Century Locri', *Ancient World* 25-26, 205–223.
Maeder, F. 2008, 'Sea-silk in Aquincum. First Production Proof in Antiquity', in Alfaro, C. & Karali, L. (eds.), *Purpureae vestes II. Vestidos, textiles y tintos. Estudios sobre la producción de los bienes de consume en la antigüedad*, Valencia, 109–118.

Maeder, F. 2009, 'Die Edle Steckmuschel und ihr Faserbart: Eine kleine Kulturgeschichte der Muschelseide', *Mitteilungen der Naturforschenden Gesellschaften beider Basel* 11, 15–26.
Maeder, F. 2016, 'Byssus und Muschelseide. Ein sprachliches Problem und seine Folgen', in Harich-Schwarzbauer, H. (ed.), *Weben und Gewebe in der Antike. Materialität - Repräsentation - Episteme - Metapoetik*, Oxford, 1–20.
Mair, A. W. 1955, *Hymns of Callimachus*. Translated by A. W. Mair, London.
Makaronas, C. 1963, 'Τάφοι παρα το Δερβένι Θεσσαλονίκης', *ArchDelt* 18, 193–196.
Mansfield, J. 1985, *The Robe of Athena and the Panathenaic Peplos*, Berkeley.
Mantis, A. 1990, Προβλήματα της εικονογραφίας των ιερέων στην αρχαία ελληνική τέχνη, Athens.
Margariti, C. & Kinti, M. 2014, 'The Conservation of a 5th-Century BC Excavated Textile Find from the Kerameikos Cemetery at Athens', in Harlow, M. & Nosch, M.-L. (eds.), *Greek and Roman Textiles and Dress. An Interdisciplinary Anthology*, Oxford, 130–149.
Margariti, C., Protopapas, S. & Orphanou, V. 2011, 'Recent Analyses of the Excavated Textile Find from Grave 35 HTR73, Kerameikos Cemetery, Athens, Greece', *Journal of Archaeological Science* 38, 3, 522–527.
Marinatos, S. 1967, *Archaeologia Homerica. Kleidung, Haar und Barttracht*, Göttingen.
Mark, I. S. 1987, 'The Ancient Image and Naiskos of Athena Polias: The Ritual Setting on a Late Fifth-Century Acropolis Relief', *AJA* 91, 2, 287–288.
Mark, I. S. 1993, *The Sanctuary of Athena Nike in Athens. Architectural Stages and Chronology*, Hesperia Supplement 26.
Marshall, F. H. 1911, *Catalogue of the Jewellery, Greek, Etruscan, and Roman in the Department of Antiquities, British Museum*, London.
Marx, P. A. 2001, 'Acropolis 625 (Endoios Athena) and the Rediscovery of Its Findspot', *Hesperia* 70, 2, 221–254.
Mauss, M. 2011, *The Gift. Forms and Functions of Exchange in Archaic Societies*. Translated by Ian Cunnison, Mansfield Centre.
Mayrhofer, M. 1992–2001, *Etymologisches Wörterbuch des Altindoarischen*, Heidelberg.
McCauley, R. N. & Lawson, E. T. 2007, 'Cognition, Religious Ritual, and Archaeology', in Kyriakidis, E. (ed.), *The Archaeology of Ritual*, Los Angeles, 209–254.
McGregor, R. 2010, 'Dressing the ka'ba from Cairo: The Aesthetics of Pilgrimage to Mecca', in Morgan, D. (ed.), *Religion and Material Culture. The Matter of Belief*, New York, 247–261.
Merkelbach, R. 1971, 'Gefesselte Götter', *Antaios* 12, 549–565.
Merkelbach, R. 1978, 'The Girl in the Rosebush: A Turkish Tale and its Roots in Ancient Ritual', *Harvard Studies in Classical Philology* 82, 1–15.
Merker, G. S. 2000, *The Sanctuary of Demeter and Kore. Terracotta Figurines of the Classical, Hellenistic, and Roman Periods, Corinth* vol. XVIII, part IV, Princeton.
Meuli, K. 1975, 'Die gefesselten Götter', in Meuli, K. (ed.), *Gesammelte Schriften*, Basel, 1035–1081.
Meyboom, P. G. P. 1995, *The Nile Mosaic of Palestrina*, Leiden.
Michel, C. & Nosch, M.-L. 2010, 'Textile Terminologies', Michel, C. & Nosch, M.-L. (eds.), *Textile Terminologies. The Ancient Near East and Mediterranean from the Third to the First Millennia BC*. Ancient Textiles Series 8, Oxford, ix–xix.
Milanezi, S. 2005, 'Beauty in Rags. On *rhakos* in Aristophanic theatre', in Cleland, L., Harlow, M. & Llewellyn-Jones, Ll. (eds.), *The Clothed Body in the Ancient World*, Oxford, 75–86.
Milgrom, J. 1983, 'Of Hems and Tassels', *Biblical Archaeology Review* 9, 3, 61–75.
Miller, D. 2005, 'Introduction', in Küchler, S. & Miller, D. (eds.), *Clothing as Material Culture*, New York, 1–20.
Miller, M. 1989, 'The Ependytes in Classical Athens', *Hesperia* 58, 3, 313–329.
Miller, M. C. 1997, *Athens and Persia in the Fifth Century BC. A Study in Cultural Reciprocity*, Cambridge.
Mills, H. 1984, 'Greek Clothing Regulations: Sacred and Profane?', *ZPE* 55, 255–265.
Mitchell, J. P. 2010, 'Performing Statues', in Morgan, D. (ed.), *Religion and Material Culture. The Matter of Belief*, New York, 262–276.
Mitsopoulou-Leon, V. 2009, 'Votive offerings for Artemis Hemera (Lousoi) and their Significance', in Prêtre, C. (ed.), *Le donateur, l'offrande et la déesse. Systèmes votifs dans les sanctuaires de déesses du monde*

grec. Actes du 31e colloque international organisé par l'UMR Halma-Ipel (Université Charles-de-Gaule, Lille 3, 13-15 décembre 2007), Kernos Supplement 23, Liège, 255–271.

Mommsen, A. 1899, 'Rhakos auf attischen Inschriften', *Philologus* 58, 343–347.

Moorman, E. M. 2011, *Divine Interiors. Mural Paintings in Greek and Roman Sanctuaries*, Amsterdam.

Moraitou, G. 2007, 'The Funeral Pyre Textile from Royal Tomb II in Vergina. Report on the 1997 Documentation, Treatment and Display', *Archaeological Textiles Newsletter* 44, 5–10.

Morgan, D. 2010, 'Introduction: the Matter of Belief', in Morgan, D. (ed.), *Religion and Material Culture. The Matter of Belief*, London, 1–18.

Morgan, J. 2007, 'Women, Religion, and the Home', in Ogden, D. (ed.), *A Companion to Greek Religion*, 2007, 297–310.

Morris, B. & Brooks, M. M. 2007, 'Jewish Ceremonial Textiles and the Torah: Exploring Conservation Practices in Relation to Ritual Textiles Associated with Holy Texts', in Hayward, M. & Kramer, E. (eds.), *Textiles and Text: Re-establishing the Links between Archival and Object-based Research*, London, 244–248.

Morrow, K. D., 1985, *Greek Footwear and the Dating of Sculpture*, Madison.

Moulhérat, C. & Spantidaki, G. 2009a, 'Cloth from Kastelli Chania', *Arachne* 3, 8–15.

Moulhérat, C. & Spantidaki, G. 2009b, 'Archaeological Textiles from Salamis: A Preliminary Presentation', *Arachne* 3, 16–29.

Murray, A. T. 1919, *Homer. The Odyssey*, London.

Muscarella, O. W. 1964, 'Ancient Safety-pins. Their Function and Significance', *Expedition* 6, 2, 34–40.

Mylonopoulos, J. 2010, 'Divine Images versus Cult Images. An Endless Story about Theories, Methods and Terminologies', in Mylonopoulos, J. (ed.), *Divine Images and Human Imaginations in Ancient Greece and Rome*, Leiden, 1–19.

Mylonopoulos, J. 2013, 'Commemorating Pious Service: Images in Honour of Male and Female Priestly Officers in Asia Minor and the Eastern Aegean in Hellenistic and Roman Times', in Horster, M. & Klöckner, A. (eds.), *Cities and Priests. Cult Personnel in Asia Minor and the Aegean Islands from the Hellenistic to the Imperial Period*, Göttingen, 121–153.

Nagy, B. 1984, 'The *Peplotheke*: What Was It?', in Boegehold, A. L. (ed.), *Studies Presented to Sterling Dow on his Eightieth Birthday*, Durham, 227–232.

Neer, R. 2010, *The Emergence of the Classical Style in Greek Sculpture*, Chicago.

Neils, J. 1992, 'Panathenaic Amphoras: Their Meaning, Makers, and Markets', in Neils, J. (ed.), *Goddess and Polis*, Hanover, NH, 29–52.

Neils, J. 2000, 'Others Within the Other: An Intimate Look at Hetairai and Maenads', in Cohen, B. (ed.), *Not the Classical Ideal: Athens and the Construction of the Other in Greek Art*, Leiden, 203–226.

Neils, J. 2001, *The Parthenon Frieze*, Cambridge.

Neils, J. 2003, 'Children and Greek Religion', in Neils, J. & Oakley, J. H. (eds.), *Coming of Age in Ancient Greece. Images of Childhood from the Classical Past*, Hanover, NH, 139–162.

Neils, J. 2008, 'Adonia to Thesmophoria: Women and Athenian Festivals', in Kaltsas, N. & Shapiro, A. (eds.), *Worshipping Women. Ritual and Reality in Classical Athens*, New York, 242–265.

Neils, J. 2009, 'Textile Dedications to Female Deities: The Case of the Peplos', in Prêtre, C. (ed.), *Le donateur, l'offrande et la déesse. Systèmes votifs dans les sanctuaires de déesses du monde grec. Actes du 31e colloque international organisé par l'UMR Halma-Ipel (Université Charles-de-Gaule, Lille 3, 13-15 décembre 2007)*, Kernos Supplement 23, Liège, 135–147.

Nick, G. 2002, *Die Athena Parthenos. Studien zum griechischen Kultbild und seiner Rezeption*, Mainz 2002.

Nielsen, I. 2009, 'The sanctuary of Artemis Brauronia. Can architecture and iconography help to locate the settings of the rituals?', in Fischer-Hansen, T. & Poulsen, B. (eds.), *From Artemis to Diana. The goddess of man and beast*, Copenhagen, 83–116.

Nilsson, M. P. 1957, *The Dionysiac mysteries of the Hellenistic and Roman Age*, Lund.

North, J. A. 1996, 'Pollution and Purification at Selinous', *SCI* 15, 293–301.

Nosch, M.-L. 2000, 'Schafherden unter dem Namenspatronat von Potnia und Hermes in Knossos', in Blakholmer, F. (ed.), *Österreichische Forschungen zur Ägäischen Bronzezeit 1998. Akten der Tagung am Institut für Klassische Archäologie der Universität Wien 2.-3. Mai 1998*, Vienna, 211–215.

Nosch, M.-L. 2007, 'The History of the Homeric Priestess Theano: A View from her Past', in Larsson Lovén, L. & Strömberg, A. (eds.), *Public Roles and Personal Status. Men and Women in Antiquity. Proceedings of the Third Nordic Symposium on Gender and Women's History in Antiquity, Copenhagen 3-5 October 2003*, Sävedalen, 165–183.

Nosch, M.-L. 2014, 'The Sources of Linen Textiles and Flax and Their Trade in Classical Greece', in Droß-Krüpe, K. (ed.), *Textile Trade and Textile Distribution in Antiquity*, Philippika 73, 17–42.

Nosch, M.-L. 2016, 'The Loom and the Ship in Ancient Greece. Shared Knowledge, Shared Terminology, Cross-Crafts, or Cognitive Maritime-textile Archaeology?', in Harich-Schwartzbauer, H. (ed.), *Weben und Gewebe in der Antike. Materialität - Repräsentation - Episteme - Metapoetik*, Oxford, 109–132.

Nosch, M.-L. Forthcoming, 'Les textiles des royautés en Grèce', in Werlings, M.-J. & Zurbach, J. (eds.), *Mélanges Pierre Carlier*.

Nosch, M.-L., & Perna, M. 2001, 'Cloth in the Cult', in Laffineur, R. & Hägg, R. (eds.), *Potnia. Deities and Religion in the Aegean Bronze Age. Proceedings of the 8th International Aegean Conference Göteborg, Göteborg University, 12-15 April 2000*. Aegaeum 22, Liege, 471–477.

Oakley, J. H. 2004, *Picturing Death in Classical Athens*, Cambridge.

Oakley, J. H. & Sinos, R. H. 1993, *The Wedding in Ancient Athens*, Madison.

O'Brien, J. V. 1993, *The Transformation of Hera: A Study of Ritual, Hero, and the Goddess in the Iliad*, Lanham 1993.

Ogden, D. 2002, 'Controlling Women's Dress: Gynaikonomoi', in Llewellyn-Jones, Ll. (ed.), *Womens' Dress in the Ancient World*, London & Swansea, 203–225.

Ohly, D. 1953, 'Die Gottin und ihre basis', *Athenische Mitteilungen* 68, 24–50.

O'Neill, E. 1938, *Aristophanes. Knights. The Complete Greek Drama*, Vol. 2, New York.

Osborne, R. 1985, *Demos: The Discovery of Classical Attika*, Cambridge.

Osborne, R. 2004, 'Hoards, Votives, Offerings: the Archaeology of the Dedicated Object', *World Archaeology* 36, 1, 1–10.

Osborne, R. 2011, 'Greek Inscriptions as Historical Writing', in Feldherr, A. & Hardy, G. (eds.), *The Oxford History of Historical Writing* Vol. 1, Oxford, 97–121.

Pace, B. 1923, 'Diana Pergaea', in Ramsay, W. M. et al. (eds.), *Anatolian Studies Presented to Sir William Mitchell Ramsay*, Manchester, 297–314.

Palagia, O. 1980, *Euphranor*, Leiden.

Palagia, O. 2008, 'Women in the Cult of Athena', in Kaltsas, N. & Shapiro, A. (eds.), *Worshipping Women. Ritual and Reality in Classical Athens*, New York, 30–77.

Panagiotakopulu, E., Buckland, P.C., Day, P.M., Doumas, C., Sarpaki, A. & Skidmore, P. 1997, 'A Lepidopterous Cocoon from Thera and Evidence for Silk in the Aegean Bronze Age', *Antiquity* 71, 420–429.

Papadimitriou, J. 1956, 'Brauron', *Ergon 1956*, 25–31.

Papadimitriou, J. 1957, 'Brauron', *Ergon 1957*, 20–24.

Papadimitriou, J. 1961a, 'Brauron', *Ergon 1961*, 20–37.

Papadimitriou, J. 1961b, 'Ἀνασκαφαὶ ἐν Βραυρῶνι', *Praktika tes en Athenais Archaiologikes Hetaireias 1956*, 73–89.

Papadimitriou, J. 1962, 'Brauron', *Ergon 1962*, 25–39.

Papadopoulou, M. forthcoming, 'Head-dress for Success: Cultic Uses of the Hellenistic Mitra', in C. Brøns & Nosch, M.-L. (eds.), *Textiles and Cult in the Mediterranean Area*, Oxford.

Papalexandrou. N. 2012, 'Review of Despinis, G. Ἄρτεμις Βραυρωνία: Λατρευτικά ἀγάλματα καί ἀναθήματα ἀπό τά ἱερά τῆς θεᾶς στή Βραυρώνα καί τήν Ἀκρόπολη τῆς Ἀθήνας, Library of the Archaeological Society of Athens 268. Archaeological Society of Athens, Athens 2010', *AJA* 116, 4, www.ajaonline.org. Open Access: Book review.

Parker, R. 1983, *Miasma. Pollution and Purification in Early Greek Religion*, Oxford.

Parker, R. 2007, *Polytheism and Society at Athens*, Oxford.

Parker, R. & Obbink, D. 2000, 'Sales of Priesthoods on Cos I', *Chiron* 30, 415–449.

Pasinli, A. 2010, *Istanbul Archaeological Museums*, Istanbul.

Pastoureau, M. 2001, *Blue: The History of a Colour*, Princeton.

Patera, M. 2012, 'Ritual Dress Regulations in Greek Inscriptions of the Hellenistic and Roman Periods', in Schrenk, S., Vössing, K. & Tellenbach, M. (eds.), *Kleidung und Identität in religiösen Kontexten der römischen Kaiserzeit*, Regensburg, 35–46.

Paton, W. R. 1969, *The Greek Anthology*, Vols. I–V. Translated by W. R. Paton, Cambridge, MA & London.

Patton, K. C. 1990, 'Gods who Sacrifice: A Paradox of Attic Iconography', *AJA* 94, 2, 326.

Paul, S. 2013, 'Roles of Civic Priests in Hellenistic Cos', in Horster, M & Klöckner, A. (eds.), *Cities and Priests. Cult Personnel in Asia Minor and the Aegean Islands from the Hellenistic to the Imperial Period*, Göttingen, 247–278.

Payne, H. 1940, *Perachora I. The sanctuaries of Hera Akraia and Limenia. Excavations of the British School of Archaeology at Athens 1930-1933*, Oxford.

Payne, H. 1962, *Perachora II. The sanctuaries of Hera Akraia and Limenia. Excavations of the British School of Archaeology at Athens 1930-1933*, Oxford.

Pelletier-Michaud, L. 2016, *Évolution du sens des termes de couleur et de leur traitement poétique. L'élégie romaine et ses modèles grecs*. PhD dissertation, Laval University.

Peppas-Delmousou, D. 1988, 'Autour des inventaires de Brauron', Knoepfler, D. (ed.), *Comptes et inventaires dans la cité Grecque. Actes du colloque international d'epigraphie tenu á Neuchâtel du 23 au 26 septembre 1986 en l'honneur de Jacques Tréheux*, Geneva, 323–346.

Perrin, B. 1920, *Plutarch's Lives*. Translated by Bernadotte Perrin, Cambridge, MA & London.

Petridou, G. 2009, '*Artemidi to Ichnos*: Divine Feet and Hereditary Priesthood in Pisidian Pogla', *Anatolian Studies* 59, 81–93.

Petrovic, I. 2010, 'The Life Story of a Cult Statue as an Allegory: Kallimachos' Hermes Perpheraios', in Mylonopoulos, J. (ed.), *Divine Images and Human Imaginations in Ancient Greece and Rome*, Leiden, 205–224.

Petrovic, I. & Petrovic, A. 2003, 'Stop and Smell the Statues: Callimachus' Epigram 51 Pf. Reconsidered (Four Times)', *Materiali e discussion per l'analisi dei testi classici* 51, 179–208.

Petrovic, I. & Petrovic, A. 2006, '"Look who is talking now!": Speaker and Communication in Greek Metrical Sacred Regulations', in Stavrianopoulou, E. (ed.), *Ritual and Communication in the Graeco-Roman World*, Liege, 151–180.

Petterson, M. 1992, *The Cults of Apollo at Sparta: The Hyakinthia, the Gymnopaidiai, and the Karneia*, Stockholm.

Pfuhl, E. 1923, *Malerei und Zeichnung der Griechen*, Munich.

Pfuhl, E., & Möbius, H. 1977, *Die ostgriechischen Grabreliefs*, Mainz am Rhein.

Picard-Schmitter, M. T. 1955, 'Sur la chlamyde de Démétrios Poliorcétès', *RA* 46, 17–26.

Pilz, O. 2013, 'The Profits of Self-Representation: Statues of Female Cult Personnel in the Late Classical and Hellenistic Periods', in Horster, M. & Klöckner, A. (eds.), *Cities and Priests. Cult Personnel in Asia Minor and the Aegean Islands from the Hellenistic to the Imperial Period*, Göttingen, 155–175.

Pingiatoglou, S. 1981, *Eileithyia*, Würzburg.

Platt, V. 2011, *Facing the Gods. Epiphany and Representation in Graeco-Roman Art, Literature and Religion*, Cambridge.

Polinskaya, I. 2013, *A Local History of Greek Polytheism. Gods, People and the Land of Aigina, 800-400 BC*, Leiden.

Pollitt, J. J. 1986, *Art in the Hellenistic Age*, Cambridge.

Premerstein, A. von. 1912, 'Der Parthenon-Fries und die Werkstatt des panathäischen Peplos', *Jahreshefte des Österreichischen archäologischen Instituts* 15, 1–35.

Prêtre, C. 2012, 'Kosmos et Kosmema. Les offrandes de parure dans les inscriptions de Délos', *Kernos supplément* 27, Liège.

Prêtre, C. forthcoming, 'Sacred Clothes and Profane Fabrics: Offerings on Delos', in Brøns, C. & Nosch, M.-L. (eds.), *Textiles and Cult in the Mediterranean Area in the 1st Millennium BC*, Oxford.

Prückner, H. 1968, *Die Lokrischen Tonreliefs*, Mainz am Rhein.

Quercia, A. forthcoming, '"Temple key" or distaff? An ambiguous artifact from the Greek and indigenous sanctuaries of southern Italy', in Brøns, C. & Nosch, M-L. (eds.), *Textiles and Cult in the Mediterranean Area in the 1st Millennium BC*, Oxford.

Radke, G. 1936, *Die Bedeutung der weissen und schwarzen Farbe in Kult und Brauch der Griechen und Römer*, Berlin.

Rappaport, R. A. 1999, *Ritual and Religion in the Making of Humanity*, Cambridge.

Raubitschek, A. E. & Stevens, G. P. 1946, 'The Pedestal of the Athena Promachos', *Hesperia* 15, 2, 107–114.
Rayet, M. O. 1881, 'Inscription de Thebes', *BCH* 5, 264–266.
Reeder, E. D. 1995, *Pandora. Women in Classical Greece*, Princeton.
Rehm, A. 1958, *Didyma, II. Die Inschriften*, Berlin.
Reinach, T. 1899, 'Un temple élevé par les femmes de Tanagra', *REG* 12, 53–115.
Reinhold, M. 1970, *History of Purple as a Status Symbol in Antiquity*, Bruxelles.
Renfrew, C. 1985, *The Archaeology of Cult. The Sanctuary at Phylakopi*, London.
Renfrew, C. 2007, 'The Archaeology of Ritual, of Cult, and of Religion', in Kyriakidis, E. (ed.), *The Archaeology of Ritual*, Los Angeles, 109–122.
Reuthner, R. 2006, *Wer webte Athenes Gewänder? Die Arbeit von Frauen im antiken Griechenland*, Frankfurt am Main & New York.
Richter, G. M. 1929, 'Silk in Greece', *AJA* 33, 1, 27–33.
Richter, G. M. 1960, *Kouroi, Archaic Greek Youths; a Study of the Development of the Kouros Type in Greek Sculpture*, London.
Richter, G. M. 1966, *The Furniture of the Greeks, Etruscans and Romans*, London.
Richter, G. M. 1968, *Korai, Archaic Greek Maidens*, London & New York.
Ridgway, B. S. 1984, *Roman Copies of Greek Sculpture*, Ann Arbor.
Ridgway, B. S. 1992, 'Images of Athena on the Akropolis', in Neils, J. (ed.), *Goddess and Polis. The Panathenaic Festival in Ancient Athens*, Hanover, 119–142.
Ridgway, B. S. 2000, *Hellenistic Sculpture II. The styles of ca. 200-100 BC*, London.
Ridgway, B. S. 2005, '"Periklean" cult images and their media', in Barringer, J. M. & Hurwit, J. M. (eds.), *Periklean Athens and its Legacy. Problems and Perspectives*, Austin, 111–118.
Riis, P. J. 1948-1949, 'The Syrian Astarte Plaques and Their Western Connections', *Berytus* 9, 69–90.
Riis, P. J. 1993, 'Ancient Types of Garments. Prolegomena to the Study of Greek and Roman Clothing', *Acta Archaeologica* 64, 1, 149–182.
Roach, M. E. & Eicher, J. B. 1973, 'The Language of Personal Adornment', in Cordwell, J. M. & Schwarz, R. A. (eds.), *The Fabrics of Culture*, London, 7–21.
Robertson, M. & Frantz, A. 1975, *The Parthenon Frieze*, London.
Robertson, N. 1983, 'The Riddle of the Arrephoria at Athens', *HSCP* 87, 241–288.
Robertson, N. 1996, 'Athena's Shrines and Festivals', in Neils, J. (ed.), *Worshipping Athena: Panathenaia and Parthenon*, Madison & London, 27–77.
Robertson, N. 2004, 'The Praxiergidae Decree (*IG* I³ 7) and the Dressing of Athena's Statue with the Peplos', *GRBS* 44, 111–161.
Robertson, N. 2013, 'The Concept of Purity in Greek Sacred Laws', in Frevel, C. & Nihan, C. (eds.), *Purity and the Forming of Religious Traditions in the Ancient Mediterranean World and Ancient Judaism*, Leiden, 195–243.
Robinson, E. S. G. 1936, 'A Find of Archaic Coins from South West Asia Minor', *Numismatic Chronicle* 16, 265–280.
Roccos, L. J. 1995, 'The Kanephoros and Her Festival Mantle in Greek Art', *AJA* 99, 4, 641–666.
Roesch, P. 1985, 'Les femmes et la fortune en Béotie', in Vérilhac, A.-M., et al. (eds.), *La femme dans le monde Méditerranéen*, Lyon, 71–84.
Roller, D. W. 1989, *Tanagran Studies I. Sources and Documents on Tanagra in Boiotia*, Amsterdam.
Roller, D. W. 2007, 'The Inscriptions of Tanagra. A Preliminary Analysis', in Fittschen, K. (ed.), *Historische Landeskunde und Epigraphik in Griechenland. Akten des Symposiums veranstaltet aud Anlass des 100. Todestages von H. G. Lolling (1848-1894) in Athen vom 28.-30.9.1994*, Münster, 151–156.
Romano, I. B. 1980, *Early Greek cult images*. Unpublished PhD dissertation, University of Pennsylvania.
Romano, I. B. 1982, 'Early Greek Idols. Their Appearance and Significance in the Geometric, Orientalizing and Archaic Periods', *Expedition* 24, 3, 3–13.
Romano, I. B. 1985, 'The Archaic Images of Hera and Zeus in the Heraion at Olympia', *AJA* 89, 2, 348.
Romano, I. B. 1988, 'Early Greek Cult Images and Cult Practices', in Hägg, R. et al. (eds.), *Early Greek Cult Practice. Proceedings of the Fifth International Symposium at the Swedish Institute at Athens, 26-29 June, 1986*, Stockholm, 127–133.

Rosenberger, V. 2008, 'Gifts and Oracles: Aspects of Religious Communication', in Rasmussen, A. H. et al. (eds.), *Religion and Society. Rituals, Resources and Identity in the Ancient Graeco-Roman World. The BOMOS-Conferences 2002-2005*, Rome, 91–106.

Rosenthal-Heginbottom, R. 2009, 'The Curtain (Parochet) in Jewish and Samaritan Synagogues' in De Moor, A. & Fluck, C. (eds.), *Clothing the House. Furnishing Textiles of the 1st Millennium AD from Egypt and Neighbouring Countries. Proceedings of the 5th Conference of the Research Group 'Textiles from the Nile Valley' Antwerp, 6-7 October 2007*, Tielt, 155–169.

Rosenzweig, R. 2004, *Worshipping Aphrodite. Art and Cult in Classical Athens*, Ann Arbor.

Rostad, A. 2006, *Human Transgression – Divine Retribution. A Study of Religious Transgression and Punishments in Greek Cultic Regulations and Lydian-Phrygian Reconciliation Inscriptions*. Doctoral thesis, Department of Classics, University of Bergen.

Rouse, W. H. D. 1902, *Greek Votive Offerings*, Cambridge.

Rudy, K. M. 2007, 'Introduction: Miraculous Textiles in Exempla and Images from the Low Countries', in Rudy, K. M. & Baert, B. (eds.), *Weaving, Veiling, and Dressing: Textiles and their Metaphors in the Late Middle Ages*, Turnhout, 1–36.

Rutherford, I. 1998, 'The Amphikleidai of Sicilian Naxos: Pilgrimage and Genos in the Temple Inventories of Delos', *ZPE* 122, 81–89.

Rutkowski, B. 1973, 'Cult Images in the Aegean World', *Archaeologia Polona* 14, 53–57.

Ryan, F. X. 2008, 'Breadth and Depth in the Account of the Dedications to Athana Lindia', *Studi ellenistici* 20, 455–470.

Sabetai, V. 2008 'Women's Ritual Roles in the Cycle of Life', in Kaltsas, N. & Shapiro, A. (eds.), *Worshipping Women. Ritual and Reality in Classical Athens*, New York, 289–333.

Şahin, S. 1999, *Die Inschriften von Perge*, Bonn.

Salviat, F. 1959, 'Décrets pour Épié fille de Dionysos: Déesses et sanctuaries thasiens', *BCH* 83, 362–379.

Santamaria, U., Morresi, F., Delle Rose, M. 2004, 'Indagini scientifiche dei pigmenti e leganti delle lastre marmoree dipinte dell'Aula del Colosso del Foro di Augusto', in P. Liverani (eds.), *I colori del bianco. Policromia nella scultura antica, Musei Vaticani. Collana di Studi e Documentazione*, Rome, 281–289.

Sapouna-Sakellarakis, E. 1978, *Die Fibeln der griechischen Inseln*, Prähistorische Bronzefunde (PBF) XIV, 4.

Şare-Ağtürk, T. 2014, 'Arakhnes's Loom: Luxurious Textile Production in Ancient Western Anatolia', *Olba* XXII, 251–280.

Satlow, M. L. 2013, 'Introduction', in Satlow, M. L. (ed.), *The Gift in Antiquity. The Ancient World: Comparative Histories*, Malden, MA & Oxford, 1–11.

Schachter, A. 1997, 'Reflections on an Inscription from Tanagra', in Bintliff, J. (ed.), *Recent Developments in the History and Archaeology of Central Greece. Proceedings of the 6th International Boeotian Conference*, Oxford, 277–286.

Schachter, A. 2003, 'Evolutions of a Mystery Cult. The Theban Kabiroi', in Cosmopoulou, M. B. (ed.), *Greek Mysteries. The Archaeology and Ritual of Ancient Greek Secret Cults*, London, 112–142.

Schauenburg, K. 1977, 'Zu Götterstatuen auf unteritalischen Vasen', *Archäologischer Anzeiger*, 4, 285–297.

Schefold, K. 1937, *Statuen auf Vasenbildern*, Berlin.

Scheid, J. & Svenbro, J. 1996, *The Craft of Zeus: Myths of Weaving and Fabric*, Cambridge, MA.

Schnapp, A. 1988, 'Why Did the Greeks Need Images?', in Christiansen, J. & Melander, T. (eds.), *Proceedings of the 3rd Symposium on Ancient Greek and Related Pottery. Copenhagen August 31-September 1987*, Copenhagen, 568–574.

Schneider, J. 2006, 'Cloth and Clothing', in Tilley, C. et al. (eds.), *Handbook of Material Culture*, London, 203–220.

Schneider, J. & Weiner, A. B. 1991, 'Introduction', in Schneider, J. & Weiner, A. B. (eds.), *Cloth and Human Experience*, Washington & London, 1–29.

Schneider-Herrmann, G. 1972, 'Kultstatue im Tempel auf italienischen Vasenbildern', *BABesch* 47, 31–42.

Schrenk, S. 2009 '(Wall-) Hangings Depicted in Late Antique Works of Art? The Question of Function', in De Moor, A. & Fluck, C. (eds.), *Clothing the House. Furnishing Textiles of the 1st millennium AD from Egypt and Neighbouring Countries. Proceedings of the 5th Conference of the Research Group Textiles from the Nile Valley Antwerp, 6-7 October 2007*, Tielt, 146–154.

Scott, M. 2011, 'Displaying Lists of What is (not) on Display. The Uses of Inventories in Greek Sanctuaries', in Haysom, M. & Wallensten, J. (eds.), *Current Approaches to Religion in Ancient Greece. Papers Presented at a Symposium at the Swedish Institute at Athens, 17-19 April 2008*, Stockholm, 239-252.
Seaton, R. C. 1912, *Apollonius Rhodius, Argonautica*, London.
Sebesta, J. L. 2001, 'Tunica Ralla, Tunica Spissa: The Colors and Textiles of Roman Costume', in Sebesta, J. L. & Bonfante, L. (eds.), *The World of Roman Costume*, Madison, 65-76.
Sebesta, J. L. 2002, 'Visions of Gleaming Textiles and a Clay Core: Textiles, Greek Women, and Pandora', in Llewellyn-Jones, Ll. (ed.), *Women's Dress in the Ancient Greek World*, London, 125-142.
Segre, M. 1993, *Iscrizioni di Cos*, Rome.
Seiterle, G. 1979, 'Artemis Die grosse Gottin von Ephesos', *Antike Welt* 10, 3, 3-16.
Seiterle, G. 1999, 'Ephesische Wollbinden. Attribut der Göttin, Zeichen des Stieropfers', in Frisinger, H. & Krinzinger, F. (eds.), *100 Jahre Österreichische Forschungen in Ephesos*, Wien, 251-254.
Senff, R. 1993, *Das Apollonheiligtum von Idalion. Architektur und Statuenausstattung eines zyprischen Heiligtums*, Jonsered.
Shamir, O. 2008, 'Organic Materials', in Ariel, D. T. et al. (eds.), *The Dead Sea Scrolls*, Jerusalem, 116-133.
Shamir, O. forthcoming, 'The High Priest Sha'atnez garments compared to textiles found in the Land of Israel', in Brøns, C. & Nosch, M.-L. (eds.), *Textiles and Cult in the Mediterranean Area in the 1st Millennium BC*, Oxford.
Shapiro, H. A. 1989, *Art and Cult under the Tyrants in Athens*, Mainz am Rhein.
Shaya, J. 2002, *The Lindos Stele and the lost Treasures of Athena: Catalogs, Collections, and Local History*, PhD dissertation, University of Michigan.
Shaya, J. 2005, 'The Greek Temple as Museum: The Case of the Legendary Treasure of Athena from Lindos', *AJA* 190, 3, 423-442.
Shear, T. L. 1978, *Kallias of Sphettos and the revolt of Athens in 286 B.C.*, Hesperia Supplement 17.
Shipley, G. 1987, *A History of Samos 800-188 BC*, Oxford.
Simon, C. G. 1986, *The Archaic Votive Offerings and Cults of Ionia*, Ann Arbor.
Simon, E. 1983, *Festivals of Attica. An Archaeological Commentary*, Madison.
Sinn, U. 1993, 'Greek Sanctuaries as Places of Refuge', in Marinatos, N. & Hägg, R. (eds.), *Greek Sanctuaries. New Approaches*, New York, 88-109.
Skovmøller, A. 2014, 'Where Marble Meets Colour: Surface Texturing of Hair, Skin and Dress on Roman Marble Portraits as Support for Painted Polychromy', in Harlow, M. & Nosch, M.-L. (eds.), *Greek and Roman Textiles and Dress. An interdisciplinary anthology*, Oxford, 279-297.
Smyth, H. W. 1926, *Aeschylus. Eumenides*. Translation by H. W. Smyth, Cambridge, MA.
Sokolowski, F. 1955, *Lois sacrées de l'Asie Mineure*, Paris.
Sokolowski, F. 1962, *Lois sacrées des cités grecques. Supplément*, Paris.
Sokolowski, F. 1965, 'A new testimony on the cult of Artemis of Ephesus', *HarvTheolR* 58, 427-431.
Sokolowski, F. 1969, *Lois sacrées des cités grecques*. Paris.
Sourvinou-Inwood, C. 1995, 'Male and Female, Public and Private, Ancient and Modern', in Reeder, E. D. (ed.), *Pandora: Women in Classical Greece*, Princeton, 111-120.
Sourvinou-Inwood, C. 2011, *Athenian Myths & Festivals. Aglauros, Erechtheus, Plynteria, Panathenaia, Dionysia*, Oxford.
Spantidaki, S. 2014, 'Embellishment Techniques of Classical Greek Textiles', in Harlow, M. & Nosch, M.-L. (eds.), *Greek and Roman Textiles and Dress: An Interdisciplinary Anthology*, Oxford, 34-45.
Spantidaki, Y. & Moulherat, C. 2012, 'Greece', in Gleba, M. & Mannering, U. (eds.), *Textiles and Textile Production in Europe*. Ancient Textiles Series 11, Oxford, 185-202.
Spiro, M. E. 1966, 'Religion: Problems of Definition and Explanation', in Banton, M. (ed.), *Anthropological Approaches to the Study of Religion*, London, 85-126.
Stafford, E. J. 2005, 'Viewing and Obscuring the Female Breast: Glimpses of the Ancient Bra', in Cleland, L., Harlow, M. & Llewellyn-Jones, Ll. (eds.), *The Clothed Body in the Ancient World*, Oxford, 96-110.
Stafford, E. J. 2007, 'Personification in Greek Religious Thought and Practice', in Ogden, D. (ed.), *A Companion to Greek Religion*, Oxford, 71-85.

Stavrianopoulou, E. 2006, 'Introduction', in Stavrianopoulou, E. (ed.), *Ritual and Communication in the Graeco-Roman World*, Liege, 7–22.
Stehle, E. 1997, *Performance and Gender in Ancient Greece*, Princeton.
Steingräber, S. 1986, *Etruscan Painting*, New York.
Stephani, L. 1881, *Compte-Rendu de la Commission Imperiale Archeologique (1978-1879)*, St Petersburg.
Stephenson, J. W. 2014, 'Veiling the Late Roman House', *Textile History* 45, 3–31.
Stewart, A. 1990, *Greek Sculpture. An Exploration*, New Haven.
Stig Sørensen, M.-L. 2000, *Gender Archaeology*, Cambridge.
Stocking, C. 2010, 'Return to Sender: Failed Reciprocities in Ancient Greek Religion', abstract for a lecture given in March 2010 v. Classical Association of the Middle West and South, Oklahoma City.
Strøm, I. 1995, 'The early sanctuary of the Argive Heraion and its external relations (8th to 6th centuries BC). Bronze Imports and Archaic Greek Bronzes' *Proceedings of the Danish Institute at Athens* 1, 37–127.
Stubbe Østergaard, J. & Nielsen, A. M. 2014, *Transformations. Classical Sculpture in Colour*, Copenhagen.
Stubbings, J. M. 1962, 'The Ivories', in T. J. Dunbabin (ed.), *Perachora. The Sanctuaries of Hera Akraia and Limenia. Volume two: Pottery, Scarabs, and Other Objects from the Votive Deposit of Hera Limenia*, 403–451.
Studniczka, F. 1886, *Beiträge zur Geschichte der altgriechischen Tracht. Abhandlung des archäologisch-epigraphischen Seminars der Universität Wien*, Vienna.
Stulz, H. 1990, *Die Farbe Purpur im frühen Griechentum*, Stuttgart.
Suhr, E. G. 1958, *Venus de Milo. The Spinner*, New York.
Surtees, L. 2014 'Loomweights', in Schaus, G. P. (ed.), *Stymphalos. The Acropolis Sanctuary*. Vol. 1, Toronto, 236–247.
Tagalidou, E. 1993, *Weihreliefs an Herakles aus Klassischer Zeit*, Jonsered.
Takács, S. A. 2005, 'Divine and Human Feet: Records of Pilgrims Honouring Isis' in Elsner, J. & Rutherford, I. (eds.), *Pilgrimage in Graeco-Roman and Early Christian Antiquity: Seeing the Gods*, Oxford, 353–369.
Tanner, J. 2001, 'Nature, Culture and the Body in Classical Greek Religious Art', *World Archaeology* 33, 2, 257–276.
Tellenbach, M., Schulz, R. & Wieczorek, A. (eds.) 2013, *Die Macht der Toga. Dresscode im Römischen Weltreich*, Regensburg.
Testart, A. 1998, 'Uncertainties of the "Obligation to Reciprocate": A Critique of Mauss', in James, W. & Allen, N. J. (eds.), *Marcel Mauss: A Centenary Tribute*, Providence, 97–110.
Theissen, G. 1983, *The Miracle Stories of the Early Christian Tradition*, Philadelphia.
Thelwall, S. 1969, *The Writings of Quintus Sept. Flor. Tertullianus*, Edinburgh.
Themelis, P. G. 1971, *Brauron: Guide to the Site and to the Museum*, Athens.
Themelis, P. G. 1994, 'Artemis Ortheia at Messene: The Epigraphical and Archaeological Evidence', in Hägg, R. (ed.), *Ancient Greek Cult Practice from the Epigraphical Evidence*, Stockholm, 101–122.
Themelis, P. G. 2002, 'Contributions to the Topography of the Sanctuary of Brauron', in Gentili, B, & Perusino, F. (eds.), *Le orse di Brauron: Un rituale di iniziazione femminile nel santuario di Artemide*, Pisa, 103–116.
Themelis, P. G. 2004, ''Ἀνασκαφὴ Μεσσήνης', *Praktika 2001*, 63–96.
Themelis, P. G. 2007, 'Τα Κάρνεια καὶ ἡ Ἀνδανία', in E. Simantoni-Bournia, E., Laimou, A. A., Mendoni, L. G. & Kourou, N. (eds.), *Ἀμύμονα ἔργα. Τιμητικός τόμος για τον καθηγητή Βασίλη Κ. Λαμπρινουδάκη*, Athens, 509–528.
Thiersch, H. 1936, *Ependytes und Ephod. Gottesbild und Priesterkleid im alten Vorderasien*, Stuttgart.
Thompson, G. 1965, 'Iranian Dress in the Achaemenian Period: Problems Concerning the Kandys and Other Garments', *Iran* 3, 121–126.
Thompson, W. E. 1982, 'Weaving: A Man's Work', *The Classical World* 75, 4, 217–222.
Thompson, W. E. 1964, 'A Pronaos Inventory', *Hesperia* 33, 86–87.
Touchais, G. 1984, 'Chronique des fouilles et découvertes archéologiques en Grèce en 1983'. *BCH* 108, 2, 735–843.
Trinkl, E. 2014, 'The Wool Basket: Function, Depiction and Meaning of the *Kalathos*', in Harlow, M. & Nosch, M.-L. (eds.), *Greek and Roman Textiles and Dress. An Interdisciplinary Anthology*. Ancient Textiles Series 19, Oxford, 190–206.

Tsigarida, E.-B. 1997, 'Jewellery from the Geometric Period to Late Antiquity (9th c. BC–4th c. AD)', in Kypraiou, E. (ed.), *Greek Jewellery. 6000 Years of Tradition*, Athens, 61–150.
Tuck, A. 2009, 'Stories at the Loom: Patterned Textiles and the Recitation of Myth in Euripides', *Arethusa* 42, 2, 151–159.
Turner, J. A. 1983, *Hiereiai: Acquisition of Feminine Priesthoods in Ancient Greece*, Dissertation University of California, Santa Barbara.
Turner, V. 1969, *The Ritual Process: Structure and Anti-Structure*, London.
Ungaro, L. 2004, 'Il rivestimento dipinto dell' "Aula del Colosso" nel Foro di Augusto', in Liverani, P. (eds.), *I colori del bianco. Policromia nella scultura antica, Musei Vaticani. Collana di Studi e Documentazione*, Rome, 275–280.
Ungaro, L. 2007, 'Roma, Foro di Augusto, Aula del Colosso. Il rivestimento parietale in marmo dipinto: analisi di laboratorio e ricostruzione', in Angelelli, C. (eds.), *Atti del XII Colloquio dell'Associazione Italiana per lo Studio e la Conservazione del Mosaico, Padova-Brescia 2006*, Tivoli, 231–240.
Ungaro, L., Vitali, M. L. 2004, 'Die bemalte Wandverkleidung der "Aula del colosso" im Augustusforum', in *Bunte Götter. Die Farbigkeit antiker Skulptur. Eine Ausstellung der Staatlichen Antikensammlungen und Glyptothek München, 16. Dezember 2003 bis 29. Februar 2004, 15. Juni-5. September 2004*. München, 216–218.
Van Gennep, A. 1909, *Les Rites de Passage*, Paris.
Van Straten, F. T. 1981, 'Gift for the Gods', in Versnel, H. S (ed.), *Faith, Hope and Worship. Aspects of Religious Mentality in the Ancient World*, Leiden, 65–151.
Van Straten, F. T. 1995, *Hiera Kala: Images of Animal Sacrifice in Archaic and Classical Greece*, Leiden.
Van Wees, H. 2005, 'Trailing Tunics and Sheepskin Coats: Dress and status in early Greece', in Cleland, L., Harlow, M. & Llewellyn-Jones, Ll. (eds.), *The Clothed Body in the Ancient World*, Oxford, 41–51.
Vaulina, M. & Wasowicz, A. 1974, *Bois grecs et romains de l'Ermitage*, Wroclaw.
Vickers, M. 1999, *Images on Textiles. The Weave of Fifth-Century Athenian Art and Society*, Konstanz.
Vidman, L. 1969, *Sylloge inscriptionum religionis isiacae et sarapicae*, Berlin.
Vidman, L. 1970, *Isis und Sarapis bei den Griechen und Römern*, Berlin.
Vikela, E. 2008, 'The worship of Artemis in Attica: cult places, rites, iconography', in Kaltsas, N. & Shapiro, A. (eds.), *Worshipping Women. Ritual and Reality in Classical Athens*, New York, 79–105.
Villing, A. 1998, 'Athena as Ergane and Promachos. The iconography of Athena in archaic east Greece', in Fisher, N. & van Wees, H. (eds.), *Archaic Greece: New Approaches and New Evidence*, London, 147–168.
Von den Hoff, R. 2008, 'Images and Prestige of Cult Personnel in Athens between the Sixth and First Centuries BC', in Dignas, B. & Trampedach, K. (eds.), *Practitioners of the Divine: Greek Priests and Religious Officials from Homer to Heliodorus*, Cambridge, 107–141.
Wace, A. J. B. 1952, 'The Cloaks of Zeuxis and Demetrius', *JOAI* 39, 111–118.
Wachsmann, S. 2012, 'Panathenaic Ships: The Iconographic Evidence', *Hesperia* 81, 237–266.
Waetzoldt, H. 2013, 'The Colours and Variety of Fabrics from Mesopotamia during the Ur III Period (2050 BC)', in Michel, C. & Nosch, M.-L. (eds.), *Textile Terminologies in the Ancient Near East and Mediterranean from the Third to the First Millennia BC*. Ancient Textiles Series 8, Oxford, 201–209.
Wagner-Hasel, B. 2013, 'Marriage Gifts in Ancient Greece', in Satlow, M. L. (ed.), *The Gift in Antiquity. The Ancient World: Comparative Histories*, Malden, MA & Oxford, 158–171.
Wagner-Hasel, B. 2016, '"Canusiner Gewand, das trübem Honigwein sehr gleicht [...]". Wollqualitäten und Luxusdiskurs in der Antike', in Harich-Schwarzbauer, H. (ed.), *Weben und Gewebe in der Antike. Materialität - Repräsentation - Episteme - Metapoetik*. Ancient Textiles Series 23, Oxford, 37–66.
Waldstein, C. 1905, *The Argive Heraeum*, Boston & New York.
Wankel, H. 1987, 'Zu den chitonen für die Ephesische Artemis', *Epigraphica Anatolica* 9, 1987, 79–80.
Warr, C. 2010, *Dressing for Heaven. Religious Clothing in Italy, 1215-1545*, Manchester & New York.
Watzinger, C. 1905, *Griechische Holzsarkophage aus der Zeit Alexanders der Grossen*, Leipzig.
Weber, M. 2006, 'Die Kultbilder der Aphrodite Urania der zweiten Hälfte des 5. Jhs. v. Chr. in Athen-Attika und das Bürgerrechtsgesetz von 451/0 v. Chr', *AM* 121, 165–223.
Weddle, P. 2010, *Touching the Gods. Physical Interaction with Cult Statues in the Roman World*. PhD dissertation, Durham University.
Weiner, A. B. 1985, 'Inalienable Wealth', *American Ethnologist* 12, 2, 210–227.

Williams, D. & Ogden, J. 1994, *Greek Gold. Jewellery of the Classical World*, London.
Whitehead, A. 2013, *Religious Statues and Personhood: Testing the Role of Materiality*, London.
Wild, J. P. 2003, 'The Eastern Mediterranean, 323 BC–AD 350', in Jenkins, D. (ed.), *The Cambridge History of Western Textiles* I, Cambridge, 102–117.
Wolters, P. & Bruns, G. 1940, *Das Kabirenheiligtum bei Theben*, Berlin.
Wortmann, D. 1968, 'Die Sandale der Hekate-Persephone-Selene', *ZPE* 2, 155–160.
Wächter, T. 1910, *Reinheitsvorschriften im griechischen Kult*, Giessen.
Yates, J. 1843, *Textrinum Antiquorum: An Account of the Art of Weaving Among the Ancients*, London.